The Cognitive Basis of Science

The Cognitive Basis of Science concerns the question 'What makes science possible?' Specifically, what features of the human mind and of human culture and cognitive development permit and facilitate the conduct of science? The essays in this volume address these questions, which are inherently interdisciplinary, requiring co-operation between philosophers, psychologists, and others in the social and cognitive sciences. They concern the cognitive, social, and motivational underpinnings of scientific reasoning in children and lay persons as well as in professional scientists. The editors' introduction lays out the background to the debates, and the volume includes a consolidated bibliography that will be a valuable reference resource for all those interested in this area. The volume will be of great importance to all researchers and students interested in the philosophy or psychology of scientific reasoning, as well as those, more generally, who are interested in the nature of the human mind.

The cognitive basis of science

edited by

Peter Carruthers

Professor and Chair, Department of Philosophy, University of Maryland College Park

Stephen Stich

Board of Governors Professor, Department of Philosophy and Center for Cognitive Science, Rutgers University

and

Michael Siegal

Professor of Psychology, University of Sheffield

Published in association with the Hang Seng Centre for Cognitive Studies, University of Sheffield, and the Evolution and Higher Cognition Research Group, Rutgers University

CAMBRIDGE UNIVERSITY PRESS
Cambridge, New York, Melbourne, Madrid, Cape Town, Singapore,
São Paulo, Delhi, Dubai, Tokyo, Mexico City

Cambridge University Press
The Edinburgh Building, Cambridge CB2 8RU, UK

Published in the United States of America by Cambridge University Press, New York

www.cambridge.org
Information on this title: www.cambridge.org/9780521011778

First published 2002

A catalogue record for this publication is available from the British Library

Library of Congress Cataloguing in Publication data

The cognitive basis of science/edited by Peter Carruthers, Stephen Stich and
Michael Siegal.

p. cm.

Includes bibliographical references and index.

ISBN 0-521-81229-1 – ISBN 0-521-01177-9 (pb.)

1. Science – Philosophy. 2. Science – Psychological aspects. I. Carruthers,
Peter, 1952– II. Stich, Stephen P. III. Siegal, Michael.

Q175 .C6127 2002
501 – dc21 2001043131

ISBN 978-0-521-81229-0 Hardback
ISBN 978-0-521-01177-8 Paperback

for
Sir Q.W. Lee
with thanks again

Contents

Contributors

SCOTT ATRAN, Anthropology, CNRS Paris and University of Michigan at Ann Arbor

PETER CARRUTHERS, Department of Philosophy, University of Maryland at College Park

KEVIN N. DUNBAR, Department of Education and Department of Psychological and Brain Sciences, Dartmouth College

JONATHAN ST B. T. EVANS, Department of Psychology, University of Plymouth

LUC FAUCHER, Department of Philosophy, University of Quebec at Montreal

RONALD GIERE, Department of Philosophy, University of Minnesota

CLARK GLYMOUR, Department of Philosophy, Carnegie-Mellon University

ALISON GOPNIK, Department of Psychology, University of California at Berkeley

PAUL L. HARRIS, Graduate School of Education, Harvard University

DENIS HILTON, Department of Psychology, Université Toulouse-II

CHRISTOPHER HOOKWAY, Department of Philosophy, University of Sheffield

PHILIP KITCHER, Department of Philosophy, Columbia University

BARBARA KOSLOWSKI, Department of Human Development, Cornell University

RON MALLON, Department of Philosophy, University of Utah

STEVEN MITHEN, Department of Archaeology, University of Reading

DANIEL NAZER, Department of Philosophy, Rutgers University

NANCY J. NERSESSIAN, Department of Philosophy, Georgia Tech

SHAUN NICHOLS, Department of Philosophy, College of Charleston

AARON RUBY, Department of Philosophy, Rutgers University

MICHAEL SIEGAL, Department of Psychology, University of Sheffield

STEPHEN STICH, Department of Philosophy, Rutgers University

PAUL THAGARD, Department of Philosophy, University of Waterloo

STEPHANIE THOMPSON, Department of Human Development, Cornell University

ROSEMARY VARLEY, Department of Human Communication Sciences, University of Sheffield

JONATHAN WEINBERG, Department of Philosophy, Indiana University

Preface

This volume is the culmination of the fourth project undertaken by Sheffield University's Hang Seng Centre for Cognitive Studies. (The first project resulted in *Theories of Theories of Mind* (1996), edited by Peter Carruthers and Peter K. Smith; the second resulted in *Language and Thought* (1998), edited by Peter Carruthers and Jill Boucher; and the third produced *Evolution and the Human Mind* (2000), edited by Peter Carruthers and Andrew Chamberlain; all three volumes were published by Cambridge University Press.) For the first time, however, the present project also involved co-operation with the Evolution and Higher Cognition Research Group at Rutgers University, led by Stephen Stich. Four inter-disciplinary workshops were held in Sheffield over the period 1998–2000, and two conferences took place – one at Rutgers in November 1999, and one at Sheffield in June 2000. This collaboration has enabled us to assemble a wider and more international field of contributors than would otherwise have been possible.

The intention behind the project was to bring together a select group of philosophers, psychologists and others in the cognitive sciences to address such questions as the following: What is it about human cognition which either enables us, or fits us, to do science? Do scientific abilities have some sort of distinctive innate basis? Or are scientific abilities socially constructed out of general-learning mechanisms? How do different elements of our cognition fit together to underpin scientific reasoning? To what extent are there continuities between the cognitive processes involved in child development, those engaged in by hunter–gatherer communities, and those which are distinctive of scientific enquiry? To what extent do the well-known biases in human reasoning impact upon science, and what place do the emotions have in an adequate account of scientific activity? How important is the social dimension of science for our understanding of science and scientific cognition?

We have selected seventeen of the best, most focused, contributions from among those delivered at the two conferences. These have been re-written in the light of editorial comment and advice from the referees. In addition, drafts of these papers were made available to the other participants via a restricted-access web-site. The result, we believe, is a highly integrated volume of original inter-disciplinary essays which will make a significant contribution to our understanding of science, and thereby to our understanding of the human mind.

We would very much like to thank all those who participated in the workshop series and/or one or another of the conferences, whose comments and contributions to discussion did much to sharpen ideas; and special thanks should go to all those who (for one reason or another) presented a paper at one of these venues without contributing a chapter to this volume. Special thanks go to: Alex Barber, George Botterill, Pascal Boyer, George Butterworth, Susan Carey, Nick Chater, Gregory Currie, Mark Day, Merideth Gattis, Rochel Gelman, James Hampton, Stephen Laurence, Peter Lipton, Howard Margolis, Richard Nisbett, David Over, David Papineau, Ian Ravenscroft, Elizabeth Spelke, Eugene Subbotsky and Peter Todd. Thanks also go to Tom Simpson for his help in preparing the bibliography and indexes.

Finally, this project would not have been possible without financial assistance from the British Academy and the Center for Cognitive Science and the Research Group on Evolution and Cognition at Rutgers University, as well as the original endowment from the Hang Seng Bank of Hong Kong. We are grateful to these institutions, and especially the Chairman of the latter (now retired), Sir Q. W. Lee, for his far-sighted support.

1 Introduction: what makes science possible?

Peter Carruthers, Stephen Stich and Michael Siegal

In this brief opening chapter we briskly review some of the recent debates within philosophy and psychology which set the stage for the present collection of essays. We then introduce the essays themselves, stressing the inter-linking themes and cross-connections between them.

1 Introduction

The central position of science in our contemporary world needs no emphasis. Without science (broadly construed, to include all forms of technical innovation) we would still be roaming the savannahs of Africa like our *Homo habilis* ancestors, digging up tubers and scavenging scraps of meat. And without science (construed narrowly, as involving the application of an experimental method) we would have seen none of the advances in knowledge, technology and accumulation of wealth which have transformed the world and most of its people in just the last four centuries or so. Science now touches every aspect of our lives, from cradle (indeed, sometimes from conception) to grave. Given the manifest importance of science, the search for a scientific understanding of scientific thought and activity itself should need no further motivating. But in fact, the attempt to explain scientific cognition is not only extremely hard, but raises a whole host of fascinating and puzzling questions about the nature, development and operations of the human mind, and its interactions with culture.

This book is about the question: what makes science possible? Specifically, what features of the human mind, of human cognitive development and of human social arrangements permit and facilitate the conduct of science? These questions are inherently inter-disciplinary, requiring co-operation between philosophers, psychologists and others in the social and cognitive sciences. And they are, it should be stressed, questions which are as much about the psychological underpinnings of science as they are about science itself. That is, they concern what it is about our minds and/or mind-guided social interactions which make science possible, and how these factors relate to other things which we can do, either as adults or children. Indeed, one of the important themes of the book is the broad-scale architecture of the mind. For in order to understand how science is possible we have to understand how our capacity for scientific theorizing fits into the structure of the mind, and what consequences that might have for the practice of science.

Steven Pinker in his well-known book *How the Mind Works* (1997) raises questions to which this volume is a partial answer. Having argued that cognitive science is approaching an understanding of many different aspects of the mind – vision, language, memory, and so on – he then lists various factors about human beings which (he says) we *cannot* yet begin to explain. One of these is consciousness. Another is science. According to Pinker, we don't really have any grip on how human beings can be capable of scientific thinking and reasoning. That is a striking and challenging claim. Taking up and answering that challenge, as this book begins to do, is something which should be of interest, not only to psychologists and philosophers of psychology, but to all those interested in understanding either the nature of the human mind, or the nature of science, or both.

2 Philosophy of science: a very short recent history

As the title of this section suggests, we here take the reader through a *very* brisk tour of recent developments in the philosophy of science.

2.1 Positivism and beyond

In the beginning of our story there was logical positivism, which dominated much of the middle part of the twentieth century (Ayer, 1946; Carnap, 1950, 1967; Hempel, 1965). The logical positivists were heirs to the classical empiricist tradition in philosophy of science and theory of knowledge – believing that all enquiry should be grounded in observation, and that the scientific task is essentially that of accommodating existing observations while correctly predicting new ones. They also believed that one central paradigm of scientific enquiry is enumerative induction. A pattern is discerned in our observations (each raven so far observed has been black), and then generalized into a universal law (all ravens are black). Much intellectual effort was expended in attempts to justify our inductive practices, and in discussion of the problem of underdetermination of theory by data (there are always, in principle, infinitely many distinct theories – that is, generalizations of the data – consistent with any finite data-set).

One stalwart and long-standing critic of logical positivism was Popper (1935, 1963, 1972). Popper pointed out that induction can only *generalize from* observation, whereas science characteristically goes *beyond* experience, by postulating theoretical entities such as electrons and X-rays, for example, which might explain it. The method of science, Popper argued, is not observation and induction, but rather non-algorithmic theory-construction followed by testing. Scientists invent theories to *explain* their data, using imagination, analogy, intuition and any other resources which may happen to come to hand. (And since explanation isn't easy, scientists will generally be satisfied if they can construct just *one* explanatory theory.) But then, having devised a theory, they subject it to rigorous testing – deriving from it predictions concerning the observations which might be expected if the theory were true, and attempting to falsify the theory by making those observations which seem least likely. A theory is justified when it escapes falsification, on this view.

Up to this point, philosophy of science had been conducted in a relatively *a priori* fashion – with some reference to real scientific examples, admittedly, but mostly with philosophers of science just thinking about what scientists *ought* to do, rather than about what they actually *do* do. This all began to change in the 1960s and 1970s, when philosophy of science took its so-called 'historical turn', through the work of Kuhn (1962), Feyerabend (1970, 1975) and Lakatos (1970).

2.2 *Historical and naturalistic turns*

As Kuhn and others noticed, when one studies the history of science one discovers that the behaviour of actual scientists often fails to conform to the norms of scientific method laid down by philosophers of science. In particular, when scientists know of data inconsistent with their theories, they do not immediately abandon those theories and start again, as Popper would have had them do. Sometimes they try to explain away the recalcitrant data while continuing to hold onto their theory; but as often as not they simply ignore it, and get on with the business of developing their favoured theoretical approach. This gives rise to a dilemma for *a priori* philosophers of science. Either they can claim that the actual practice of scientists has been irrational, or at least inappropriate – in which case the immense success of science is rendered utterly mysterious. (How *can* science be so successful if scientists have mostly been doing it all wrong?) Or they can take the historical results as a refutation of their proposed methodologies. Almost all philosophers of science converged on the latter course – and wisely so, surely.

At about the same time as, and not unrelated to, the historical turn in philosophy of science, much of philosophy was undergoing a 'naturalistic turn'. This took place in epistemology and philosophy of mind generally, as well as in the philosophy of science in particular. Most philosophers started to accept as a serious constraint on their theorizing, that both human mental processes and human modes of acquiring knowledge are *natural*, happening in accordance with causal laws as do all other events in nature; and that philosophical attempts to achieve an understanding of the nature of these processes should be seen as continuous with scientific enquiry. This resulted in a plethora of causal theories of mental and epistemic phenomena – with the development of causal theories of reference (Kripke, 1972; Putnam, 1975), of memory (Martin and Deutscher, 1966; Locke, 1971), of perception (Grice, 1961; Dretske, 1969; Goldman, 1976), of knowledge (Armstrong, 1973; Dretske, 1981; Goldman, 1986), and of justification (Goldman, 1979, 1986). Indeed, the dominant theory of the overall nature of the mind, which rose into ascendancy during this period, was functionalism, which sees mental states and events as individuated by their characteristic causal roles (Putnam, 1960, 1967; Lewis, 1966; Armstrong, 1968 – see section 3 below).

It became important, then, to see science, too, as a natural phenomenon, somehow recruiting a variety of natural processes and mechanisms – both cognitive and social – to achieve its results. Philosophers of science began to look, not just to history, but also to cognitive psychology in their search for an understanding of scientific activity (Nersessian, 1984b, 1992a; Giere, 1988, 1999a; Thagard, 1992). This trend is continued into the present volume, with a number of its philosophical authors appealing to psychological models or data in their chapters (e.g. Carruthers, chapter 4; Faucher *et al.*, chapter 18; Giere, chapter15; Nersessian, chapter 7).

2.3 Science and the social

Our story so far has mostly been one of good news – with philosophy of science in the last century, like science itself, arguably progressing and/or getting somewhat closer to the truth. But one out-growth of the historical turn in philosophy of science was a form of social constructivism and relativism about science (Bloor, 1976; Rorty, 1979; Latour and Woolgar, 1986; Shapin, 1994). On this view, scientific theories are proposed, accepted and rejected in accordance with a variety of political and social forces which needn't have any connection with truth or reality, nor with reliable standards of evidence and rationality. Indeed, on this account the very idea of 'reality' as something set over and against our various socially constructed perspectival representations is unintelligible. The only real sense which can be made out of one theory being better or worse than another is in terms of its political–social influence.

While social constructivism has not found wide acceptance among philosophers of science generally (nor among the contributors to this volume in particular), it has perhaps played a useful role in emphasizing the social dimension of science and scientific activity. And one characteristic of recent work has been to combine elements from the social constructivist position with an overall realism about science – allowing that science progresses through varied social interactions among scientists, while arguing that those interactions can still be such as to facilitate increases in knowledge of the world (Kitcher, 1993, chapter 14 in this volume; Goldman, 1999; Thagard, 1999).

If one had to characterize the current state of play in contemporary philosophy of science – using very broad brush-strokes, of course – it would be that it is naturalistic and broadly realist in orientation; interested in both descriptive and explanatory accounts of scientific activity (historical, psychological and sociological); but also concerned with normative issues concerning the probity and/or reliability of various scientific practices. That picture pretty much represents, too, the orientation of most of the contributions to this volume.

3 Philosophy of mind: another short recent history

Since this book is as much about the mind as it is about science, in this section we provide a brief overview of the main developments to have taken place in the philosophy of mind over the last half-century or so.

3.1 Behaviourism and beyond

If our story of recent philosophy of science began with logical positivism, then our account of recent philosophy of mind has to begin with logical behaviourism, which also dominated much of the middle part of the twentieth century (Ryle, 1949; Wittgenstein, 1953). The leading idea of behaviourism is that it is a mistake to treat talk about the mental as talk about inner causes of overt behaviour. To think in this way, according to Ryle, is to commit a kind of *category-mistake*. Talk about the mental is not talk about mysterious inner causes of behaviour, but is rather just a way of talking about dispositions to behave and patterns of behaviour.

Behaviourism did have some attractions. It allowed humans to be included smoothly within the natural order by avoiding postulation of anything 'ghostly' inside the organic machinery of the body. It was thus able to reject any sort of ontological dualism, between non-physical minds and physical bodies. For the main objection to such a dualism has always been the problem of explaining how there can be any sort of causal commerce between the states and events of a non-physical mind and those of a physical brain or body.

The deficiencies of logical behaviourism were even more apparent, however. There were two main problems. One is that it seems quite implausible that knowledge of one's own mind could consist in knowledge of one's behavioural dispositions, since this hardly leaves any room for the idea of first-person authority about, or any kind of privileged access to, one's own thoughts and feelings. (Hence the old joke about the two behaviourists who meet in the street – 'You're feeling fine', says one, 'But how am I?')

The other major deficiency is this: logical behaviourism was offered as a piece of conceptual analysis. It was supposed to be an account of what had all along been the import of our psychological discourse. That being the Rylean stance, a serious criticism of logical behaviourism is that it fails on its own terms, as an exercise in analysis. According to behaviourism what look like imputations of internal mental events or states should actually be construed as 'iffy' or conditional statements about people's actual and possible behaviour. The objection to the pretensions of behaviourist conceptual analysis, then, is that nobody has ever actually produced a single completed example of the behavioural content of such an analysis.

Indeed, there are principled reasons why no such behavioural analysis can be provided. For as Davidson (1970) pointed out, a particular belief or desire only issues in action together with, and under the influence of, other intentional states of the agent. There is no way, therefore, of saying what someone who holds a certain belief will do in a given situation, without also specifying what other beliefs and desires that agent holds. So analysis of a belief or a desire as a behavioural disposition requires invoking other beliefs and desires. This point has convinced practically everyone that Ryle was wrong. A belief or a desire does not just consist in a disposition to certain sorts of behaviour. On the contrary, our common-sense psychology construes these states as internal states of the agent which play a causal role – never singly, but always at least jointly – in *producing* behaviour.

3.2 Physicalism and functionalism

With dualism and logical behaviourism firmly rejected, attempts since the 1960s to give a philosophical account of the status of the mind have centred on some combination of physicalist *identity theory* with *functionalism* of one or another sort.

There are two distinct versions of identity theory which have been the focus of philosophical debate – *type-identity* theory and *token-identity* theory. According to the former, each type of mental state is identical with some type of brain state – for example, pain is the firing of C-fibres. According to token-identity theory, in contrast, each particular mental state or event (a 'token' being a datable particular rather than a type) is identical with some brain state or event, but it allows that individual instances of the same mental type may be instances of different physical types.

Type-identity theory was first advocated as a hypothesis about correlations between sensations and brain processes which would be discovered by neuroscience (Place, 1956; Smart, 1959; Armstrong, 1968). Its proponents claimed that the identity of mental states with brain states was supported by correlations which were just starting to be established by neuroscience, and that this constituted a scientific discovery akin to other type-identities, such as *heat is molecular motion*, *lightning is electrical discharge* and *water is* H_2O.

Most philosophers rapidly came to think that the early confidence in such type-correlations was misplaced, however. For consider a sensation type, such as pain. It might be that whenever *humans* feel pain, there is always a certain neurophysiological process going on (for example, C-fibres firing). But creatures of many different Earthly species can feel pain, and it is also possible that there are life-forms on other planets which feel pain, even though they are not closely similar in their physiology to any terrestrial species. So, quite likely, a given type of sensation is correlated with lots of different types of neurophysiological state. Much the same can be argued in the case of beliefs, desires and other mental kinds.

The conclusion drawn from these considerations was that type-identity theory is unsatisfactory, because it is founded on an assumption that there will be one-one correlations between mental state types and physical state types. Rather, we should expect mental state types to be *multiply-realized* in physical states. This is just what the thesis of token-identity affirms: each token mental state is identical with some token physical state; but instances of the same mental state type can be identical with instances of different physical types.

At about the same time, and connected with these debates concerning mind–brain identity, *analytic functionalism* was proposed as an account of the manner in which we conceive of mental states. The guiding idea behind functionalism is that some concepts classify things by what they *do*. So transmitters transmit, while aerials are objects positioned to receive air-borne signals; and wings are limbs for flying with, while eyes are light-sensitive organs for seeing with, and genes are biological structures which control development. Similarly, then, it was proposed that mental concepts are concepts of states or processes with a certain *function*, or distinctive causal role (Putnam, 1960, 1967; Lewis, 1966).

Functionalism seemed to be the answer to several philosophical prayers. It could account for the multiple realizability of mental states, since physiological states of a number of distinct types could nevertheless share the same causal role. And it had obvious advantages over behaviourism, since it accords much better with ordinary intuitions about causal relations – it allows mental states to interact and influence each other, rather than being directly tied to behavioural dispositions. Finally, it remains explicable that dualism should ever have seemed an option. For although (according to functionalists) we conceptualize mental states in terms of causal roles, it can be a contingent matter what actually *occupies* those causal roles; and it was a conceptual possibility that the role-occupiers might have turned out to be composed of some sort of *mind-stuff*.

There were two main problems with analytical functionalism, however. One is that it is committed to the analytic–synthetic distinction, which many philosophers think (after Quine, 1951) to be unviable. And it is certainly hard to decide quite *which* truisms concerning the causal role of a mental state should count as analytic (true in virtue of meaning), rather than just obviously true. (Consider examples such as: that *belief* is the sort of state which is apt to be induced through perceptual experience and liable to combine with *desire* to generate action; that *pain* is an experience frequently caused by bodily injury or organic malfunction, liable to cause characteristic behavioural manifestations such as groaning, wincing and screaming; and so on.)

Another commonly voiced objection to functionalism was that it is incapable of capturing the felt nature of conscious experience (Block and Fodor, 1972; Nagel 1974). Objectors have urged that one could know everything about the functional role of a mental state and yet still have no inkling as to *what it is like to be in that state* – its so-called *quale* or subjective *feel*. Moreover, some mental states seem to be conceptualized purely in terms of feel; at any rate, with beliefs about causal role taking a secondary position. For example, it seems to be just the feel of pain which is essential to it (Kripke, 1972). We seem to be able to imagine pains which occupy some other causal role; and we can imagine states having the causal role of pain which are not pains (which lack the appropriate kind of feel).

3.3 Theory-theory

In response to such difficulties, many have urged that a better variant of functionalism is *theory-theory* (Lewis, 1970, 1980; Churchland, 1981; Stich, 1983; Fodor 1987). According to this view, mental state concepts (like theoretical concepts in science) get their life and sense from their position in a substantive *theory* of the causal structure and functioning of the mind. To know what a belief is (to grasp the concept of belief) is to know sufficiently much of the theory of mind within which that concept is embedded. All the benefits of analytic functionalism are preserved. But there need be no commitment to the viability of an analytic–synthetic distinction.

What of the point that some mental states can be conceptualized purely or primarily in terms of feel? A theory-theorist can allow that we have *recognitional capacities* for some of the theoretical entities characterized by the theory. (Compare the diagnostician who can recognize a cancer – immediately and without inference – in the blur of an X-ray photograph.) But it can be claimed that the concepts employed in such capacities are also partly characterized by their place in the theory – it is a *recognitional* application of a *theoretical* concept. Moreover, once someone possesses a recognitional concept, there can be nothing to stop them prizing it apart from its surrounding beliefs and theories, to form a concept which is *barely* recognitional. Our hypothesis can be that this is what takes place when people say that it is conceptually possible that there should be pains with quite different causal roles.

Some or other version of theory-theory is now the dominant position in the philosophy of mind (which is not to say that there are no difficulties, and no dissenting voices, of course). And in many of its forms, theory-theory is of-a-piece with the sort of naturalism in philosophy which holds that philosophical and scientific enquiries are continuous with one another. From this perspective, both philosophy of mind and cognitive psychology are engaged in fundamentally the same enterprise – to characterize the nature and operations of the human mind.

4 Developments in developmental psychology

In this section we once again provide a very brisk recent history, this time in respect of developments in developmental psychology.

4.1 The Piagetian account

Piaget claimed that children's initial knowledge of relations of cause and effect is limited to what they see (in this respect his position was close to that of some logical positivists). In his early work he characterized their ideas about causality as restricted by 'syncretism' – in a word, by the tendency to connect everything with everything else (Piaget, 1928, p. 4). If asked to complete the beginning of a sentence such as, 'The man fell off his bicycle because . . . ', children under five or six years will respond with, 'Because he broke his arm' rather than, say, 'Because he lost his balance.' On this basis, Piaget denied that young children are able to detect causal relations. In later work with Inhelder (Inhelder and Piaget, 1958), Piaget reiterated his view that children's scientific cognition is muddled and chaotic, and that their beliefs about events are juxtaposed together instead of causally linked.

Not only did Piaget judge young children to be incapable of identifying causal relations clearly, but he contended that they assign internal states and motives to inanimate objects. For example, they believe that inanimate objects – especially those that move – can possess consciousness and have sensations and emotions just as persons do. Piaget (1928, 1929) interpreted children's answers to questions about the movement and feelings of objects to indicate that their notions of causality are primitive, and reflect an inability to reason about the physical world.

According to the Piagetian analysis, moreover, the causal understanding of young children points to a suite of domain-general processes which underpin the nature of early cognitive development. Children's understanding across domains as diverse as geometry, physics and biology constitute a 'structured whole' in the sense that they share common properties in reasoning and problem-solving. What children know is tied to appearances rather than involving underlying transformations and causal mechanisms which aren't visible to the eye. In conservation experiments, for example, young children typically believe that when water is poured from a short, wide glass into a tall, narrow one, the amount has changed even though nothing has been added or subtracted. Such children are also supposed not to understand the difference between animate and inanimate objects – believing, for example, that a shadow is a substance emanating from an object but participating with the night; and they attribute animistic qualities to leaves in projecting shadows. Only with increasing age do children know how shadows are projected and deny that objects cast shadows at night. Similarly, Piaget proposed that young children misunderstand the nature of dreams; they believe that dreams originate from outside and remain external to the self.

In later childhood, after the age of seven years, children can use transformations and invisible mechanisms in their causal reasoning. However, not until they achieve a formal operational understanding in early adolescence do they systematically test hypotheses, on a Piagetian approach. While much of this theoretical apparatus has been rejected in later work, many developmental psychologists continue to share Piaget's view that children's understanding undergoes radical conceptual change over the course of development (Carey, 1985; Wellman, 1990; Perner, 1991; Gopnik and Meltzoff, 1997).

4.2 Modern evidence for early causal understanding

We now know that Piaget significantly underestimated children's capacity for causal understanding. Even on the most sympathetic evidence, children's knowledge should be seen as variable rather than as constrained by domain-general stages (Siegler, 1994). It is now well documented that, although no single factor can fully explain children's inability to conserve, their responses on conservation tasks have much to do with the child's perception of the relevance and purpose of the context in which questions are asked (Donaldson, 1978; Siegal, 1997, 1999). Similarly, young children seem surprised when inanimate objects appear to move by themselves, or when they unaccountably appear or disappear (see the reviews by Carey, 2000b, and Rakison and Poulin-Dubois, 2001). Moreover, children as young as three years old have been shown to be very adept in distinguishing real from pretend objects and events in a simplified testing procedure where they are asked to sort items into those which are real (can be seen, touched and acted upon) and those which are not (Wellman and Estes, 1986; Leslie, 1994a).

Young children can also use causal knowledge in the fundamental process of naming and classifying artefacts. In a recent demonstration, Gelman and Bloom (2000) asked preschool children to name a series of simple objects. In one condition, the objects were described as purposefully created (for example, 'Jane went and got a newspaper. Then she carefully bent it and folded it until it was just right. Then she was done. This is what it looked like.'); in another, the objects were described as having been created accidentally ('Jane was holding a newspaper. Then she dropped it by accident, and it fell under a car. She ran to get it and picked it up.'). Even three year-olds were more likely to provide artefact names (e.g. 'hat') when they believed that the objects were intentionally created, and material-based descriptions (e.g. 'newspaper') when they believed that the objects were created accidentally. All of this work points to an early capacity for causal understanding underlying children's ability to classify objects in the physical world.

4.3 Childhood science?

How, then, can we characterize children's causal understanding of the world – that is, children's 'science'? One proposal is that the child begins with naïve theories of the world (different in different domains of activity) which undergo transformation when encountering contrary evidence. This 'theory-theory' account claims that the growth of children's scientific understanding is not dissimilar from that of adult scientists who revise their theories in the light of contrary evidence (Gopnik and Meltzoff, 1997). As Gopnik and Glymour (chapter 6 in this volume) put it, 'The assumption behind this work is that there are common cognitive structures and processes, common representations and rules, that underlie both everyday knowledge and scientific knowledge.' On this account, children's scientific knowledge is built out of initial innate theories, which are then altered and fundamentally restructured in the light of observational evidence, by means of processes similar to those employed in scientific reasoning.

Note, however, that this 'theory-theory' account of childhood cognitive development is not the same as the 'theory-theory' account of our understanding of mental-state concepts, discussed in section 3 above. Philosophical theory-theorists, while agreeing with these developmentalists that the end-point of normal development is a theory of the workings and causal structure of the human mind, are not committed to any particular mechanism for reaching that end-point. On the contrary, they can be modularists (see below), or they can believe that the fundamental developmental process is one of *simulation* of the minds of others (Goldman, 1993 – note, however, Goldman himself isn't a theory-theorist about the end-point of development), or whatever.

Similarly, 'modular' accounts of development, too, propose that children have mechanisms to detect causal relations from the start (Scholl and Tremoulet, 2000). These form the rudiments of scientific understanding. Thus children have core knowledge in areas such as number and syntax which provide a structural foundation for the development of scientific understanding (Macnamara, 1982; Bloom, 2000; Wynn, 2000). But the process which carries children from their initial knowledge to their later understanding, in these domains, is not (or not fundamentally) one of theorizing, but rather one of biological maturation, on a modularist account.

Related considerations have to do with the unevenness of the process of development. For theory-theorists agree with Piaget, at least to the extent of thinking that the basic engine of cognitive development is domain-general, even if that engine has to work with materials of varying sophistication and complexity in different domains. Modularists, in contrast, will hold that the factors influencing cognitive development may vary widely across domains.

Nevertheless, on either theory-theory or modular accounts, children have a head-start in their causal understanding of the world. The questions are what – if anything – needs to be added in order to get genuine scientific thinking emerging; and how much of the initial cognitive beliefs and reasoning processes which are at work in childhood survive into mature scientific thought.

5 Recent psychology of reasoning

In this section we briefly review some of the recent psychological work on reasoning – and especially scientific reasoning, or reasoning related to science – in adults.

5.1 Stage one: three traditions

During the heyday of behaviourism, from the 1930s until about 1960, there was relatively little work done on human reasoning. In the decades after 1960, reasoning and problem-solving became increasingly important topics for experimental investigation, leading to the rise of three main research traditions, as follows.

(1) Simon, Newell and their many students looked at problem-solving in tasks like the Tower of Hanoi and 'Crypt-arithmetic' (Newell and Simon, 1972). (Sample problem: Assign numbers to letters so that SATURN + URANUS = PLANETS.) The normative theory which guided this work was based on Newell and Simon's 'General Problem Solver' (GPS) – a computer program for searching the problem space. The empirical work in this tradition seemed to indicate that people had the basic idea of the GPS, although they could easily get lost in the book-keeping details. Simon labelled the increasingly elaborate normative theory which emerged from this work 'the logic of discovery', and in later work he and his collaborators tried to simulate various episodes in the history of science (Langley *et al.*, 1987; Kulkarni and Simon, 1988).

(2) In the UK, Wason (and later both Wason and Johnson-Laird), focused on the selection task and related problems (Wason, 1960; Wason and Johnson-Laird, 1970; Evans, Newstead and Byrne, 1993). The normative theory guiding this work was Popper's falsificationism. The empirical results showed that ordinary subjects were quite bad at using the strategy of seeking evidence which would falsify the hypothesis at hand.
(3) Starting a bit later (in the late 1960s and early 1970s), Kahneman and Tversky began looking at various sorts of probabilistic reasoning (Kahneman, Slovic and Tversky, 1982). The normative theory they relied on was Bayesian. They made a great splash by finding a significant number of problems (base-rate problems, 'conjunction fallacy' problems, overconfidence problems and a host of others) on which subjects did very poorly indeed.

5.2 *Reactions*

The reactions and criticisms to these three lines of research on human reasoning were very different.
(1) A number of people argued that the GPS strategy was much too simple for explaining how science actually works (e.g. Giere, 1989). But since the normative theory was open-ended, Simon and his collaborators often took the criticisms on board and sought to construct more complex problem-solving programs which could do a better job of simulating discoveries in the history of science. The approach has remained steadfastly individualistic, however, and thus it is often criticized by historians and sociologists of science for neglecting the social aspects of scientific discovery.
(2) In response to the Wason and Johnson-Laird tradition, reactions were of two sorts. First, there was a great deal of experimental work looking at variations on the theme of the original selection-task experiments. Perhaps the best-known findings indicated that there are quite massive 'content effects' – there are *some* selection-task problems on which people do quite well. Many different explanations have been offered in an attempt to explain these content effects. The best known of these are the 'pragmatic reasoning schemas' proposed by Cheng and Holyoak (1985) and the social contract account first put forward by Cosmides (1989). (In addition, Sperber and colleagues have proposed a pragmatic explanation, based on relevance theory – see Sperber, Cara and Girotto, 1995.) Cosmides' work has been widely discussed in recent years because it is embedded within the theoretical framework of evolutionary psychology. The versions of the selection-task on which people do well, Cosmides argues, are just those that trigger special-purpose reasoning mechanisms which were designed

by natural selection to handle problems that would have been important for our hunter–gatherer forebears.

The second sort of reaction to the Wason/Johnson-Laird tradition was to challenge the Popper-inspired normative theory which had been assumed in analysing the original studies. Anderson's 'rational analysis' account was an influential first move along these lines (Anderson, 1990), and more recently, Chater, Oaksford and others have followed his lead (Oaksford and Chater, 1994). Moreover, Koslowski (1996) has argued in detail that once people's performance in causal and scientific reasoning tasks is interpreted in the light of a scientific realist perspective (particularly with the latter's commitment to underlying mechanisms), then much of that performance can be seen as normatively sensible and appropriate.

(3) Perhaps because some of the major figures in the Kahneman and Tversky 'heuristics and biases' tradition made a point of stressing that their findings had 'bleak implications' for human reasoning, this work attracted the most attention and provoked a number of different lines of criticism. Some critics focused on alleged shortcomings in the experiments themselves – noting, for example, that the problems might be interpreted in ways that the experimenters did not intend (Adler, 1984). Cohen argued, in contrast, that the experiments could not possibly show that humans had an irrational reasoning 'competence', and thus that the results were at best the result of performance errors (Cohen, 1981).

Gigerenzer and others have mounted a sustained and very interesting attack on the normative assumptions that the heuristics and biases experimenters make (Gigerenzer, 2000). In case after case, he has argued that (for various reasons) it is far from clear *what* the 'right' or rational answer is to the questions posed to subjects. One theme in this critique has been that many statisticians favour a frequentist interpretation of probability; and on that interpretation, many of the heuristics and biases problems have *no* correct answer because they ask about the probabilities of single events not relativized to a reference class. Gigerenzer and his collaborators have carried out elegant experiments to demonstrate that reformulating some of the heuristics and biases problems in terms of frequencies rather than single event probabilities can dramatically improve performance.

In recent years, Gigerenzer has put an evolutionary spin on these results, claiming that our minds evolved to deal with probabilistic information presented in the form of 'natural frequencies' (Gigerenzer, Todd and the ABC Research Group, 1999; Gigerenzer, 2000). He has joined forces with Cosmides and Tooby and other evolutionary psychologists to argue that our performance on many of the heuristics and biases problems can be explained by the fact that they are not the sorts of problems that our minds evolved to deal with.

5.3 *An emerging synthesis?*

In the last few years, a number of people, notably Evans and colleagues (Wason and Evans, 1975; Evans and Over, 1996), and more recently Stanovich (1999), have proposed 'dual-processing theories' which may accommodate the findings (and many of the theoretical arguments) from both those within the heuristics and biases tradition, and their critics. On this account, reasoning is subserved by two quite different sorts of system. One system is fast, holistic, automatic, largely unconscious and requires relatively little cognitive capacity. The other is relatively slow, rule-based, more readily controlled and requires significantly more cognitive capacity. Stanovich speculates that the former system is largely innate and that, as evolutionary psychologists suggest, it has been shaped by natural selection to do a good job on problems similar to those which would have been important to our hominid forebears. The latter system, by contrast, is more heavily influenced by culture and formal education, and is more adept at dealing with the problems posed by a modern, technologically advanced and highly bureaucratized society. This new, slow system is largely responsible for scientific reasoning. Stanovich also argues that much of the individual variation seen in heuristics and biases tasks can be explained by differences in cognitive capacity (more of which is required for the second system), and by differences in cognitive style which lead to different levels of inclination to *use* the second system.

Important questions for our understanding of science to emerge out of this new synthesis include the extent to which scientific reasoning is influenced by the implicit system, and how scientific practices can control for or modify such influences. (See Evans, chapter 10 in this volume, for a discussion of these issues.)

6 Themes and connections: a guide through the volume

In this section we will say just a few words about each of the chapters in this volume, picking up and identifying a number of recurring themes and issues as we go along.

6.1 *Part one*

The first three chapters in the main body of the book all relate in one way or another to the question of the innate basis of scientific reasoning.

Mithen (chapter 2) distinguishes some of the key elements of scientific thinking, and then traces the evidence of their emergence from the early hominines of some 4 million years ago to the first human farming communities of 11,500 years ago. Mithen emphasizes that the cognitive components of scientific reasoning are multiple (cf. Dunbar, chapter 8), and he sketches how they probably appeared at different stages and for different reasons in the course of human evolution.

Atran (chapter 3) explores the universal cognitive bases of biological taxonomy and taxonomic inference, drawing on cross-cultural work with urbanized Americans and forest-dwelling Maya Indians. His claim is that there is a universal, essentialist and innately channelled appreciation of species as the causal foundation for the taxonomic arrangement of life forms, and for inferences about the distribution of causally-related properties of living beings.

Carruthers (chapter 4) examines the extent to which there are continuities between the cognitive processes and epistemic practices engaged in by human hunter–gatherers, on the one hand, and those which are distinctive of science, on the other. He argues that the innately channelled architecture of human cognition provides all the materials necessary for basic forms of scientific reasoning in older children and adults, needing only the appropriate sorts of external support, social context and background beliefs and skills in order for science to begin its advance. This contradicts the claims of those who maintain that 'massive reprogramming' of the human mind was necessary for science to become possible (Dennett, 1991).

6.2 Part two

The seven chapters which make up part two of the book examine aspects of contemporary scientific cognition in both children and adults.

Varley (chapter 5) reports the results of a series of experiments testing the folk-scientific abilities of two severely a-grammatic aphasic men. Since the capacities of one of the two men were near-normal, Varley argues that core components of scientific reasoning, at least – specifically, hypothesis generation and testing, and reasoning about unseen causes and mechanisms – must be independent of language. This is an important result, since it is often claimed that scientific thinking is dependent upon language (Dennett, 1991; Bickerton, 1995).

Gopnik and Glymour (chapter 6) take their start from 'theory-theory' accounts of child development, according to which children propose and revise theories pretty much in the manner of adult scientists. They argue that the theories in question are best understood as 'causal maps', and that the recent development of computational 'Bayes nets' may provide the resources for us to understand their formation and change, in both science and child development.

Nersessian (chapter 7) makes a strong plea for enquiries into historical episodes in science to be seen as one important source of understanding of the cognitive basis of science. Her focus is on the cognitive basis of the model-based reasoning practices employed in creative thinking, and on the way in which these can lead to representational change (or 'conceptual innovation') across the sciences.

Dunbar (chapter 8) reports and discusses the results of a series of *In Vivo* studies of science, in which the reasoning processes of scientists were observed and recorded 'on line'. He argues that these results support a pluralist conception of scientific activity, in which types of cognitive process which are used elsewhere in human activity are deployed in distinctive patterns and sequences in the service of particular goals, and with different such patterns occurring in different sciences and in different aspects of scientific activity. Dunbar also points to evidence of cross-cultural variation in scientific reasoning practices (cf. Faucher *et al.*, chapter 18).

Koslowski and Thompson (chapter 9) emphasize the important role of collateral information ('background knowledge') in scientific reasoning, both in proposing and in testing scientific theories. This role has generally been ignored (and often explicitly 'factored out') in psychological studies apparently demonstrating that naïve subjects are poor scientists. With the role of collateral information properly understood, Koslowski and Thompson present evidence that school-age children are able to reason quite appropriately in experimental contexts.

Evans, too (chapter 10), is concerned with the effects of background belief on reasoning, but from a different perspective. He reviews the experimental evidence of various forms of 'belief bias' in people's reasoning. He discusses the implications of this data for scientific practice, drawing conclusions which enable him (like Koslowski and Thompson) to be generally sanguine about the prospects for a positive assessment.

Finally in part two of the book, Hilton (chapter 11) is also concerned with the psychological evidence of irrationality in people's reasoning, this time in their reasoning about causality. While somewhat less optimistic in his approach than the previous two chapters, he does think that a significant proportion of this data can be explained in terms of (perfectly sensible and appropriate) pragmatic factors, and that those irrationalities which remain can be mitigated by appropriate social arrangements (a point which is also made by Evans, and which is further emphasized by Kitcher, chapter 14).

6.3 Part three

The three chapters making up part three of the book all share a concern with the place of motivation and emotion within science. Traditionally, science (like human reasoning generally) has been seen as a passionless enterprise, in which emotion can only interfere. Scientists are supposed to be *dispassionate enquirers*. The separation of reason from emotion has come under vigorous attack (Damasio, 1994). And all three of the contributions in part three of the book continue that critique, focusing on scientific reasoning especially.

Thagard (chapter 12) argues against prevailing models in the philosophy of science by claiming that emotional reactions are an integral and ineliminable part of scientific practice, in all three of the main domains of *enquiry*, *discovery* and *justification*. The emotions which mostly concern him are broadly truth-directed ones such as *interest*, *surprise* and an aesthetic–intellectual response to *beauty* ('elegance', 'simplicity', etc.).

Hookway (chapter 13) explores some of the ways in which epistemic emotions such as doubt and dogmatism may help as well as hinder the pursuit of knowledge, especially in the way that they help guide enquiry by making certain questions salient for us while leading us to ignore others.

Kitcher (chapter 14) is concerned with the influence of rather more mundane and materialistic motivations – such as a desire for fame or for prestigious prizes – on scientific practice. He sketches a research programme for investigating the impact of different social arrangements and inducements on the conduct of science, and a framework for thinking about how such arrangements should be assessed within democratic societies such as our own.

6.4 Part four

The four chapters in part four of the book are, in various ways, about the social dimension of scientific cognition. (This is an important theme in a number of other chapters as well, including those by: Carruthers, chapter 4; Dunbar, chapter 8; Evans, chapter 10; Hilton, chapter 11; and Kitcher, chapter 14.)

Giere (chapter 15) argues that science can be better understood if we notice the extent to which scientific cognition is *distributed*, incorporating many factors outside the minds of individual scientists. These would include scientific instruments, libraries, calculations conducted with the aid of computers and a variety of forms of social arrangement and social structure.

Siegal (chapter 16) critically examines the idea proposed by Carey (1985), that childhood development may involve stages of 'strong conceptual change' analogous to revolutionary periods in science. He concentrates on the domain of biology in particular (and there are a number of connections here with the ideas of Atran, chapter 3). Siegal suggests that for key aspects of biology, the evidence for conceptual change in the strong sense is inconclusive, and that children's understanding is both domain-specific and highly sensitive to the information they receive from their surrounding culture.

Harris (chapter 17) makes a powerful case for recognizing the importance of *testimony* in the development of childhood beliefs, contrasting this position with a view of the child as a 'stubborn autodictat' (a view quite common among developmental psychologists), which sees the child as a lone little scientist, gathering data and forming and testing hypotheses for itself.

Faucher *et al.*, too (chapter 18), mount an attack on the sort of 'child-as-scientist' models proposed by Gopnik and others, arguing that even if they are right about childhood they are wrong about science. And, like Harris, Faucher *et al.*, too, emphasize what children acquire from the surrounding culture during development. But Faucher *et al.*'s emphasis is also on the acquisition of *norms* from surrounding culture, especially norms concerning the gathering of evidence and the assessment of theories. They also discuss recent data suggesting that culture-specific norms can have a deep and lasting impact on the operations of our cognition.

7 Conclusion

These are exciting times, not only for science and the scientific understanding of science, but also for our understanding of the human mind. The philosophy and psychology of science and scientific practice – like science itself – continues to make progress, and to raise challenging new questions. It is our view that the chapters in this volume should contribute substantially to that continued advance.

Part one

Science and innateness

2 Human evolution and the cognitive basis of science

Steven Mithen

The capacity for science is likely to have a biological basis that evolved in a piecemeal fashion during the course of human evolution. By examining the anatomy and activities of human ancestors from the earliest *Homo* at 2.5 mya to anatomically modern humans engaged in farming at 8,000 BP, this chapter attempts to identify when, and in which order, several of the key facets of scientific practice emerged. While it recognizes that to emerge as a recognizable entity science required a particular suite of social and economic circumstances, that emergence can be fully understood only by examining both the abilities and the constraints of the evolved human mind.

1 Introduction

The cognitive basis of science needs to be explored from at least two different perspectives – the developmental and the evolutionary. These may be intimately connected. A nativist stance to the development of mind in the child argues that emergent modes of thought are significantly shaped by our evolutionary past (e.g. Pinker, 1997). This stance requires any study of the origin of science in the child to involve a concern for the thought, behaviour and environments of our human ancestors. That concern should involve going beyond the guess-work and just-so stories favoured by many evolutionary psychologists to examining the evidence itself – the fossil and archaeological records (Mithen, 1996).

Even if a nativist stance was thought quite inappropriate, the study of early prehistory remains an essential undertaking if one wishes to understand the cognitive basis of science. This is especially the case if one wishes to avoid the narrow definitions of science that restrict this to practices undertaken only from the end of the Middle Ages (e.g. Butterfield, 1957), or even to no more than the last 250 years (e.g. Cunningham and Williams, 1993). If we wish to search for the origins of either science or its cognitive foundations, we must look much further back into history, indeed into the time before written records. That is my intention in this chapter. But rather than looking back in time I will start with the earliest traces of humanity and rapidly move forward, identifying what might be critical evolutionary steps towards a scientific mode of thought and practice.

My key argument is that the 'capacity' to undertake science arose gradually during the course of human evolution, finally appearing as an emergent property from a suite of cognitive processes that had evolved independently within human ancestors. As such, science cannot be wholly understood as a cultural construct; neither can the capacity for science be entirely reserved for modern humans – several of our ancestors employed, and were perhaps reliant on, key elements of scientific thinking for their survival.

2 Key elements of science

To make this survey through time we need to have some idea of what we are searching for in terms of what constitutes the critical elements of scientific thought. As I would not be so presumptuous as to attempt a definition of science myself, the following might best be characterized as a 'popularist' notion of science rather than one with any philosophical basis. With that caveat, the following appear to be essential:

- Detailed and extensive observation of the natural world.
- The generation of hypotheses which have the potential for falsification.
- A concern with causation.
- The use of tools to extend human perception and cognition, tools which may be no more complex than a pencil and paper. Material culture facilitates – often enables – science in three ways: it can record, store and transmit information; it can be used as an adjunct to the human brain as in an aide memoire, a calculator or computer; it can extend perception as in the use of a telescope or a microscope.
- The accumulation of knowledge and advancement of understanding through time.
- The use of metaphor and analogy to facilitate scientific thought and to aid communication (Kuhn, 1979).

I must emphasize that I am not suggesting that all of the above features are necessarily essential to all types of scientific thought and practice. Moreover, I suspect that there are further dimensions of science (or some types of science) that I have not mentioned above. But rather than strive for a comprehensive definition I wish to focus on these key elements and examine when they may have appeared in the course of human evolution. To do so, I will group our human ancestors and relatives into four broadly consecutive categories: Early hominines, Early Humans, The Neanderthals and anatomically modern humans. I will divide the latter into two, late Pleistocene hunter–gatherers and early Holocene farmers. My discussion of each group will draw on analogies with either non-human primates or historically documented hunter–gatherers where appropriate.

3 Early hominines

By this group I am referring to the australopithecines and earliest members of *Homo* that are found in Africa between 4.5 mya and c.1.8 mya. Some members of this group, such as the *Paranthropus* species continue in the fossil record until 1.0 mya, but by that time our concern has shifted to species that are most likely direct ancestors of *Homo sapiens*. A general feature of this group – as far as can be ascertained with a fragmentary fossil record – is that they all possessed a substantial degree of bipedalism, although they are unlikely to have been habitual or dedicated bipeds (Johanson and Edgar, 1996). *A.afarensis* at 3.5 mya appears to have a joint arboreal and terrestrial adaptation, while also retaining anatomical features of a knuckle walking ancestor (Richmond and Strait, 2000).

The fossils of these early hominines display immense morphological diversity. They are broadly divided into robust and gracile forms, with the former often classified into their own genus of *Paranthropus*. Brain size varies between c.400 and c.750cc. Exactly how many species we are dealing with, and which specimen should belong to which species, are matters of substantial debate. Classification is made problematic by the often fragmentary nature of the fossil sample, the unknown extent of sexual dimorphism and other dimensions of intra-species variation, and a coarse grained chronology. Similarly, whether any of these fossils should be placed into the genus *Homo* can be questioned: those traditionally classified as *H. habilis* and *H. rudolfensis* might be more appropriately placed within the Australopithecine genus (Wood and Collard, 1999).

Debates also exist about the lifestyle of the early hominines. The robust forms appear to have evolved as specialized plant eaters with their massive jaws functioning as grinding machines. The gracile forms were more likely omnivorous, but the extent of meat within the diet, and how that meat was acquired, are highly contentious issues (Isaac, 1978; Binford, 1981, 1986; Bunn and Kroll, 1986). Meticulous studies of the animal bones found on Plio-Pleistocene archaeological sites have demonstrated some degree of hunting, but this is likely to have been a supplement to scavenging from carnivore kills. On the slim anatomical evidence that exists, all members of this group are likely to have been able to manufacture stone tools of an Oldowan type (Susman, 1991), but whether all had the cognitive capacity to do so, or the ecological needs for such tools, remains unclear. Traditionally tool making has been assumed to have been restricted to *H. habilis*, but there is no substantial academic basis for that claim. Indeed, Oldowan tools appear earlier in the archaeological record than the first traces of the fossils traditionally attributed to *Homo* (Wood, 1997).

The cognitive capacities of the early hominines are normally discussed in the context of those of the great apes. Indeed they are most effectively thought of as bipedal apes, behaving and thinking in manners far closer to modern chimpanzees and gorillas than to modern humans. Nevertheless, there are significant differences which may have implications for the eventual emergence of scientific thought in *Homo*.

The extent to which Oldowan technology differs from chimpanzee technology remains unclear. Some argue that there are no significant differences (Wynn and McGrew, 1989), while the failure of Kanzi to acquire the necessary skills to make Oldowan-like choppers (Toth *et al.*, 1993; Schick *et al.*, 1999) argues otherwise. The fact that early hominines carried either stone nodules or tools over distances an order of magnitude greater than that covered by anvil-carrying Tai chimpanzees (Boesch and Boesch, 1984) is, I think, important. Moreover, making stone flakes and choppers appears to involve hand–eye coordination and motor actions quite different to those used in other activities, whereas chimpanzee tool manufacture appears to be on a continuum with non-tool-assisted feeding activities (Mithen, 1996).

Flaking stone is fundamentally different to breaking or biting twigs, owing to its far greater degree of unpredictability about the end result. In this regard, striking one stone against another is a form of hypothesis testing – the hypothesis being the type of flake that will be removed. This is certainly the case with modern stone knappers who engage in the replication of pre-historic artefacts. When wishing to shape an artefact they will prepare a striking platform, select a hammer stone of a specific weight and density, strike the artefact at a specific angle and with a specific force, and then often inspect the flake and the remnant flake-scar. They will judge the extent to which it matched their expectations, and hence successfully shaped the artefact; differences between the expected and realized flake will be noted and will serve to modify their future knapping actions. This process is quite different to trial and error learning due to the predictive element, and one must surmise that it was also engaged in by pre-historic stone knappers. The consequences of working without this hypothesis testing approach can perhaps be seen in Kanzi's trial and error efforts at making stone artefacts (Toth *et al.*, 1993).

Hypothesis testing might also be an appropriate term to apply to the foraging activities of the early hominines. A key aspect of this after 2 mya appears to have been searching for animal carcasses in relatively open savannas, placing the hominines under severe risk of predation themselves (Isaac, 1984). Predator risk would have been minimized by using time efficiently, which would mean a non-random search for carcasses by predicting where in the landscape they would most likely be found. Potts (1988) has argued that caches of stones were made so that they would be readily available if a carcass needed butchering – in other words, the hominines were predicting where carcasses would be found and anticipating the level of predator risk.

Such behaviour may not be significantly different to the foraging tactics of many animal species. Indeed Dunbar (1995) has argued that all foraging animals essentially deploy hypothesis testing to acquire knowledge about the world, and that this key element of scientific thinking is widespread in the animal world. Similarly, the detailed observations that the early hominines are likely to have made about the natural world may not be different in kind to the observations that the great apes make about their world to maintain their detailed knowledge of their environment, especially the location and availability of edible plants.

One cause for caution, however, comes from the studies of vervet monkeys by Cheney and Seyfarth (1990) in their seminal volume *How Monkeys see the World*. They were able to show that the monkeys were quite ignorant or uninterested in what seem to humans as very obvious and important features of the natural world. Even though snakes and leopards are key predators, the monkeys were quite unaware of the implications of snake tracks, or of a fresh animal carcass in a tree left by a leopard. Hence we must be cautious about attributing hominines with the same understanding of 'signs' in the landscape that we, or modern hunter–gatherers, would possess. The latter certainly have an immense knowledge of the natural world, and are constantly making new observations. But that activity is often substantially underwritten by language in the sense that new observations, such as of an animal presence, ripe berries, or animal tracks are a major source of talk and used for social bonding (Mithen, 1990). Without the capacity for spoken language the drive for the same extent of observation may be lacking.

In summary, the early hominines most likely had some cognitive foundations for science, but these may have been no greater than those found in the great apes of today. I would attribute them with some capacity to make detailed observations of the natural world, to draw generalizations and inferences from these and to engage in prediction and hypothesis testing.

4 Early humans

Like the early hominines, this is a highly diverse group of human ancestors and relatives within which there is likely to be considerable variation of cognitive capacities. This group ranges from the 1.6 mya *H.ergaster*, as notably represented by WT-15000, the Nariokotome Boy (Walker and Leakey, 1993), to the late specimens of *H.heidelbergensis* as represented by those from the 'Pit of Bones' at Atapuerca dating to 300,000 years ago (Arsuaga *et al.*, 1993). I group these together for practical purposes alone, and will treat all these species as essentially pre-linguistic, even though some of the later members of this group may have evolved vocalizations far more complex than that found within the great apes today.

There are several general features to this group that distinguish them from the early hominines (Johansen and Edgar, 1996). They were habitually bipedal and had significantly larger brain sizes, ranging from 900cc to 1300cc. These two features may be evolutionarily related to each other, together with an increased extent of meat eating and more sophisticated tool manufacture. Increasing aridity in east Africa after 2 million years BP may have led to the spread of savannah and the selection of bipedalism as a means of thermal regulation (Wheeler, 1994). This in turn may have made possible the exploitation of a scavenging niche, while a higher quality of diet may have allowed a reduction in gut size and release of metabolic energy to fuel an expansion of the brain (Aiello and Wheeler, 1995). The selective pressures for brain enlargement are most likely to have come from increased sociality, another requirement for living in more open environments.

A further factor common to this group – and indeed all later humans – is that a distinctively modern human pattern of life-history is likely to have appeared (Key, 2000). The important differences between this and the life-history likely to have been experienced by the early hominines is an extended post-menopausal life span, secondary altriciality, and the appearance of a distinct period of childhood. By the latter is meant a period post-weaning when there is nevertheless a dependency on food provisioning by adults. This, in turn, will have required significant changes in social organization. This may have involved the protection and provisioning of females and their infants by males, and perhaps the development of pair bonding. Alternatively, an intensification of female kin networks has been proposed by O'Connell, Hawkes and Blurton-Jones (1999), in the form of the 'grand-mothering hypothesis'.

This transition to a human life-history pattern is likely to have occurred by the time of *H.ergaster*, and before the significant increase in brain size that occurs after 600,000 years ago. In this regard, the changes in social behaviour which provided the possibility for caring for relatively helpless infants appears to have been a necessary pre-condition for the expansion of the brain and a further extension of secondary altriciality.

The emergence of childhood is evidently a key development for the eventual appearance of scientific thought. This period of development, allowing for play, experiment and learning is essential in modern humans for acquiring culture, which may involve a scientific mode of thought. So the fact that a childhood phase of life-history was now present meant that a crucial threshold to the later emergence of science had been crossed.

This period of development may be related to the increased complexity of technology that emerges after 1.4 mya. As from that date bifacial technology, often in the form of handaxes, appears in the archaeological record. These tools require a significantly greater degree of planning and technical skill to manufacture than Oldowan choppers (Gowlett, 1984) and demonstrate the imposition of a pre-conceived form. Two other major behaviour developments must be noted. First is the dispersal from Africa. This appears to have occurred very soon after the appearance of *H.ergaster* in light of well-dated skulls from Dmanisi, Georgia, at 1.7 mya (Gabunia *et al.*, 2000). The *H.erectus* fossils from Java have been dated to 1.8/1.6 mya (Swisher *et al.*, 1994), although some contend that a date of 1.0 mya is more accurate. *H.heidelbergensis* was in Europe at 780,000 years ago as represented by the TD6 locality at Atapuerca (Carbonell *et al.*, 1995). These early dispersals from Africa appear to have occurred with the use of an Oldowan-like technology.

The second key behavioural development may be related to the dispersal into northern latitudes. This is big game hunting which has been demonstrated by the Schöningen spears, dating to 400,000 years ago (Thieme, 1997), and the faunal remains from Boxgrove at 500,000 years ago (Roberts, 1997). Although big game hunting may have occurred prior to these dates, the evidence from these sites provides the first direct and largely unambiguous evidence, and may be a reflection of new cognitive skills.

With regard to the evolution of the cognitive basis for science, the most significant development within the Early Humans is most likely to be a theory of mind and its resultant impact on their concern with causation. The extent to which a theory of mind is present within the great apes is highly contentious (Whiten, 1996; Povinelli, 2000), and even with much larger-brained early humans any evidence for this must be highly circumstantial. But as I have argued elsewhere (Mithen, 2000) the increased sociality, complex technology and co-operative hunting undertaken by the early humans imply a significantly enhanced theory of mind over that found within earlier members of *Homo* or the australopithecines. This was most likely an essential pre-condition for the evolution of a full language capacity.

Theory of mind is concerned with inferring people's beliefs and desires, which may be quite different to one's own. It is fundamentally related to understanding causation in terms of why another person acted in one way rather than another. Whether such concerns extended to the non-social world at this time is a moot point. The proposal (Mithen, 1996) that the early human mind was constituted by multiple, relatively isolated intelligences, suggests that it wasn't. And hence whereas one of these early humans may have asked questions about why another person behaved in one way or another, I doubt if questions were asked about natural events – why the sun rises, why it rains, why there is thunder, and what any of these might mean for the future. In other words, any concern for causation was restricted to the social world alone. Its extension beyond that was a critical step for the emergence of scientific thought and did not, I contend, take place within the early humans.

This argument is not incompatible with the further development of observation and hypothesis testing about the natural world that may have occurred among early humans. When hunting, these humans may have been interested in the location and behaviour of animals without asking exactly why they were found there or why they behaved in their species-specific ways – as any scientist knows, finding correlations is very useful but quite different from understanding causation. The evidence for big game hunting suggests the tracking of animals using traces such as footprints, scats, marks on vegetation and a multitude of other signs (see Mithen, 1990, for a review of information gathering by hunter–gatherers). As Carruthers (chapter 4 in this volume) has argued, this activity might be seen as being very close to a scientific mode of thought in terms of piecing together fragmentary bits of information, drawing inferences, developing hypotheses and testing these. We must be extremely careful, however, not to impose ethnographically documented tracking behaviour by *H.sapiens* on to pre-modern humans who may, as I have argued, have a quite different mentality. Only with the anatomically modern humans of the Upper Palaeolithic can we find direct evidence for such tracking behaviour contained with the imagery of their cave art (Mithen, 1988).

5 The Neanderthals

Homo neanderthalensis most likely evolved from *H.heidelbergensis* in Europe by 250,000 years ago. This species survived until a mere 28,000 years ago in Europe and western Asia, with its demise traditionally attributed to the competitive exclusion caused by the dispersal of anatomically modern humans (see reviews in Stringer and Gamble, 1993; Mellars 1996).

The Neanderthals lived through substantial climatic change including the last interglacial (OIS 5). Study of their anatomical remains have suggested a physiology that is distinctly cold-adapted and a lifestyle that involved considerable physical exertion. There are several anatomical indications that the Neanderthals had a vocal tract not significantly different to that of modern humans (Schepartz, 1993). As their brain size also covers the same range as modern humans it is difficult to sustain an argument that they lacked vocal skills of a level of complexity that should be described as language. Whether it was a language with an extensive lexicon and a set of grammatical rules remains unknown and subject to much debate.

A key element of their subsistence was the exploitation of larger herbivores such as bovids, horse and deer (Mellars, 1996). Meticulous studies of the faunal remains from sites such as Combe Grenal and the caves of west central Italy has documented hunting of these animals, most probably with the use of short thrusting spears – although the Schöningen discoveries (Thieme, 1997) suggest that throwing spears may also have been deployed. Some faunal assemblages suggest that scavenging took place. Chemical analysis of Neanderthal bones indicate a high meat content in the diet which accords with the predominantly tundra-like landscapes they exploited.

A wide variety of knapping techniques were employed by the Neanderthals, including several variants of the levallois technique. By making extensive use of re-fitting, Boëda (1988) has shown that this work often involved substantial skill, far exceeding that required to make bi-faces and no less complex than that employed by anatomically modern humans. Neanderthal tool making sometimes involved the production of long blades of a type once thought restricted to the Upper Palaeolithic, although these were manufactured using a quite different technique and appear to have been less standardized (Mellars, 1996). Evidence for the manufacture of tools from bone, or even the use of unmodified pieces, are extremely scarce and as many faunal assemblages are well preserved this cannot be accounted for by taphonomy alone.

As among the Early Humans any evidence for activities such as art, ritual and symbolism is both scarce and highly ambiguous. A series of incised bones from Bilzingsleben dating to c.300,000 BP provide the most persuasive examples of non-utilitarian marking (Mania and Mania, 1983), together with the supposed 'figurine' from Berekhat Ram of c.250,000 BP (D'Errico and Nowell, 2000). The latter is a 3 cm long stone nodule which has been incised using a flint tool, but whether its superficial resemblance to a female form was intentional, accidental or is one that exists merely in the eye of (certain) archaeologists alone remains unclear.

The later Neanderthals were certainly engaging in burial, although no more than a handful of convincing examples are known (Gargett, 1989). Unlike the first burials of anatomically modern humans, none of these appear to contain grave-goods and there is no evidence for graveside ritual. While burial may imply belief in supernatural beings and an afterlife, it may be no more than a reflection of strong social bonds and an extension of the evident care delivered to injured or diseased individuals during their lifetime.

The phenomenon of the Chatelperronian, c.35,000–30,000 BP, has been discussed and debated in immense detail (D'Errico *et al.*, 1998). This is an industry which evidently emerged from the Mousterian but has strong Upper Palaeolithic (i.e. anatomically modern human) affinities in both its lithic technology and the use of organic materials to make various types of beads and implements. The most widespread opinion is that it was manufactured by Neanderthals after they had come into contact with modern humans – it was a product of acculturation. D'Errico *et al.* (1998), however, argue that this interpretation cannot be sustained by the stratigraphic evidence and absolute dates. They propose that the Neanderthals independently invented both an Upper Palaeolithic-like technology and the use of body decoration, immediately before modern humans had arrived in Europe – a coincidence too great for many archaeologists to countenance.

With regard to the cognitive foundations for science, the Neanderthals provide further evidence to substantiate the claims made for Early Humans. Their tool making methods, hunting patterns and circumstantial evidence for social behaviour implies planning, hypothesis testing and meticulous observation of the natural world.

Language facilitated communication, especially of those beliefs and desires that may be difficult to transmit through gesture alone. Such verbal communication may have been essential to sustaining the complex tool making techniques in terms of transmitting these across generations. Verbal instruction, however, appears to play a minor role in craft apprenticeship among modern humans (Wynn, 1991) and may have played no role in the cultural transmission of Neanderthal technology. A persuasive argument can be made that language evolved to facilitate social interaction (Dunbar, 1996), and this may have remained its fundamental role among the Neanderthals (Mithen, 1996).

Perhaps the most significant feature of the Neanderthal archaeological record is its relative stability through time. Although there is substantial temporal and spatial variation in tool making techniques and subsistence behaviour no directional change through time can be identified. Some patterning with major climatic changes is apparent (e.g. Turq, 1992) but there seems to be no gradual accumulation of knowledge through time – one of the essential requirements for the practice of science. Hence those Neanderthals living at 50,000 years ago appear to have no greater store of knowledge or understanding of the world than those living at 250,000 years ago. As a contrast, I will shortly compare this to the dramatic changes in the knowledge of modern humans across not the 200,000- but the 10,000- (or even mere 2,000-) year period that occurred at the end of the last ice age 11,500 years ago – an environmental change equivalent to that which the Neanderthals experienced at 125,000 years ago.

There is one set of artefacts that might reflect the use of material culture as a form of recording device, a use that seems essential for the development of science. These are the incised bones from Bilzingsleben, one of which had psuedo-parallel lines cut into the edge. Unlike many other examples, it is difficult to imagine how these could have been created unintentionally (as in the course of cutting plants on a bone support) or for some utilitarian purpose. Not dissimilar artefacts from the Upper Palaeolithic are readily accepted as means for recording information – what D'Errico (1995) has described as 'Artificial Memory Systems'. Precisely what is being recorded (the passage of time, the number of animals killed, or people attending a ceremony) is generally thought to be of less significance than the fact that material culture is being used in this manner. The Bilzingsleben artefacts may have a merely fortuitous resemblance to these Upper Palaeolithic examples, however, most of which are rather more complex in design.

6 *Homo sapiens*

Genetic and fossil evidence indicates that the first anatomically modern humans had evolved in Africa by 130,000 years ago. This coincides with OI stage 6, a period of aridity which may have created the conditions for the speciation event creating *H.sapiens* (Lahr and Foley, 1998). The Omo Kibish specimen dates to 130,000 years ago, while fragmentary fossil remains from the lower levels of Klaisies River Mouth at the southern tip of South Africa date to c.125,000. By 100,000 *H.sapiens* are found in Western Asia at Qafzeh Cave, but that population seems unlikely to be the source of *H.sapiens* in Eurasia. Genetic evidence suggests a major demographic expansion at c.70,000 years ago which may also be related to climatic events and most likely resulted in wave of dispersal out of Africa. By 60,000 years ago *H.sapiens* had reached Australia, an achievement which involved a major sea crossing. Europe was colonized by *H.sapiens* at 40,000 years ago – their late arrival seeming to mimic the relatively late arrival of earlier hominids in that continent compared to elsewhere in the Old World (Gamble, 1993). The date at which the Americas were colonized is a major topic of debate. Monte Verde in southern Chile is the earliest authenticated settlement with a date of c.15,000 years ago. To arrive at Monte Verde at that time it seems likely that people arrived in North America by a coastal route, but whether that merely took them south of the ice sheets in California, or all the way to Panama or even further south is quite unclear (Dixon, 2001). It is not until c.13,500 years ago that archaeological sites become widespread throughout the Americas; when this happens people are evidently occupying a very wide range of environments, from the Amazonian rainforest to northern tundra. Some archaeologists take this to imply that the Americas had in fact originally been colonized prior to 20,000 years ago.

One consequence of the dispersal throughout the Old World is that pre-existing species of *Homo* were pushed into extinction, notably the Neanderthals in Europe and *H.erectus* in S.E. Asia. As those species had been occupying their regions for many thousands of years and appear to have had specialized physiological adaptations to their environments, the competitive success of *H.sapiens* with its own adaptations to equatorial environments is remarkable. This persuades many archaeologists that modern humans had not just a cultural or reproductive advantage over other types of humans, but a cognitive advantage which may have resided in a more complex form of language or a quite different type of mentality (Mithen, 1996). Support for the latter is readily evident from the dramatic developments that occur in the archaeological record relating to new ways of behaving and thinking by modern humans. Within this, we can see that further foundations have been laid for the eventual emergence of a scientific mode of thought.

The major developments in the archaeological record include the manufacture of a new range of stone tools, a greater use of organic materials, the manufacture of body decoration and the production of art objects. Such developments are used to define the Late Stone Age in Africa and the Upper Palaeolithic in Eurasia, and first appear at c.50,000 years ago in Africa. Bar-Yosef (1998) believes that these amount to a 'revolution' similar to the invention of agriculture and that a spread of this new culture can be monitored into Asia and then Europe – one possibly carried by a new population. Some believe that the developments at c.50,000 years ago are so dramatic that this must reflect a major cognitive change at that date – a position with which I have some sympathy – rather than coinciding with the emergence of *H.sapiens* at 130,000 years ago.

Evidence for cultural complexity among the very earliest *H.sapiens* is gradually appearing. The Middle Stone Age caves of South Africa contain a considerable amount of red ochre. This was most likely used for body painting, as no decorated cave walls or objects have been found (Watts, 1999). Some pieces of ochre have incised designs, as do small fragments of bone. In this light a symbolic culture appears to be present in South Africa by 70,000 years ago at least, and was probably coincident with the first *H.sapiens*. Moreover, bone harpoons have been recovered from Katanda, Zaire, and are believed to date to 90,000 years ago (Yellen *et al.*, 1995). These are similar in form to those found 60,000 years later in the Upper Palaeolithic and suggest that the first *H.sapiens* had the intellectual abilities to invent and use such technology.

The fact that such artefacts and evidence for symbolic behaviour remains so scarce until 50,000 years ago may be explained by two factors. First, by no more than poor preservation and the lack of discovery. Many parts of Africa where the first modern humans existed have received far less archaeological attention than regions of Europe. Secondly, as Shennan (2001) has recently argued, human population levels may have been too sparse and fragmented to sustain the degree of cultural transmission required for new ideas to spread and new artefacts be made in sufficient numbers to have an impact on the archaeological record. And hence the explosion in cultural behaviour after 50,000 years ago might be explained by human demography alone and its knock-on effects on cultural transmission.

I remain cautious about such explanations owing to the continuation of a distinctively Middle Palaeolithic stone tool technology, one little different to that of the Neanderthals, in the tool kits made by modern humans of south Africa and western Asia until 50,000 years ago at least. A major cognitive change may have occurred at that time, with this being as much a consequence as a cause of the cultural shifts stimulated by demographic change. (See Carruthers, forthcoming, for one proposal.)

However this issue of the emergence of modern behaviour and thought is eventually resolved, it is quite evident that by the time of the Upper Palaeolithic there had been at least three major developments relevant to the eventual emergence of science.

The first is the most important – the use of material culture in a manner to extend human cognition. During the earlier Palaeolithic, artefacts functioned to extend human physical capabilities. Stone tools provided sharp edges for cutting or a solid robust mass for crushing, hides provided clothing, stone or brushwood walls provided windbreaks. By such means humans were able to live in environments which would have otherwise been quite impossible for them.

By the Upper Palaeolithic, material culture had also begun to extend human cognitive abilities. The most evident example is the use of artificial memory systems (D'Errico, 1995). These are the pieces of bone and stone that have been incised with an ordered series of marks, often made with different types of tools and possibly over a long period of time. Certain artefacts from the Upper Palaeolithic have a striking resemblance to the calendar sticks of recently documented hunter–gatherers. Others appear to use a range of marks in spatial patterns that have a striking resemblance to early written texts – it appears that a system of notation is being utilized.

This, therefore, is the first unambiguous evidence that material culture is being used as a means for recording, storing and transmitting information – an essential role of material culture in scientific practice. To use material culture in this manner appears second nature to us; nevertheless this is a recent development (in evolutionary terms) and lies behind not only the flourishing of art during the Upper Palaeolithic but also the ultimate emergence of science.

This was not the only new role that material culture began to play. As I have argued elsewhere (Mithen, 1998) items of material culture, particularly those which we categorize as art, may have acted as mental anchors for ideas which were inherently difficult to grasp and store within the evolved mind. The clearest examples would be ideas about religious beings who could flout the laws of nature, laws that the evolved mind had come to take for granted (Mithen, 1998). In this regard the paintings and sculptures representing such beings – which must, I think be present in the oeuvre of Palaeolithic art – play essentially the same role as visual images and models in modern science. A key example would be the model of the double helix of DNA – if this was lacking the idea of DNA and how it behaves would be inherently more difficult to grasp (perhaps impossible for many). Both types of representation are means of extending the mind into the material world as a means to overcome limitations imposed by its evolutionary history – one that did not require people to hold mental representations of part-human part-animal figures or the structure of biological molecules.

The art of the Upper Palaeolithic is also indicative that another critical foundation for scientific thought had emerged by at least 30,000 years ago – possibly significantly earlier with the first modern humans. This is the use of analogy and metaphor, which is recognized as a key feature, not only for the effective communication of scientific ideas, but for the original development of those ideas (Kuhn, 1979; Nersessian, chapter 7 in this volume). Our evidence from Palaeolithic art is circumstantial – images of combined human and animal figures (Mithen, 1996). The most telling image is the lion-man carving from Hohlenstein-Stadel in Germany which dates to c.30,000 years ago. This is an 11cm carving of a mammoth tusk into a figure which has a lion's head and a human body. With such evidence it becomes unavoidable to think that the Upper Palaeolithic people were attributing animal qualities to humans ('brave as a lion') and vice versa. Further indications of this are found in the use of mammoth ivory to carve beads that mimic sea-shells, objects that are presumed to have had prestige value (White, 1989). As I have argued elsewhere (Mithen, 1996) the emergence of metaphorical thought might be as significant for cognitive and cultural evolution as that of language itself. I suspect it relates to a major re-organization of human mentality away from one of a domain-specific nature to one that is cognitively fluid.

The third critical development that appears restricted to anatomically modern humans, and may not have become fully established until after 50,000 years ago, is the accumulation of knowledge. Although pre-modern humans certainly transmitted a substantial amount of knowledge from generation to generation, as evident from the robust traditions of tool making, there appears to have been very limited – if any – accumulation of knowledge through time.

To build upon the ideas, techniques, theories and practices of a previous generation seems quite normal to us. It is part of our everyday experience as we are frequently asked to think back as to how we lived a mere 10 or 20 years ago, let alone to the last century. Such accumulation and development of knowledge is evidently a critical feature of science. It is only after the appearance of modern humans that we can see this beginning – the technology used by a Neanderthal living 30,000 years ago is no more advanced than that used by one who had lived 300,000 years ago.

We see the accumulation of knowledge most evidently in the development of technology during the course of the last ice age, particularly in the stratified sequences from south west Europe where cultural developments can be related to environmental change. Two phenomena are present. One is how technology is adapted to the prevailing environmental conditions. Hence the tools and hunting strategies used on the tundras of southern France during the height of the last glacial maximum are quite different to those used when thick woodland spread across the landscape after 11,500 years ago. Each can be seen as an appropriate solution to particular hunting and gathering problems posed by those environments (Zvelebil, 1984). The technology of pre-modern humans had not been used to adapt to prevailing environmental conditions in the same manner.

A second feature is the accumulation of knowledge through time. This is harder to document during the final stages of the ice age, and has been of limited recent interest to Palaeolithic archaeologists who have wished to avoid invoking notions of 'progress' and the separation of hunter–gatherers into advanced and 'primitive' forms. Yet there can be little question, I believe, that as from the start of the Upper Palaeolithic there was an accumulation of knowledge. This is perhaps best seen in what Straus (1990) has described as a Palaeolithic 'arms race' when there was selection between contrasting types of hunting technology (microlithic armatures or antler spears) during the Magdalenian period. The fact that by 12,000 years ago an almost identical microlithic technology had emerged in several different parts of the world (e.g. Europe, the Near East, Africa) suggests that this was the most efficient solution to foraging with a stone age technology – one reached independently by different people travelling along different cultural trajectories in different parts of the world.

To summarize, by the start of the Upper Palaeolithic three further foundations of scientific thought and practice were present within the human mind/society – the use of material culture, metaphorical and analogical thought and the accumulation of knowledge through time. These added to the existing foundations that had been inherited from pre-modern ancestors – detailed observation, hypothesis testing and an interest in causation. In this regard the complete foundations for the emergence of science were most likely in place within those *H.sapiens* hunter–gatherer societies of the late Pleistocene.

Whether or not we should describe such societies as undertaking 'science' is a moot point, one that depends upon how the term is defined. While the mixing of pigments, construction of dwellings, planning of foraging activities and the reading of tracks may have involved scientific modes of thought, these were most likely also intimately tied into religious beliefs which many would consider the precise opposite of scientific thinking. The development of science itself is likely to have required a very particular set of social and economic circumstances leading to an isolation of these two modes of thought from one another. Such circumstances are unlikely to have arisen within hunter–gatherer communities. It is appropriate to conclude this chapter, therefore, with some brief remarks about the origin of agriculture and its contribution to the eventual development of scientific thought and practice.

7 The first farming communities

Agriculture was independently invented in at least seven areas of the world following the end of the last ice age, the most notable being the Near East, Central Mexico, China and the Andes (Smith, 1995). My comments will be restricted to the Near East where farming villages with fields of wheat, barley, lentils and peas, flocks of sheep and goats, and the use of cattle and pigs had arisen by 8,500 years ago. Exactly how and why the transition from hunting and gathering to farming occurred remains a matter of considerable archaeological debate and is not my concern in this chapter. Banning (1998) provides a comprehensive review of the early Neolithic, providing descriptions of the sites and developments referred to below.

With regard to the social and economic foundations for science the presence of a farming economy allowed two key developments. First, the size of human communities could enlarge very considerably – some of the first towns such as 'Ain Ghazal are likely to have reached 2,000 inhabitants. Moreover, with the origin of farming there was a substantial increase in trade and exchange between communities in both staples and luxury goods. With both more permanent and more transient members of these towns the possibility for new ideas to arise, to spread and to merge with others must have increased by several orders of magnitude. And hence it is not surprising that we witness a great number of technological developments, in addition to those relating to ideology and religious practices.

A second consequence of a farming economy is that it becomes possible to use surpluses to support people who make no contribution to food production. Exactly how food production and distribution was organized with the early farming settlements remains unclear; some settlements such as Beidha appear to have grain stores which may have been under the control of a centralized authority. But it is quite evident that these settlements housed craft specialists, as seen by workshops in the corridor buildings at Beidha, and may have had a priestly caste.

Whether or not they were essential, large populations and the presence of specialists are likely to have greatly facilitated scientific thought and practices. As within the hunter–gatherer communities, anything approaching science within the early farming settlements is likely to have remained entwined with religious thought and may, therefore, not be eligible for the title of 'scientific' thought. But it is worthwhile to conclude this chapter by noting four major cultural developments associated with the first Near Eastern farming towns which seem to require thought and practice which might be described as scientific in nature.

First, the understanding of the reproduction and growth of animals and plants must have increased very substantially. One must be cautious here, as archaeologists are now aware that modern – and most likely pre-historic – hunter–gatherers have very extensive botanical and zoological knowledge (e.g. Blurton-Jones and Konner, 1976). The idea that agriculture arose because people first came to understand how plants and animals grow and reproduce has long been rejected. Nevertheless, that knowledge is likely to have been enhanced as people came to depend upon a small number of species and needed to maximize yields to feed burgeoning populations, especially as soil exhaustion and erosion set in.

Once farming economies had been established there was a very substantial development of architecture, dramatically illustrated by the change from the small circular dwellings of the Pre-Pottery Neolithic A to the rectangular buildings of the Pre-Pottery Neolithic B period at around 9,000 years ago (Banning, 1998). The latter were built from either mud-brick or stone and frequently involved two storeys and a variety of complex architectural features. Such architecture required the development of completely new building techniques. The fact that some of these buildings remain standing to a substantial extent today testifies to the development of that knowledge and practice.

The construction of PPNB houses involved the laying of thick plaster floors. A basic understanding of how to make plaster emerged within the preceding PPNA period, but saw a considerable development within the early farming settlements. The manufacture of plaster involves applying heat to transform limestone into lime and then into plaster, which when hardened is like an artificial limestone with properties akin to concrete. Kingery, Vandiver and Pickett (1988) have shown that this was a very sophisticated technology that required burning the limestone at temperatures of 750–850°C. This was used to create hard, easily cleaned and in some cases decorative floors. It was also used for producing sculpture, notably in the plastered skulls that have come from towns such as Jericho.

The fourth development within the early farming communities that appears closely aligned to a scientific mode of thought is the substantial development of textiles. Previous hunter–gatherers are most likely to have made extensive use of plant fibres when making clothing and equipment. But as with plaster, this technology received a substantial boost within the early farming towns. The finds from Nahal Hemar Cave in Israel (Schick, 1988), have provided our best insight into the new textiles, showing how several techniques were used to make yarn, cloth and twine. In later Neolithic sites spinning is evident from the presence of spindle whorls. Although sheep had been domesticated, it remains unclear whether sheep wool was spun and woven in the Neolithic. All the direct evidence concerns vegetable fibres, notably linen, but increased pastoralism in the Late Neolithic may suggest a shift to the use of wool (Banning, 1998).

In addition to these four developments within the early farming communities – agriculture itself, building, pyrotechnology and textiles – there was a substantial growth in other domains such as basketry, brewing and, eventually, pottery. Such an explosion of craft activity may not qualify as 'science'. But it demonstrates how the new social and economic circumstances provided by settled farming lifestyles provided a context in which the evolved propensities of the human mind could be used to create completely new bodies of knowledge. The cognitive foundations for these new craft activities were put in place while *H.sapiens* lived as a hunter–gatherer during the Pleistocene. The same foundations most likely enabled the development of science at a much later date in human history.

8 Conclusion

The human mind is a product of a long evolutionary history. Rather than trying to identify any one particular time or one particular species when the cognitive foundations for science emerged, this chapter has argued that these came about piecemeal over at least 5 million years. Our common ancestor with the great apes most likely already engaged in hypothesis testing and made acute observations about the natural and social worlds. Both are likely to have become further enhanced among the earliest *Homo*, together with a interest in causation arising from the emergence (or further development) of a theory of mind within species such as *H.ergaster* and *H.heidelbergensis*. Language is also likely to have evolved within this species and the presence of this is likely to have enhanced these other mental attributes. With the first *H.sapiens*, and especially those after 50,000 years ago, we find evidence for three further cognitive foundations for science – the use of material culture to extend human perception and cognition, the accumulation of knowledge through time, and the use of metaphor and analogy. By the end of the last ice age the complete cognitive foundations for science appear to have been in place. The potency of the mind that had evolved is evident from the substantial development of craft activities requiring sophisticated and extensive technological knowledge in the first farming communities. A scientific mode of thought appears to have been a necessary component of such activities, but the emergence of science as a discrete domain of behaviour is likely to have required a suite of social, historic and economic circumstances that had not yet arisen in human history.

3 Modular and cultural factors in biological understanding: an experimental approach to the cognitive basis of science

Scott Atran

> The experience that shaped the course of evolution offers no hint of the problems to be faced in the sciences, and the ability to solve these problems could hardly have been a factor in evolution.
>
> – Noam Chomsky, *Language and Problems of Knowledge*, 1988, p. 158

This chapter explores the universal cognitive bases of biological taxonomy and taxonomic inference using cross-cultural experimental work with urbanized Americans and forest-dwelling Maya Indians. The claim is that there is a universal, essentialist appreciation of generic species as the causal foundation for the taxonomic arrangement of biodiversity, and for inference about the distribution of causally related properties that underlie biodiversity. Generic species reflect characteristics of both the scientific genus and species. A principled distinction between genus and species is not pertinent to knowledge of local environments, nor to the history of science until after the Renaissance. Universal folk-biological taxonomy is domain-specific; that is, its structure does not spontaneously or invariably arise in other cognitive domains, such as the domains of substances, artefacts or persons. It is plausibly an innately determined evolutionary adaptation to relevant and recurrent aspects of ancestral hominid environments, such as the need to recognize, locate, react to and profit from many ambient species. Folk biology also plays a special role in cultural evolution in general, and particularly in the development of Western biological science. Although the theory of evolution may ultimately dispense with core concepts of folk biology, including *species*, *taxonomy* and *teleology*, in practice these may remain indispensable to doing scientific work.

1 Introduction

What follows is a discussion of three sets of experimental results that deal with various aspects of universal biological understanding among American and Maya children and adults. The first set of experiments shows that by the

The comparative studies reported here were co-directed with Douglas Medin and funded by NSF (SBR 97-07761, SES 99-81762) and the Russell Sage Foundation.

age of four–five years (the earliest age tested in this regard) urban American and Yukatek Maya children employ a concept of innate species potential, or underlying essence, as an inferential framework for understanding the affiliation of an organism to a biological species, and for projecting known and unknown biological properties to organisms in the face of uncertainty. The second set of experiments shows that the youngest Maya children do not have an anthropocentric understanding of the biological world. Children do not initially need to reason about non-human living kinds by analogy to human kinds. The fact that American children show anthropocentric bias appears to owe more to a difference in cultural exposure to non-human biological kinds than to a basic causal understanding of folk biology *per se*. Together, the first two sets of experiments suggest that folk psychology can't be the initial source of folk biology. They also indicate that to master biological science, people must learn to inhibit activation of universal dispositions to view species essentialistically and to see humans as inherently different from other animals.

The third set of results shows that the same taxonomic rank is cognitively preferred for biological induction in two diverse populations: people raised in the Mid-western USA and Itza' Maya of the Lowland Meso-american rainforest. This is the generic species – the level of *oak* and *robin*. These findings cannot be explained by domain-general models of similarity because such models cannot account for why both cultures prefer species-like groups in making inferences about the biological world, although Americans have relatively little actual knowledge or experience at this level. In fact, general relations of perceptual similarity and expectations derived from experience produce a 'basic level' of recognition and recall for many Americans that corresponds to the superordinate life-form level of folk-biological taxonomy – the level of *tree* and *bird*. Still Americans prefer generic species for making inductions about the distribution of biological properties among organisms, and for predicting the nature of the biological world in the face of uncertainty. This supports the idea of the generic-species level as a partitioning of the ontological domains of *plant* and *animal* into mutually exclusive essences that are assumed (but not necessarily known) to have unique underlying causal natures.

The implication from these experiments is that folk biology may well represent an evolutionary design: universal taxonomic structures, centred on essence-based generic species, are arguably routine products of our 'habits of mind', which may be in part naturally selected to grasp relevant and recurrent 'habits of the world'. The science of biology is built upon these domain-specific cognitive universals: folk biology sets initial cognitive constraints on the development of any possible macro-biological theory, including the initial development of evolutionary theory. Nevertheless, the conditions of relevance under which science operates diverge from those pertinent to folk biology. For natural science, the motivating idea is to understand nature as it is 'in itself', independently of the human observer (as far as possible). From this standpoint, the species-concept, like taxonomy and teleology, may arguably be allowed to survive in science as a regulative principle that enables the mind to readily establish stable contact with the surrounding environment, rather than as an epistemic concept that guides the search for truth.

2 Four points of general correspondence between folk biology and scientific systematics

In every human society, people think about plants and animals in the same special ways. These special ways of thinking, which can be described as 'folk biology', are basically different from the ways humans ordinarily think about other things in the world, such as stones, stars, tools or even people. The science of biology also treats plants and animals as special kinds of object, but applies this treatment to humans as well. Folk biology, which is present in all cultures, and the science of biology, whose origins are particular to Western cultural tradition, have corresponding notions of living kinds. Consider four corresponding ways in which ordinary folk and biologists think of plants and animals as special.

First, people in all cultures classify plants and animals into species-like groups that biologists generally recognize as populations of inter-breeding individuals adapted to an ecological niche. We call such groups – like *redwood*, *rye*, *raccoon* or *robin* – 'generic species' for reasons that will become evident. Generic species are usually as obvious to a modern scientist as to local folk. Historically, the generic-species concept provided a pre-theoretical basis for scientific explanation of the organic world in that different theories – including evolutionary theory – have sought to account for the apparent constancy of 'common species' and the organic processes that centre on them (Wallace, 1889/1901, p. 1).

Second, there is a common-sense assumption that each generic species has an underlying causal nature, or essence, which is uniquely responsible for the typical appearance, behaviour and ecological preferences of the kind. People in diverse cultures consider this essence responsible for the organism's identity as a complex, self-preserving entity governed by dynamic internal processes that are lawful even when hidden. This hidden essence maintains the organism's integrity even as it causes the organism to grow, change form and reproduce. For example, a tadpole and frog are in a crucial sense the same animal although they look and behave very differently, and live in different places. Western philosophers, such as Aristotle and Locke, attempted to translate this common-sense notion of essence into some sort of metaphysical reality, but evolutionary biologists reject the notion of essence as such. Nevertheless, biologists have traditionally interpreted this conservation of identity under change as due to the fact that organisms have separate genotypes and phenotypes.

Third, in addition to the spontaneous division of local flora and fauna into essence-based species, such groups have 'from the remotest period in... history... been classed in groups under groups. This classification [of generic species into higher- and lower-order groups] is not arbitrary like the grouping of stars in constellations' (Darwin, 1883, p. 363).[1] The structure of these hierarchically included groups, such as *white oak/oak/tree* or *mountain robin/robin/bird*, is referred to as 'folk-biological taxonomy'. Especially in the case of animals, these non-overlapping taxonomic structures can often be scientifically interpreted in terms of speciation (related species descended from a common ancestor by splitting off from a lineage).

Fourth, such taxonomies not only organize and summarize biological information; they also provide a powerful inductive framework for making systematic inferences about the likely distribution of organic and ecological properties among organisms. For example, given the presence of a disease in robins one is 'automatically' justified in thinking that the disease is more likely present among other bird species than among non-bird species. In scientific taxonomy, which belongs to the branch of biology known as systematics, this strategy receives its strongest expression in 'the fundamental principle of systematic induction' (Warburton, 1967; Bock, 1973). On this principle, given a property found among members of any two species, the best initial hypothesis is that the property is also present among all species that are included in the smallest higher-order taxon containing the original pair of species. For example, finding that the bacteria *E-scheriehia coli* share a hitherto unknown property with robins, a biologist would be justified in testing the hypothesis that all organisms share the property. This is because *E.coli* link up with robins only at the highest level of taxonomy, which includes all organisms. This or any general-purpose system of taxonomic inference for biological kinds is grounded in a universal belief that the world naturally divides into the limited causal varieties we commonly know as (generic) species.

3 Universal folk-biological taxonomy

In all societies that have been studied in depth, folk-biological groups, or taxa, are organized into ranks, which represent an embedding of distinct levels of reality. Most folk-biological systems have between three and six ranks (Berlin, 1992). Taxa of the same rank are mutually exclusive and tend to display similar

[1] Thus, comparing constellations in the cosmologies of Ancient China, Greece and the Aztec Empire shows little commonality. By contrast, herbals like the Ancient Chinese *ERH YA*, Theophrastus' *Peri Puton Istorias*, and the Aztec *Badianus Codex*, share important features, such as the classification of generic species into tree and herb life forms (Atran, 1990, p. 276).

linguistic, biological and psychological characteristics. Ranks and taxa, whether in folk-biological or scientific classification, are of different logical orders, and confounding them is a category mistake. Biological ranks are second-order classes of groups (e.g. *species*, *family*, *kingdom*) whose elements are first-order groups (e.g. *lion*, *feline*, *animal*). Folk-biological ranks vary little across cultures as a function of theories or belief systems. Ranks are universal but not the taxa they contain. Ranks are intended to represent fundamentally different levels of reality, not convenience.[2]

3.1 Folk kingdom

The most general folk-biological rank is the folk kingdom. Examples are *plant* and *animal*. Such taxa are not always explicitly named, and represent the most fundamental divisions of the biological world. These divisions correspond to the notion of 'ontological category' in philosophy (Donnellan, 1971) and psychology (Keil, 1979). From an early age, it appears, humans cannot help but conceive of any object they see in the world as either being or not being an animal (Inagaki and Hatano, 1993) and there is evidence for an early distinction between plants and non-living things (Hatano and Inagaki, 1994).

3.2 Life form

The next rank down is that of life form. Most taxa of lesser rank fall under one or another life form. Life-form taxa often have lexically unanalysable names (simple primary lexemes), such as *tree* and *bird*, although some life-form names are analysable, such as *quadruped*. Biologically, members of a life-form taxon are diverse. Psychologically, members of a life-form taxon share a small number of perceptual diagnostics: stem aspect, skin covering and so forth (Brown, 1984). Life-form taxa may represent adaptations to broad sets of ecological conditions, such as competition among single-stem plants for sunlight and tetrapod adaptation to life in the air (Hunn, 1982; Atran, 1985). Classifying by life-form may occur early on: two-year-olds distinguish familiar kinds of quadruped (e.g. dog and horse) from sea versus air animals (Mandler, Bauer and McDonough, 1991).

[2] Generalizations across taxa of the same rank thus differ in logical type from generalizations that apply to this or that taxon. *Termite*, *pig* and *lemon tree* are not related to one another by a simple class-inclusion under a common hierarchical node, but by dint of their common rank – in this case the level of generic species. A system of rank is not simply a hierarchy, as some suggest (Rosch, 1975; Premack, 1995; Carey, 1996). Hierarchy – that is, a structure of inclusive classes – is common to many cognitive domains, including the domain of artefacts. For example, *chair* often falls under *furniture* but not *vehicle*, and *car* falls under *vehicle* but not *furniture*. But there is no ranked system of artefacts: no inferential link, or inductive framework, spans both *chair* and *car*, or *furniture* and *vehicle*, by dint of a common rank, such as the artefact *species* or the artefact *family*.

3.3 Generic species

The core of any folk taxonomy is the generic-species level. Like life-form taxa, generic species are often named by simple lexemes, such as *oak* and *robin*. Sometimes, generic species are labelled as binomial compounds, likes *hummingbird*. On other occasions, they may be optionally labelled as binomial composites, such as *oak tree*. In both cases the binomial makes the hierarchical relation apparent between generic and life form.

Generic species often correspond to scientific genera (e.g. *oak*) or species (e.g. *dog*), at least for the most phenomenally salient organisms, such as larger vertebrates and flowering plants. On occasion generic species can correspond to local fragments of biological families (e.g. *vulture*), orders (e.g. *bat*) and – especially with invertebrates – even higher-order biological taxa (Atran, 1987; Berlin, 1992). Generic species may also be the categories most easily recognized, most commonly named and most easily learned by children in small-scale societies (Stross, 1973). Indeed, ethnobiologists who otherwise differ in their views of folk taxonomy tend to agree that one level best captures discontinuities in nature and provides the fundamental constituents in all systems of folk-biological categorization, reasoning and use (Bartlett, 1940; Bulmer, 1974; Berlin, 1978; Hunn, 1982; Ellen, 1993). Ethnobiologists, historians of systematics and field biologists mostly agree 'that species come to be tolerably well defined objects . . . in any one region and at any one time' (Darwin 1883, p. 137) and that such local species of the common man are the heart of any natural system of biological classification (Diamond and Bishop, 1999).

The term 'generic species' is used here, rather than 'folk genera/folk generic' or 'folk species/folk specieme,' for two reasons:

(1) Perceptually, a principled distinction between biological genus and species is not pertinent to most people around the world. For humans, the most phenomenally salient species (including most species of large vertebrates, trees, and evolutionarily isolated groups such as palms and cacti) belong to monospecific genera in any given locale. Closely related species of a polytypic genus are often hard to distinguish locally, and no readily perceptible morphological or ecological 'gap' can be discerned between them (Diver, 1940).[3]

[3] In a comparative study of Itza' Maya and rural Michigan college students, we found that the great majority of mammal taxa in both cultures correspond to scientific species, and most also correspond to monospecific genera: 30 of 40 (75%) basic Michigan mammal terms denote biological species, of which 21 (70%, or 53% of the total) are monospecific genera; 36 of 42 (86%) basic Itza' mammal terms denote biological species, of which 25 (69%, or 60% of the total) are monospecific genera (López *et al.*, 1997). Similarly, a Guatemalan government inventory of the Itza' area of the Peten rainforest indicates that 69% (158 of 229) are monospecific (AHG/APESA, 1992) the same percentage of monospecific tree genera (40 of 58) as in our study of the Chicago area (Medin *et al.*, 1997).

(2) Historically, the distinction between genus and species did not appear until the influx of newly discovered species from around the world compelled European naturalists to sort and remember them within a worldwide system of genera built around (mainly European) species types (Atran, 1987). The original genus-concept was partially justified in terms of initially monotypic generic European species to which other species around the world might be attached (Tournefort, 1694).

People in all cultures spontaneously partition the ontological categories *animal* and *plant* into generic species in a virtually exhaustive manner. 'Virtually exhaustive' means that when an organism is encountered that is not readily identifiable as belonging to a named generic species, it is still *expected* to belong to one. The organism is often assimilated to one of the named taxa it resembles, but sometimes it is assigned an 'empty' generic-species slot pending further scrutiny (e.g. 'such-and-such a plant is some [generic-species] kind of tree', Berlin, 1999). This partitioning of ontological categories is part and parcel of the categories themselves: no plant or animal can fail in principle to belong uniquely to a generic species.

3.4 Folk-specific

Generic species may be further divided into folk-specifics. These taxa are usually labelled binomially, with secondary lexemes. Compound names, like *white oak* and *mountain robin*, make the hierarchical relation transparent between a generic species and its folk-specifics. Folk-specifics that have tradition of high cultural salience may be labelled with primary lexemes, such as *winesap* (a kind of apple tree) and *tabby* (a kind of cat). In general, whether and how a generic species is further differentiated depends on the cultural significance of the organisms involved. Occasionally, an important folk-specific taxon will be further sub-divided into contrasting folk-varietal taxa: for example, *short-haired tabby* versus *long-haired tabby*. Folk-varietals are usually labelled trinomially, with tertiary lexemes that make transparent their taxonomic relationship with superordinate folk-specifics and generic species, for example *swamp white oak*.

Thus, in addition to generic species, people everywhere tend to form groups both subordinate and superordinate to the level of preferred groups. Cultures across the world organize readily perceptible organisms into a system of hierarchical levels that are designed to represent the embedded structure of life around us, with the generic-species level being most informative. In some cultures people may develop 'theories' of life that are meant to cover all living kinds, such as Western theories of biology (Carey, 1985; Atran, 1990). But the very possibility of theorizing wouldn't exist without a universal construal of generic species to provide the trans-theoretical basis for scientific speculation about the biological world.

4 Folk biology doesn't come from folk psychology: experiment 1

One influential model of conceptual development in folk biology is Carey's (1985) notion that young children's understanding of living things is embedded in a folk-psychological, rather than folk-biological, explanatory framework, and that until age ten, it is based on their understanding of humans. Carey reports three major findings to bolster the claim that children's conceptions of the biological world are anthropocentric. First, projections from humans are stronger overall than projections from other living kinds. The other two findings are consequences of this difference in induction potential. The second result is that there are asymmetries in projection: inferences from human to mammals are stronger than from mammals to humans. Third, children violate projections according to similarity: inferences from humans to bugs are stronger than from bees to bugs. Together, these findings suggest that humans are the preferred base for children's inferences about the biological world.

This research has had a powerful impact on psychological theory and educational practice; but it suffers from a serious limitation. It has been conducted almost exclusively with individuals from North American, urban, technologically advanced populations. In the few studies that go beyond this sample (e.g. studies by Inagaki and Hatano in Japan), the focus is still on urban, majority-culture children from advanced societies. Thus, it is not clear which aspects of children's naïve biology are likely to be universal and which depend critically on cultural conceptions and conditions of learning. We are also left with little insight into how to best design science curricula for non-majority, non-urban children. Human-centred reasoning patterns might reflect lack of knowledge about non-human living things rather than a radically different construal of the biological world. Consider results of the following experiment:

Participants: Participants were 98 Yukatek Maya-speaking children and 24 Yukatek Maya-speaking adults from rural villages in southcentral Quintana Roo, Mexico. 50 four-to-five-year-olds and 48 six-to-seven-year-olds were tested and included. Equal numbers of males and females were included in each group. By and large, the four-to-five-year-olds were monolingual, the six-to-seven-year-olds had begun learning Spanish, and almost all of the adults understood Spanish as a second language. All testing was done in Yukatek Maya.

Materials: Detailed colour drawings of objects were used to represent base and target categories. Four bases were used: Human, Dog, Peccary and Bee. Targets were divided into two sets. Each set included a representative of the categories Human (man, woman), Mammal (coatimundi, deer), Bird (eagle, chachalaca), Reptile (boa, turtle), Invertebrate (worm, fly), Tree (Kanan, Gumbo Limbo), Stuff (stone, mud), Artefact (bicycle, pencil) and Sun (included in both sets). The children were tested on each set at different times, with both sets divided equally among girls and boys.

Procedure: Children were shown a picture of one of the bases and taught a new property about it. For example, the experimenter might show the dog picture, and say, 'Now, there's this stuff called andro. Andro is found inside some things. One thing that has andro inside is dogs. Now, I'm going to show you some pictures of other things, and I want you to tell me if you think they have andro inside like dogs do.' Participants were then shown each of the targets and asked: 'Does it have andro inside it, like the [base]?' Properties were unfamiliar internal substances of the form 'has X inside'. A different property was used for each base, and bases and targets were presented in random order for each participant.

4.1 Results and discussion

Each time a child projected a property from a base to a target it was scored 1, otherwise 0. Responses did not differ reliably across any target pair (e.g. coatimundi vs. deer).

There was no overall preference in projections from humans.[4] Projection from Human is least common across age groups, and inferences from Dog most common. *Post hoc* tests indicate that younger children project properties more readily than adults (older children do not differ from younger children or adults). Results show that inferences from Human are not stronger than those from any other base overall.

Another perspective on reasoning patterns is provided by analyses of how distinct patterns of generalization are for different bases. For this purpose we conducted trend analyses to look at projections as a function of similarity. We assume that birds are at distance 1 from mammals, reptiles at distance 2, insects at distance 3 and trees at distance 4. The factors were AGE GROUP and SIMILARITY (distance) and our interest was in the linear component. All age groups show clear similarity effects with dogs, peccaries and bees as bases. The generalization (similarity) gradients become sharper with age for dogs as a base. With humans as a base, four-to-five-year-olds generalize broadly in an undifferentiated manner – they show no reliable effect of similarity. In contrast, adults show characteristically sharp gradients with humans as a base.

[4] ANOVA: BASE (4) by GENDER (2) by AGE GROUP (3) revealed only a main effect of BASE: $F(3,43) = 16.2$, $p < .001$.

The six-to-seven-year-olds show a very weak similarity gradient. In short, the clearest developmental change is in determining the role of humans in the folk-taxonomic system (see figure 1a).

There was no overall asymmetry in projection, as shown by conducting an ANOVA-test on the *difference scores*: The dependent variable for each subject was their base to target (e.g. Human to Mammal) score minus their target to base (e.g. Dog to Human) score. For inferences involving the bases Human and Dog, the data are inconsistent with Carey because only adults show the asymmetry favouring Human to Mammal over Dog to Human (see figures 1a–1d).

There were no violations of similarity, again as shown by ANOVAs on difference scores: Calculated as mean projection from Human to target (e.g. Human to Mammal) minus mean projection from the target to a base in the same class as the target (e.g. Dog). Results are inconsistent with Carey because for all ages inferences from Dog to Mammal are better than inferences from Human to Mammal (see figures 1a–d).

4.2 *General discussion*

In sum, four-to-five-year-old Yukatek Maya children do not show commitment to an anthropocentric understanding of the living world. This suggests that folk psychology is not a necessary or universal source for folk biology. Further support for this position comes from the Ross *et al.* (submitted) study of biological reasoning in three age groups (five-to-six, seven-to-eight, nine-to-ten years), among urban US children and rural Native American children (Menominee). The projection task, like that used with Maya, was patterned after Carey (1985). Ross *et al.*'s sample of urban Boston children replicates major aspects of Carey's findings with Boston-area children. For example, Ross *et al.* find significant asymmetries between strength of projections from humans to categories in the same class (e.g. bear, raccoon) as the base (e.g. wolf), and the strength of projection from the non-human base (e.g. wolf) to humans for Boston five-to-eight-year-olds. For young Menominee, however, Ross *et al.* find no evidence that projections from humans are reliably higher than projections from other bases. Neither do they find violations of similarity. Ross *et al.* argue that humans are the only animal that urban children know much about and so they generalize from them. Consistent with this view, Inagaki (1990) presents evidence that experience influences children's biological reasoning. She found that kindergarten children who raised goldfish were more likely than their counterparts who did not raise goldfish to reason about a novel aquatic animal (a frog) by analogy to goldfish rather than by analogy to humans.

Thus, Carey's results with Boston-area children cannot be generalized into a claim of biology's psychological origins. These findings are difficult to reconcile with Carey's strong claim of folk biology as the product of learned experience

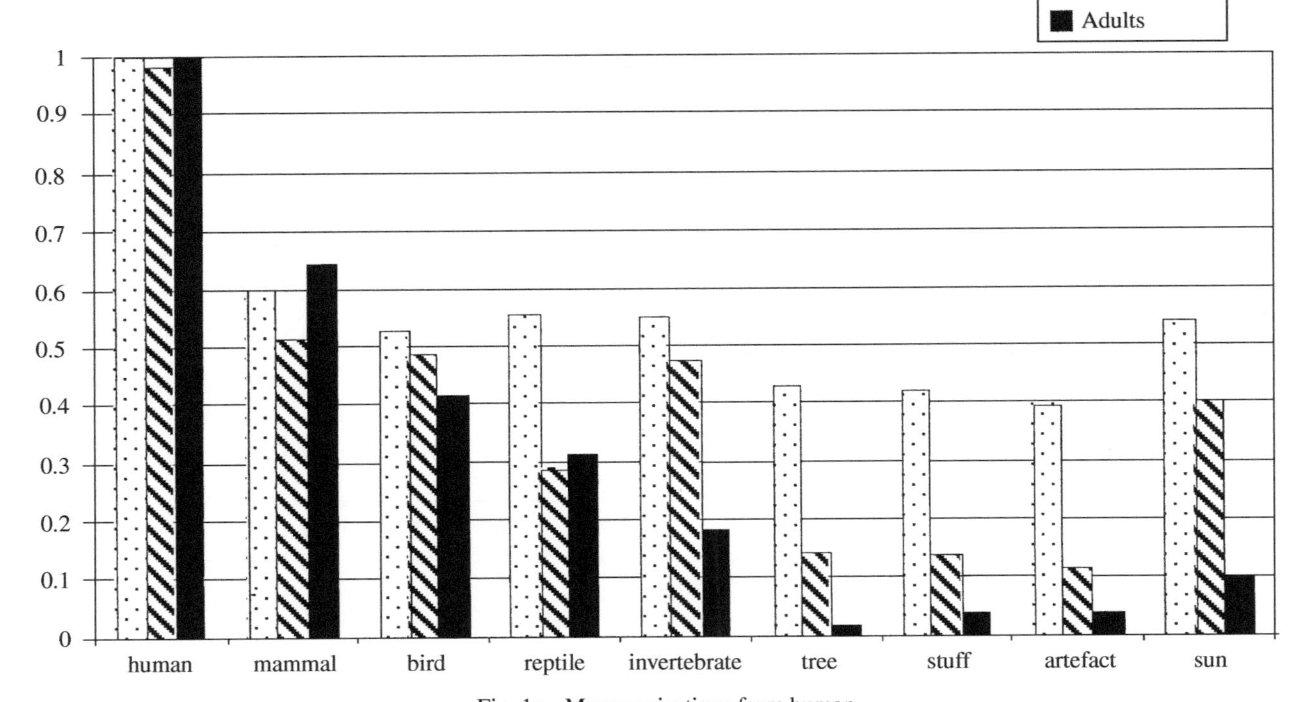

Fig. 1a Maya projections from human

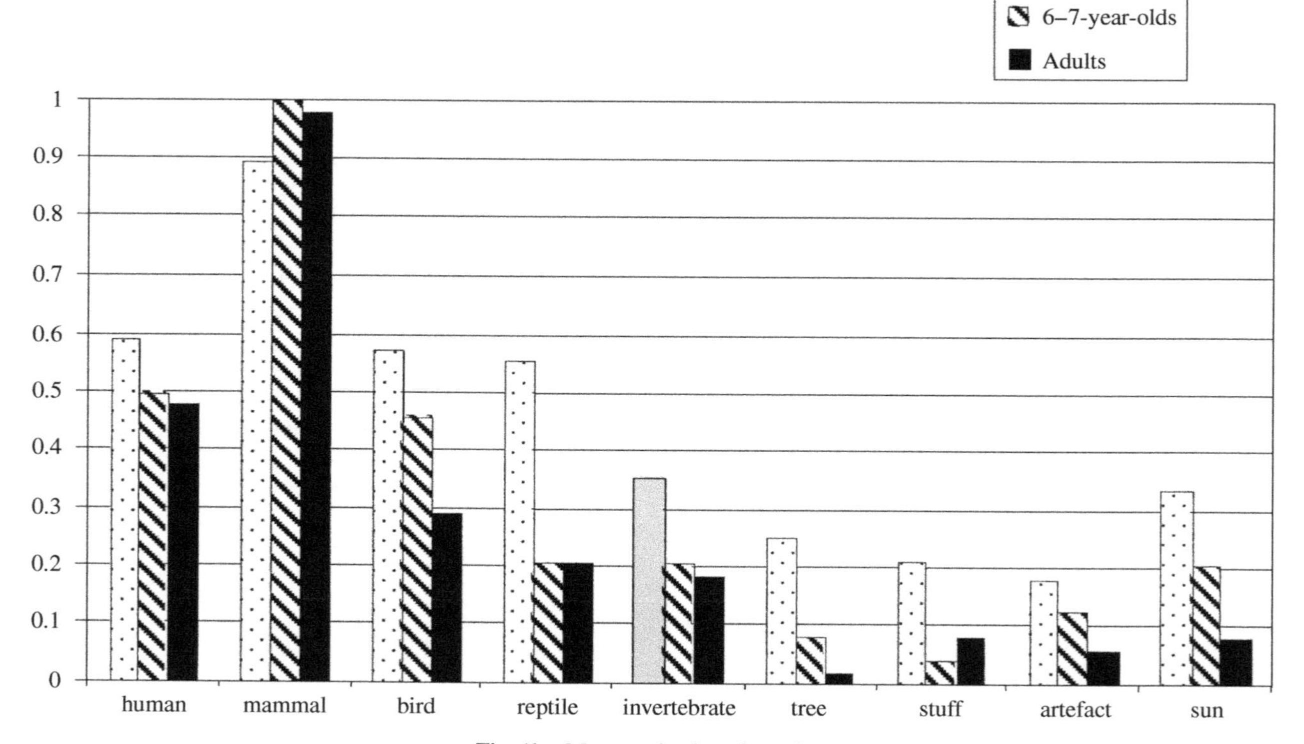

Fig. 1b Maya projections from dog

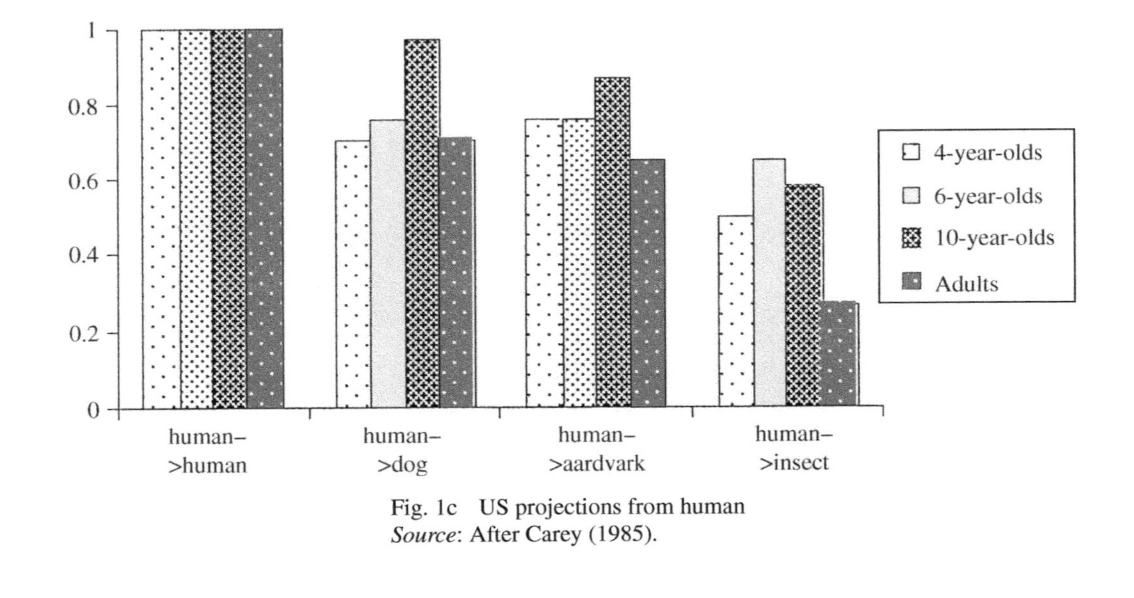

Fig. 1c US projections from human
Source: After Carey (1985).

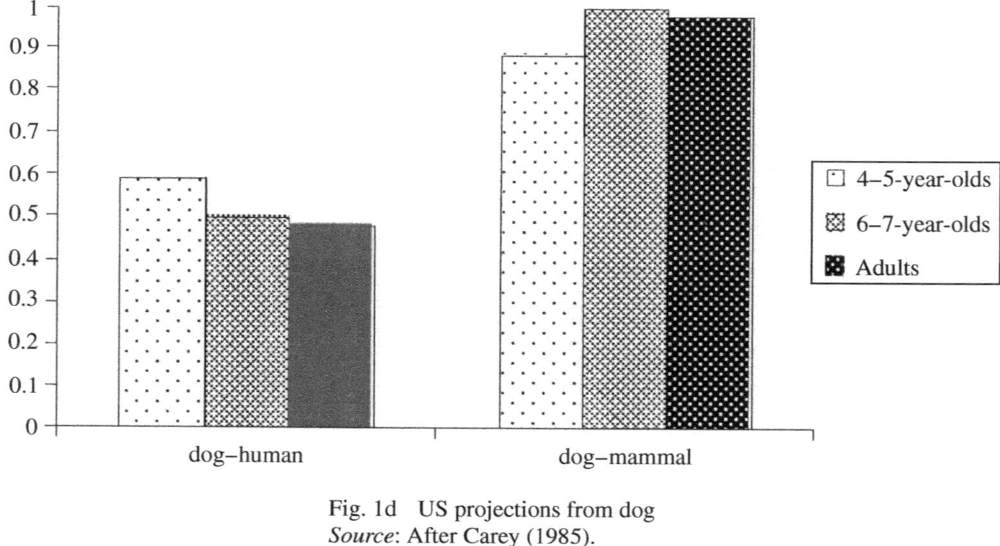

Fig. 1d US projections from dog
Source: After Carey (1985).

and radical conceptual change. If at least some basic aspects of the final or stable state of adult folk biology are pretty much the same across cultures, then there is no obvious sense to the notion that people in different cultural milieu acquire the *same* basic knowledge about folk biology through radical conceptual changes grounded in radically different learning environments. These results also undermine related claims that children acquire basic knowledge of the world as little scientists through successively novel theoretical formulations (Carey and Spelke, 1994), or that scientists come up with new theories because they manage to continue to think like children (Gopnik, Meltzoff and Kuhl, 1999a).

5 Childhood conceptions of species essences: experiment 2

Young individuals have the potential to develop certain adult characteristics before those characteristics appear. The origins of these characteristics can be explained in two broadly different ways: nature and nurture. Some characteristics seem likely to develop from birth because they are essential to the species to which the individual belongs, such as a squirrel's ability to jump from tree to tree and hide acorns. Other characteristics are determined by the environment in which the individual is reared, such as a squirrel's fear or lack of fear of human beings.

Gelman and Wellman (1991) argue that young children predict category-typical characteristics of individual animals based on the innate potential of the animal (i.e. the species of its birth-parent) rather than the environment in which it was raised (i.e. the species of its adoptive parent). Using an adoption study, they showed that four-year-old children judge that a baby cow raised by pigs will have the category-typical characteristics of cows (*moos, straight tail*) rather than pigs (*oinks, curly tail*). They interpret the results as showing that pre-school children believe that the innate potential or essence of species determines how an individual will develop, even in contrary environments.

This study is inconclusive with regard to children's assumptions about innate potential, for two reasons. First, before the children in the study predicted the adult properties of the adopted baby, they were shown a drawing of the baby animal and told its species identity. Because the experimenters told the child that the baby and mother were of the same species, it does not address the question of how the children identify to which species the baby belongs in the first place (Johnson and Solomon, 1997). Second, the study explored only known facts about species and their associated properties. It did not examine whether or not children use the concept of species-essence or biological parentage as an inferential framework for interpreting and explaining hitherto unknown facts. It may be that a child has learned from experience, and as a matter of fact, that a calf is a cow because it was born to a cow. Still, the child may not know that being a member of a certain species *causes* a cow to *be* a cow (Carey, 1996). The current study was designed to test the extent to which children's assumptions about innate species-potential govern projection of both known and unknown properties, and to avoid the problems noted above (for details, see Atran *et al.*, 2001).

Participants: Participants were 48 Yukatek Maya-speaking children and 24 Yukatek Maya-speaking adults. 24 four-to-five-year-olds and 24 six-to-seven-year-olds were tested and included. Equal numbers of males and females were included in each group. All testing was done in Yukatek Maya.

Procedure: In a forced choice task, children were asked whether an adult animal adopted at birth would resemble its adoptive parent (e.g. cow) or birth-parent (e.g. pig) on four different individual traits: known behaviours (e.g. *moo/oink*), known physical features (e.g. *straight/curly tail*), unknown behaviours (e.g. *looks for chachalacas/looks for pigeons*), and unknown physical features (e.g. *heart gets flatter/rounder when it is sleeping*). Known traits were context-free, category-typical features that the children readily associated with species, whereas unknown traits were chosen to minimize any possibility of factual or pre-learned associations of traits with categories. Each unknown trait within a set was attributed to the birth-parent for half the participants and to the adoptive parent for the other half. This assured that projection patterns of the unknown traits were not based on prior associations.

The stories were accompanied by sketches of each parent. Sketches were designed to unambiguously represent a particular species of animal with minimum detail. In addition, sketches of known physical features (e.g. a sketch of a curly or straight tail), unknown physical features (e.g. flat vs. round heart) and relevant aspects of unknown behavioural contexts (e.g. closed vs. open eyes, mahogany vs. cedar tree) were shown to participants. These sketches in no way indicated the species to which the traits belonged. Participants indicated choice of birth or adoptive parent species by pointing to the relevant parent sketch.

The story was followed by two comprehension controls: a birth control (*Who gave birth to the baby? Go ahead and point out the drawing of who gave birth to the baby*) and a nurture control (*Who did the baby grow up with?*). If the child failed either control the adoption story was repeated and a second failure in comprehension resulted in exclusion of the child from the experiment. Children then were presented with the four experimental probes. For example: *The cow mooed and the pig oinked. When the baby is all grown up will it moo like a cow or oink like a pig?* For each set, the four probes (counter-balanced in order across children) were followed by a bias control in which the participant was asked: *When the baby was growing up did it eat with animals that looked like X or animals that looked like Y?*

A final probe involved a transformation story to explore the extent to which species essences are associated with inheritance vs. vital internal properties as

Table 1 *Percent birth parent choice for each probe type for each group*

Group	Known			Unknown				Bias control (food)
	Behaviour	Phys. feat.	Mean	Behaviour	Phys. feat.	Mean	Blood	
4- to 5- year-olds	0.74**	0.68**	0.71	0.69**	0.68**	0.69	0.56	0.06***
6- to 7- year-olds	0.96***	0.97***	0.97	0.82***	0.83***	0.83	0.79**	0.01***
Adults	1.0***	0.96***	0.98	0.90***	0.93***	0.92	0.88***	0***
Mean	0.90	0.87	0.88	0.81	0.81	0.81	0.74	0.02

Note: p < 0.05*, p < 0.01**, p < 0.001***.

such (e.g. blood). Keil (1989, p. 224), found that the younger children are undecided as to whether inheritance or internal properties are primarily responsible for an animal's species identity.

5.1 *Results and discussion*

For each probe, participants were given a score of one if they chose the birth parent and zero if they chose the adoptive parent. A GENDER × AGE GROUP × SET×PROBE repeated-measures ANOVA indicated only a main effect of probe type, $F(5, 62) = 3.9$, $p < 0.01$. Each mean was tested against chance (0.5) and results appear in Table 1. Overall, the results show a systematic and robust preference for attributions from the birth-parent. This preference was observed for all age groups and for known and unknown behaviour and physical properties. The trend is somewhat stronger in older children and adults and slightly stronger for known than unknown properties. Means for all probes were significantly different from chance, except the *kind* and *blood* for the youngest children. The *kind* probe was only marginally different from chance for the young children ($p = 0.10$), possibly because of the foreign character of the Mayanized Spanish word for *kind*, 'klaasej'. Results on the *blood* probe for the youngest children might suggest genuine indecision as to whether inheritance or vital internal functioning is primarily responsible for an animal's species identity. The low mean on the bias-control probe for all groups indicates that the method of the current experiment did not bias participant responses toward the birth-parent.

5.2 *General discussion*

Results of this study indicate that Yukatek Maya children and adults assume that members of a species share an innate causal potential that largely determines category-typical behavioural and physical properties even in conflicting environments. Projection of properties to the birth-parent in the face of uncertainty and novelty implies that even young Maya children use the notion of underlying essence as an inferential framework for understanding the nature of biological species. By age seven, children have effectively attained adult competence in inferential use of the notion of innate species potential. These findings, together with Gelman and Wellman's (1991) earlier results for urban American children, suggest that such an essentialist bias in children is universal.

6 Essence (generic species) vs. appearance (basic levels) in folk biology: experiment 3

In a justly celebrated set of experiments Rosch and her colleagues set out to test the validity of the notion of a psychologically preferred taxonomic level (Rosch *et al.*, 1976). Using a broad array of converging measures they found that there is indeed a 'basic level' in category hierarchies of 'naturally occurring objects', such as 'taxonomies' of artefacts as well as living kinds. For artefact and living kind hierarchies, the basic level is where: (1) many common features are listed for categories, (2) consistent motor programs are used for the interaction with or manipulation of category exemplars, (3) category members have similar enough shapes so that it is possible to recognize an average shape for objects of the category, (4) the category name is the first name to come to mind in the presence of an object (e.g. 'table' vs. 'furniture' or 'kitchen table').

There is a problem, however: the basic level that Rosch *et al.* (1976) had hypothesized for artefacts was confirmed (e.g. *hammer*, *guitar*); however, the hypothesized basic level for living kinds (e.g. *maple*, *trout*), which Rosch initially assumed would accord with the generic-species level, was not. For example, instead of *maple* and *trout*, Rosch *et al.* found that *tree* and *fish* operated as basic-level categories for American college students. Thus, Rosch's basic level for living kinds generally corresponds to the life-form level, which is superordinate to the generic-species level (see Zubin and Köpcke, 1986, for German speakers).

To explore the apparent discrepancy between preferred taxonomic levels in small-scale and industrialized societies, and the cognitive nature of ethnobiological ranks in general, we use inductive inference. Inference allows us to test for a psychologically preferred rank that maximizes the strength of any potential induction about biologically relevant information, and whether or not this preferred rank is the same across cultures. If a preferred level carries the most information about the world, then categories at that level should favour a wide range of inferences about what is common among members. (For detailed findings under a variety of lexical and property-projection conditions, see Atran *et al.*, 1997; Coley, Medin and Atran, 1997.)

The prediction is that inferences to a preferred category (e.g. *white oak* to *oak*, *tabby* to *cat*) should be much stronger than inferences to a superordinate category (*oak* to *tree*, *cat* to *mammal*). Moreover, inferences to a subordinate category (*swamp white oak* to *white oak*, *short-haired tabby* to *tabby*) should not be much stronger than or different from inferences to a preferred category. What follows is a summary of results from one representative set of experiments in two very diverse populations: Mid-western Americans and Lowland Maya.

Participants: Participants were two sets of 12 adult Itza' Maya and two sets of 12 adult Mid-western Americans, with equal numbers of males and females. The Itza' are Maya Amerindians living in the Petén rainforest region of Guatemala. Until recently, men devoted their time to shifting agriculture, hunting and silviculture, whereas women concentrated on the myriad tasks of household maintenance. The Americans were self-identified as people raised in Michigan and recruited through an advertisement in a local newspaper.

Materials: Based on extensive fieldwork, we chose a set of Itza' folk-biological categories of the kingdom (K), life-form (L), generic-species (G), folk-specific (S), and folk-varietal (V) ranks. We selected three plant life forms (*che'* = tree, *ak'* = vine, *pok~che'* = herb/bush) and three animal life forms (*b'a'al~che' kuxi'mal* = 'walking animal', i.e. mammal, *ch'iich'* = birds including bats, *käy* = fish). Three generic-species taxa were chosen from each life form; each generic species had a subordinate folk-specific, and each folk-specific had a salient varietal.

The properties chosen for animals were diseases related to the 'heart' (*puksik'al*), 'blood' (*k'ik'el*) and 'liver' (*tamen*). For plants, diseases related to the 'roots' (*motz*), 'sap' (*itz*) and 'leaf' (*le'*). Properties were chosen according to Itza' beliefs about the essential, underlying aspects of life's functioning. Properties used for inferences had the form, 'is susceptible to a disease of the <root> called <X>'. For each question, 'X' was replaced with a phonologically appropriate nonsense name (e.g. 'eta') to minimize the task's repetitiveness.

Procedure: All participants responded to a list of over 50 questions in which they were told that all members of a category had a property (the premise) and were asked whether 'all', 'few', or 'no' members of a higher-level category (the conclusion category) also possessed that property. The premise category was at one of four levels, either life form (e.g. L = bird), generic-species (e.g. G = vulture), folk-specific (e.g. S = black vulture), or varietal (e.g. V = red-headed black vulture). The conclusion category was drawn from a higher-level category, either kingdom (e.g. K = animal), life form (L), generic-species (G), or folk-specific (S). Thus, there were ten possible combinations of premise and conclusion category levels: L→K, G→K, G→L, S→K, S→L, S→G, V→K, V→L, V→G and V→S. For example, a folk-specific-to-life-form (S→L) question might be, 'If all black vultures are susceptible to the blood disease called *eta*, are all other birds susceptible?' If a participant answered 'no', the follow-up question would be 'Are some or a few other birds susceptible to disease *eta*, or no other birds at all?'

The corresponding life forms for the Americans were: *mammal*, *bird*, *fish*, *tree*, *bush* and *flower*. (On *flower* as an American life form see Dougherty, 1979.) The properties used in questions for the Michigan participants were 'have protein X', 'have enzyme Y', and 'are susceptible to disease Z'. These were chosen to be internal, biologically based properties intrinsic to the kind in question, but abstract enough so that rather than answering what amounted to factual questions participants would be likely to make inductive inferences based on taxonomic category membership.

6.1 Results and discussion

Responses were scored in two ways. First we totalled the proportion of 'all or virtually all' responses for each kind of question (e.g. the proportion of times respondents agreed that if red oaks had a property, all or virtually all oaks would have the same property). Second, we calculated 'response scores' for each item, counting a response of 'all or virtually all' as 3, 'some or few' as 2 and 'none or virtually none' as 1. A higher score reflected more confidence in the strength of an inference.

Figure 2a summarizes the results from all Itza' informants for all life forms and diseases, and shows the proportion of 'all' responses (black), 'few' responses (chequered), and 'none' responses (white). For example, given a premise of folk-specific (S) rank (e.g. red squirrel) and a conclusion category of generic-species (G) rank (e.g. squirrel), 49% of responses indicated that 'all' squirrels, and not just 'some' or 'none', would possess a property that red squirrels have. Figure 2b summarizes results of Michigan response scores for all life forms and biological properties. Analyses used *t*-tests with significance levels adjusted to account for multiple comparisons.

Following the main diagonals of figures 2a and 2b refers to changing the levels of both the premise and conclusion categories while keeping their relative level the same (with the conclusion one level higher than the premise). Induction patterns along the main diagonal indicate a single inductively preferred level. Examining inferences from a given rank to the adjacent higher-order rank (i.e. V→S, S→G, G→L, L→K), we find a sharp decline in strength of inferences to taxa ranked higher than generic species, whereas V→S and S→G inferences are nearly equal and similarly strong. Notice that for 'all' responses, the overall Itza' and Michigan patterns are nearly identical.

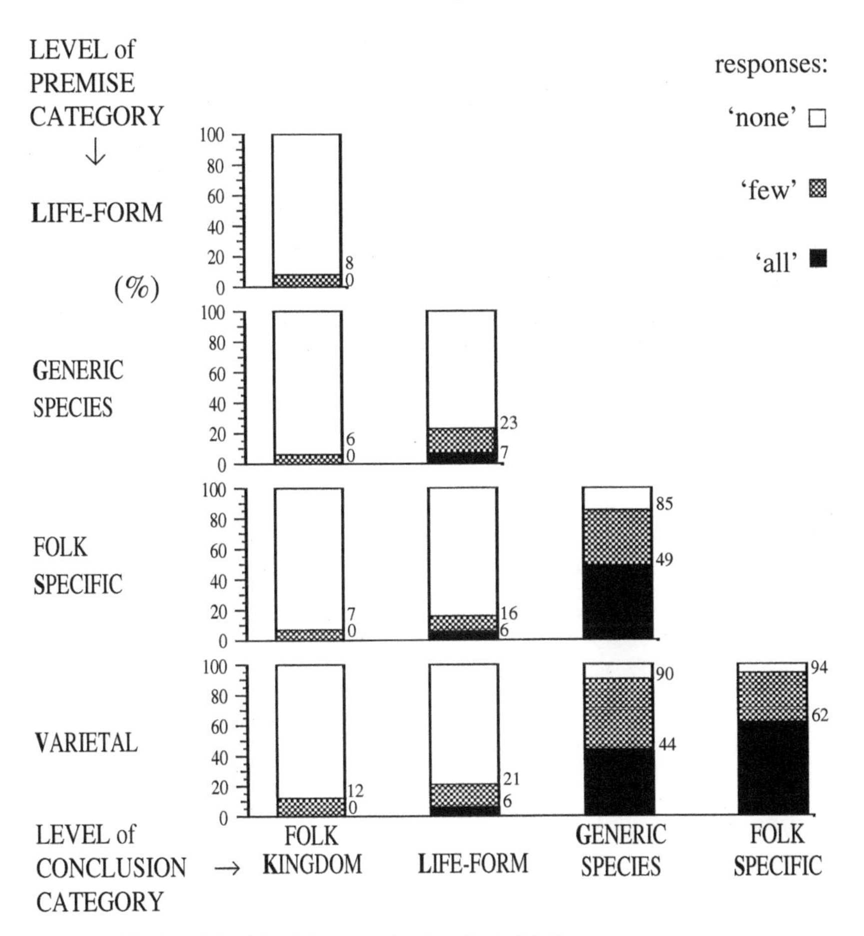

Fig. 2a Combined Itza' results for all six life forms

Moving horizontally within each graph corresponds to holding the premise category constant and varying the level of the conclusion.[5] We find the same pattern for 'all' responses for both Itza' and Americans as we did along the main diagonal. However, in the combined response scores ('all' + 'few') there is evidence of increased inductive strength for higher-order taxa among Americans

[5] Moving vertically within each graph corresponds to changing the premise while holding the conclusion category constant. This allows us to test another domain-general model of category-based reasoning: the Similarity-Coverage Model (Osherson *et al.*, 1990). In this model, the closer the premise category is to the conclusion category, the stronger induction should be. Our results show only weak evidence for this general reasoning heuristic, which fails to account for the various 'jumps' in inductive strength that indicate absolute privilege.

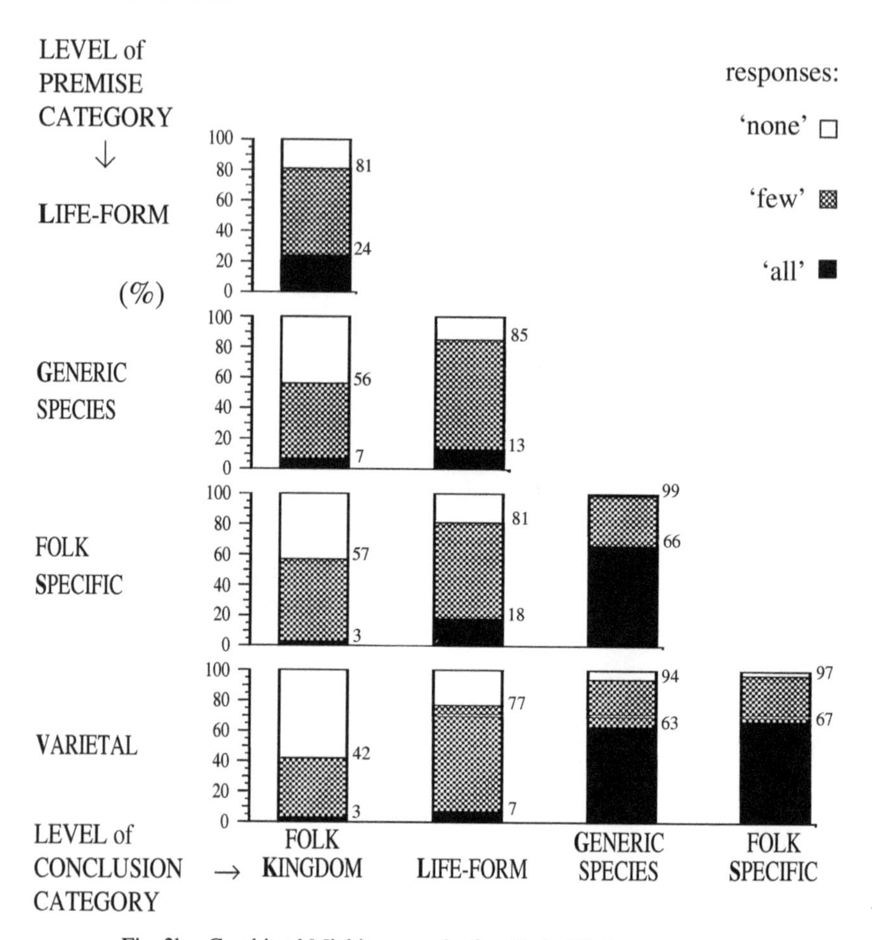

Fig. 2b Combined Michigan results for all six life forms

versus Itza'. On this analysis, both Americans and Itza' show the largest break between inferences to generic species vs. life forms. But only American subjects also show a consistent pattern of rating inferences to life-form taxa higher than to taxa at the level of folk kingdom: G→K *vs.* G→L, S→K *vs.* S→L and V→K *vs.* V→L.

Finally, moving both horizontally and along the diagonal, regression analysis reveals a small but significant difference between Itza' inductions using conclusions at the generic-species versus folk-specific levels: V→G and S→G are modestly weaker than V→S. For Michigan participants, the folk-specific level is not differentiated from the generic-species level. In fact, most of the difference between V→G and V→S inductions results from inference patterns for the Itza' tree life form. There is evidence that Itza' confer preferential status upon trees at the folk-specific level (e.g. savannah nance tree). Itza' are forest-dwelling Maya with a long tradition of agroforestry that antedates the Spanish conquest (Atran and Ucan Ek', 1999; Atran *et al.*, 1999).

6.2 *General discussion*

These results indicate that both the ecologically inexperienced Americans and the ecologically experienced Itza' prefer taxa of the generic-species rank in making biological inferences. These findings cannot be explained by appeals either to cross-domain notions of perceptual 'similarity' or to the structure of the world 'out there', as most ethnobiologists contend (Hunn, 1976; Boster, 1991; Berlin, 1992). If inferential potential were a simple function of perceptual similarity then Americans should prefer life forms for induction (in line with Rosch *et al.*). Yet Americans prefer generic species as do Maya. Unlike Itza', however, Americans perceptually discriminate life forms more readily than generic species (although one might expect that having less biodiversity in the American environment allows each species to stand out more from the rest). Despite the compelling needs established by lived experience, the Americans and Maya overwhelmingly, and in nearly equal measure, subordinate such influences to a preference for generic species.

The findings suggest that root categorization and reasoning processes in folk biology are due to domain-specific conceptual presumptions and not exclusively to domain-general, similarity-based (e.g. perceptual) heuristics. To be sure, language may signal expectation that little or poorly known generic species are more biologically informative than better known life forms for Americans (e.g. via common use of binomials, such as *oak/red oak*). Further experiments, however, still show reliable results in the absence of clear linguistic cues (e.g. *oak/white oak/swamp white oak* vs. *dog/poodle/toy poodle* – Coley, Medin and Atran, 1997).

Humans everywhere presume the biological world to be partitioned at the generic-species rank into non-overlapping kinds, each with its own unique causal essence, or inherent underlying nature, whose visible products may or may not be readily perceived.[6] People anticipate that the biological information-value of these preferred kinds is maximal whether or not there is also visible indication of maximal co-variation of perceptual attributes. This does not mean that more general perceptual cues have no inferential value when applied to the

[6] By contrast, a partitioning of artefacts (including those of organic origin, such as foods) is neither mutually exclusive nor inherent: some mugs may or may not be cups; an avocado may be a fruit or vegetable depending upon how it is served; a given object may be a bar stool or a waste bin depending on the social context or perceptual orientation of its user; and so on.

folk-biological domain. On the contrary, the findings here point to a significant role for such cues in targeting basic-level life forms as secondary foci for inferential understanding in a cultural environment where biological awareness is poor, as among many Americans. Arguably, there is an evolutionary design to a cognitive division of labour between domain-general perceptual heuristics and domain-specific learning mechanisms: the one enabling flexible adaptation to variable conditions of experience, and the other invariably steering us to those abiding aspects of biological reality that are both causally recurrent and especially relevant to the emergence of human life and cognition (Atran, 1998).

7 Science and common sense

Much of the history of systematics has involved attempts to adapt locally relevant principles of folk biology to a more global setting, such as the taxonomic embedding of biodiversity, the primacy of species, and the teleo-essentialist causality that makes sense of taxonomic diversity and the life functions of species. This process has been far from uniform (e.g. initial rejection of plant but not animal life forms, recurrent but invariably failed attempts to define essential characters for species and other taxa, intermittent attempts to reduce teleological processes to mechanics, and so forth – Atran, 1990).

Historical continuity between universal aspects of biological common sense and the science of biology should not be confounded with epistemic continuity or use of folk knowledge as a learning heuristic for scientific knowledge. Scientists have made fundamental ontological shifts away from folk understanding in the construal of species, taxonomy and underlying causality. For example, biological science today rejects fixed taxonomic ranks, the primary and essential nature of species, teleological causes 'for the sake' of species existence, and phenomenal evidence for the existence of taxa (e.g. trees do not constitute a scientifically valid superordinate plant group, but bacteria almost assuredly should).

Nevertheless, from the vantage of our own evolutionary history, it may be more important that our ordinary concepts be adaptive than true. Relative to ordinary human perceptions and awareness, evolutionary and molecular biology's concerns with vastly extended and minute dimensions of time and space has only marginal value. The ontological shift required by science may be so counter-intuitive and irrelevant to everyday life as to render inappropriate and maladaptive uses of scientific knowledge in dealing with ordinary concerns. Science can't wholly subsume or easily subvert folk-biological knowledge.

7.1 Taxonomy

Historically, taxonomy is conservative, but it can be revolutionized. Even venerable life forms, like *tree*, are no longer scientifically valid concepts because they have no genealogical unity (e.g. legumes are variously trees, vines, bushes, etc.). The same may be true of many long-standing taxa. Phylogenetic theorists question the 'reality' of zoological life forms, such as *bird* and *reptile*, and the whole taxonomic framework that made biology conceivable in the first place. Thus, if birds are descended from dinosaurs, and if crocodiles but not turtles are also directly related to dinosaurs, then: crocodiles and birds form a group that excludes turtles; or crocodiles, birds and turtles form separate groups; or all form one group. In any event, the traditional separation of *bird* and *reptile* is no longer tenable.

From Linnaeus to the present day, biological systematics has used explicit principles and organizing criteria that traditional folk might consider secondary or might not consider at all (e.g. the geometrical composition of a plant's flower and fruit structure, or the numerical breakdown of an animal's blood chemistry). Nevertheless, as with Linnaeus, the modern systematist initially depends implicitly, and crucially, on a traditional folk appreciation. As Bartlett (1936, p. 5) noted with reference to the Maya of Petén (cf. Diamond, 1966, for zoology):

> A botanist working in a new tropical area is . . . confronted with a multitude of species which are not only new to him, but which flower and fruit only at some other season than that of his visit, or perhaps so sporadically that he can hardly hope to find them fertile . . . [C]onfronted with such a situation, the botanist will find that his difficulties vanish as if by magic if he undertakes to learn the flora as the natives know it, using their plant names, their criteria for identification (which frequently neglect the fruiting parts entirely), and their terms for habitats and types of land.

As Linnaeus needed the life form *tree* and its common species to actually do his work (1738, 1751), so did Darwin need the life form *bird* and its common species. From a strictly cosmic viewpoint, the title of his great work, *On the Origins of Species*, is ironic and misleading – much as if Copernicus had entitled his attack on the geocentric universe, *On the Origins of Sunrise*. Of course, in order to attain that cosmic understanding, Darwin could no more dispense with thinking about 'common species' than Copernicus could avoid thinking about the sunrise (Wallace 1889/1901, pp. 1–2). In fact, not just species, but all levels of universal folk taxonomy served as landmarks for Darwin's awareness of the evolving pathways of diversity: from the folk-specifics and varietals whose variation humans had learned to manipulate, to intermediate-level families, and life-form classes, such as bird (Darwin 1883, pp. 353–4):

> [In the Galapagos Islands] There are twenty-six land birds; of these twenty-one or perhaps twenty-three are ranked a distinct species, and would commonly be assumed to have been here created; yet the close [family] affinity of most of these birds to American species is manifest in every character, in their habits, gestures, and tones of voice. So it is with other animals, and with a large proportion of plants . . . Facts such as these, admit of no sort of explanation on the ordinary view of creation.

Use of taxonomic hierarchies in systematics today reveals a similar point. For example, by calculating whether or not the taxonomic diversity in one group varies over time as a function of the taxonomic diversity in another group, evidence can be garnered for or against the evolutionary interdependence of the two groups. Comparisons of the relative numbers of families of insects and flowering plants reveal the surprising fact that insects were just as taxonomically diverse before the emergence of flowering plants as after. Thus, evolutionary effects of plant evolution on the adaptive radiation of insects are probably less profound than previously thought (Labandeira and Sepkoski, 1993). The heuristic value of (scientifically elaborated) folk-based strategies for cosmic inquiry is compelling, despite scientific awareness that no 'true' distinctions exist between various taxonomic levels.

7.2 The species-concept

In philosophy of biology, the current debate over the species-concept seems to centre on whether or not there is a single theoretically significant level of organization that covers all organisms (Kitcher, 1993; Sober, 1993). Accepting the primacy of evolutionary theory seems to rule out species-concepts that may be preferable on mainly pragmatic or operational grounds, such as historical primacy (Linnaean species), maximal covariance of many present and absent characters (pheneticists' basic taxonomic units), or minimally nested character–state distributions (speciation in pattern cladism).

Unfortunately, no one species-concept is presently able to simultaneously deal adequately with issues of inter-breeding (delimiting the boundaries of reproduction and gene flow), phylogenesis (fixing genealogical ascendance and descendance) and ecology (determining the geographical distribution of biodiversity) – all of which seem fundamental to the causal patterning and development of life on earth (Hull, 1997). One response has been to call for pluralism, yielding as many species-concepts as may accord with various equal, different or combined considerations from psychology, history, logic, metaphysics or the several branches of biology (Ehreshefsky, 1992; Dupré, 1993). From the perspective of a bystander eyeing the multiplication of uses and abuses pluralism seems able to generate, such an alternative could well leave not only truth, but clarity as well, in the abyss.

Perhaps the species-concept, like teleology, should be allowed to survive in science more as a regulative principle that enables the mind to establish a regular communication with the ambient environment, than as an epistemic principle that guides the search for nomological truth. Once communication is established with the world, science may discover deeper channels, or more significant overlapping networks, of causality. The persistence of a species-concept would function to ensure only that these diverse scientific explorations are never wholly disconnected or lost from one another, or from that aspect of phenomenal reality that will always remain as evident to a Maya Indian as to a modern scientist.

7.3 Teleology

Not only do taxonomic structure and species continue to agitate science – for better or worse – but so do the non-intentional and non-mechanical causal processes that people across the world assume to underlie the biological world. Vitalism is the folk belief that biological kinds – and their maintaining parts, properties and processes – are teleological, and hence not reducible to the contingent relations that govern inert matter. Its cultural expression varies (cf. Hatano and Inagaki, 1994). Within any given culture people may have varying interpretations and degrees of attachment to this belief: some who are religiously inclined may think that a 'spiritual' essence determines biological causality; others of a more scientific temperament might hold that systems of laws which suffice for physics and chemistry do not necessarily suffice for biology. Many, if not most, working biologists (including cognitive scientists) implicitly retain at least a minimal commitment to vitalism: they acknowledge that physico-chemical laws should suffice for biology, but suppose that such laws are not adequate in their current form, and must be enriched by further laws whose predicates are different from those of inert physics and chemistry.

It is not evident how complete elimination of teleological expressions (concepts defined functionally) from biological theory can be pursued without forsaking a powerful and fruitful conceptual scheme for physiology, morphology, disease and evolution. In cognitive science, a belief that biological systems, such as the mind–brain, are not wholly reducible to electronic circuitry like computers, is a pervasive attitude that implicitly drives considerable polemic, but also much creative theorizing. As Kant (1951[1790]) first intimated, even if vitalism represents a lingering folk belief that science ultimately seeks to discard, it remains an important and perhaps indispensable cognitive heuristic for regulating scientific inquiry.[7]

[7] Of course, most modern biologists no longer believe in strong vitalism, which holds that no reductive explanation is possible. Rather, many hold to a weaker form of vitalism, which disallows property or *type* reduction but allows reductive explanations of *token* biological processes. In other words, special-science processes must be realized in lower-level mechanisms on a token-for-token basis, although different token occurrences of the same higher-order type of process may have different varieties of lower-level realization. Strong vitalism is false, whereas weak vitalism might be true. My claim is only that there is a psychological bias to presume *some* sort of vitalism true.

8 Conclusion: emerging science in an evolutionary landscape

The evolutionary argument for a naturally selected cognitive disposition, such as folk biology, involves converging evidence from a number of venues: functional design (analogy), ethology (homology), universality, precocity of acquisition, independence from perceptual experience (poverty of stimulus), selective pathology (cerebral impairment), resistance to inhibition (hyperactivity) and cultural transmission. None of these criteria may be necessary, but the presence of all or some is compelling, if not conclusive.

8.1 Functional design

All organisms must function to procure energy to survive, and they also must procure (genetic) information for recombination and reproduction (Eldredge, 1986). The first requirement is primarily satisfied by other species, and an indiscriminate use of any individual of the other species (e.g. energy-wise, it does not generally matter which chicken or apple you eat). The second requirement is usually satisfied only by genetic information unique to individual conspecifics (e.g. genetically, it matters who is chosen as a mate and who is considered kin). On the one hand, humans recognize other humans by individuating them with the aid of species-specific triggering algorithms that 'automatically' co-ordinate perceptual cues (e.g. facial recognition schemata, gaze) with conceptual assumptions (e.g. intentions) (Baron-Cohen, 1995). On the other hand, people do not spontaneously individuate the members of other species in this way, but as exemplars of the (generic) species that identifies them as causally belonging to only one essential kind.

Natural selection basically accounts only for the appearance of complexly well-structured biological traits that are designed to perform important functional tasks of adaptive benefit to organisms. In general, naturally selected adaptations are structures functionally 'perfected for any given habit' (Darwin 1883, p. 140), having 'very much the appearance of design by an intelligent designer... on which the well-being and very existence of the organism depends' (Wallace 1889/1901, p. 138). Plausibly, the universal appreciation of generic species as the causal foundation for the taxonomic arrangement of biodiversity, and for taxonomic inference about the distribution of causally related properties that underlie biodiversity, is one such functional evolutionary adaptation. But a good story is not enough.[8]

[8] Although the adaptive relationship of structure to function is often manifest, as with the giraffe's neck or the rhinoceros' horns, often it is not. In such cases, evolutionary theorists adopt a strategy of 'reverse engineering'. Reverse engineering is what military analysts do when a weapon from an enemy or competitor in the arms market falls into their hands and they try to figure out exactly how it was put together and what it can do. Reverse engineering is easiest, of course, if the structure contains some signature of its function, like trying to figure out what a toaster does

8.2 *Ethology*

One hallmark of adaptation is a phylogenetic history that extends beyond the species in which the adaptation is perfected: for example, ducklings crouching in the presence of hawks, but not other kinds of birds, suggests dedicated mechanisms for something like species recognition. Some non-human species can clearly distinguish several different animal or plant species (Lorenz, 1966; Cerella, 1979; Herrnstein, 1984). Vervet monkeys even have distinct alarm calls for different predator species or groups of species: snake, leopard and cheetah, hawk and eagle, and so forth (Hauser, 2000). Chimpanzees may have rudimentary hierarchical groupings of biological groups within groups (Brown and Boysen, 2000). To be sure, the world itself is neither chaos nor flux: species are often locally self-structuring entities that are reproductively and ecologically isolated from other species through natural selection. But there is no *a priori* reason for the mind to always focus on categorizing and relating species *qua* species, unless doing so served some adaptive function (e.g. it makes little difference *which* tiger could eat a person or *which* mango a person could eat). And the adaptive functions of organisms rarely, if ever, evolve or operate in nature as all-purpose mechanisms.

8.3 *Universality*

Ever since the pioneering work of Berlin and his colleagues, evidence from ethnobiology and experimental psychology has been accumulating that all human societies have similar folk-biological structures (Berlin, Breedlove and Raven, 1973; Hunn, 1977; Hays, 1983; Brown, 1984; Atran, 1990, 1999; Berlin, 1992). These striking cross-cultural similarities suggest that a small number of organizing principles universally define folk-biological systems. Basic aspects of folk-biological structure (e.g. taxonomic ranking, primacy of generic-species) seem to vary little across cultures as a function of theories or belief systems.

given the tell-tale sign of toasted bread-crumbs left inside. But in many cases recognizing the appropriate signs already requires some prior notion of what function the structure may have served. Thus, after a century and a half of debate, it is only now that scientists clearly favour the hypothesis that bipedality was primarily selected to enhance field of view. Comparative studies of humans with bipedal birds and dinosaurs, as well as experiments comparing energy expenditure and running speed in two-footed vs. four-footed running and walking, appear to exclude the competing hypotheses that bipedality evolved for running or energy conservation. For most higher-order human cognitive faculties, however, there may be little useful comparative evidence from elsewhere in the animal kingdom. This is because of their apparent structural novelty, poor representation in the fossil record (e.g. stone tools tell little of language or theory of mind) and lack of surviving intermediate forms. The moral is that reverse engineering can be helpful, and occasionally successful, but success is by no means guaranteed even in the richest of evidentiary contexts.

8.4 *Ease of acquisition*

Acquisition studies indicate a precocious emergence of essentialist folk-biological principles in early childhood that are not applied to other domains (Keil, 1995; Hatano and Inagaki, 1999; Atran *et al.*, 2001).

8.5 *Independence from perceptual experience*

Experiments on inferential processing show that humans do not make biological inductions primarily on the basis of perceptual experience or any general similarity-based metric, but on the basis of imperceptible causal expectations of a peculiar, essentialist nature (Atran *et al.*, 1997; Coley, Medin and Atran, 1997).

8.6 *Pathology*

Cerebral impairments (Williams syndrome, brain lesions caused by certain types of herpes virus, etc.) suggest selective retention or loss of folk-biological taxonomies or of particular taxonomic ranks. Neuropsychological studies have reported a pathological performance in recognition at the life-form and generic-species levels (e.g. recognizing an item as an animal but not as a bird or robin), and dissociation at the life-form level (e.g. not recognizing items as trees). Existing studies, however, do not say anything about the generic-species rank as the preferred level of representation for reasoning, perhaps because of methodology (linked to averaging over items and failure to include sets of generic species) (Warrington and Shallice, 1984; Sartori and Job, 1988; Job and Surian, 1998).

8.7 *Inhibition and hyperactivity*

One characteristic of an evolved cognitive disposition is evident difficulty in inhibiting its operation (Hauser, 2000). Consider beliefs in biological essences. Such beliefs greatly help people explore the world by prodding them to look for regularities and to seek explanations of variation in terms of underlying patterns. This strategy may help bring order to ordinary circumstances, including those relevant to human survival. But in other circumstances, such as wanting to know what is correct or true for the cosmos at large, such intuitively ingrained concepts and beliefs may hinder more than help. For example, the essentialist bias to understand variation in terms of deviance is undoubtedly a hindrance to evolutionary thinking. In some everyday matters, the tendency to essentialize or explain variation in terms of deviation from some essential ideal or norm (e.g. people as mental or biological 'deviants') can be an effortlessly 'natural' but wrong way to think.

Because intuitive notions come to us so naturally they may be difficult to unlearn and transcend. Even students and philosophers of biology often find it difficult to abandon common-sense notions of species as classes, essences or natural kinds in favour of the concept of species as a logical individual – a genealogical branch whose endpoints are somewhat arbitrarily defined in the phyletic tree and whose status does not differ in principle for that of other smaller (variety) and larger (genus) branches. Similarly, racism – the projection of biological essences onto social groups – seems to be a cognitively facile and culturally universal tendency (Hirschfeld, 1996). Although science teaches that race is biologically incoherent, racial thinking is as notoriously difficult to suppress as it is easy to incite.

8.8 *Cultural transmission*

Human cultures favour the rapid selection and stable distribution of those ideas that: (1) readily help to solve relevant and recurrent environmental problems, (2) are easily memorized and processed by the human brain, and (3) facilitate the retention and understanding of ideas that are more variable (e.g. religion) or difficult to learn (e.g. science) but contingently useful or important. Folk-biological taxonomy readily aids humans everywhere in orienting themselves and surviving in the natural world. Its content tends to be stable within cultures (high inter-informant agreement, substantial historical continuity) and its structure isomorphic across cultures (see Boster, 1991; López *et al.*, 1997). Folk-biological taxonomy also serves as a principled basis for transmission and acquisition of more variable and extended forms of cultural knowledge.

Consider, for example, the spontaneous emergence of totemism – the correspondence of social groups with generic species – at different times and in different parts of the world. Why, as Lévi-Strauss (1963) aptly noted, are totems so 'good to think'? In part, totemism uses representations of generic species to represent groups of people; however, this pervasive meta-representational inclination arguably owes its recurrence to its ability to ride piggyback on folk-biological taxonomy. Generic species and groups of generic species are inherently well-structured, attention-arresting, memorable and readily transmissible across minds. As a result, they readily provide effective pegs on which to attach knowledge and behaviour of less intrinsically well-determined social groups. In this way totemic groups can also become memorable, attention-arresting and transmissible across minds. These are the conditions for any idea to become culturally viable. (See Sperber, 1996, for a general view of culture along the lines of an 'epidemiology of representations'.) A significant feature of totemism that enhances both memorability and its capacity to grab attention is that it violates the general behaviour of biological species: members of a totem, unlike members of a generic species, generally do not inter-breed, but mate only with members of other totems in order to create a system of social exchange. Notice that this violation of core knowledge is far from arbitrary. In fact, it is such a pointed violation of human beings' intuitive ontology that it readily mobilizes most of the assumptions people ordinarily make about biology in order to better help build societies around the world (Atran and Sperber, 1991).

In sum, folk-biological concepts are special players in cultural evolution, whose native stability derivatively attaches to more variable and difficult-to-learn representational forms, thus enhancing the latter's prospects for regularity and recurrence in transmission within and across cultures. This includes knowledge that cumulatively enriches (e.g. to produce folk expertise), overrides (e.g. to produce religious belief) or otherwise transcends (e.g. to produce science) the common-sense ontology prescribed by folk biology. The theory of evolution may ultimately dispense with core concepts of folk biology, including species, taxonomy and teleology, but in practice these may remain indispensable to scientific work. Theory-driven scientific knowledge also cannot banish folk knowledge from everyday life.

To say an evolved mental structure is 'innate' is not to say that every important aspect of its phenotypic expression is 'genetically determined'. The particular organisms observed, actual exemplars targeted and specific inferences made can vary significantly from person to person. Much as mountain rain will converge to the same mountain-valley river basin no matter where the rain falls, so each person's knowledge will converge on the same cognitive 'drainage basin' (Waddington, 1959; Sperber, 1996). This is because: (1) inputs naturally cluster in causally redundant ways inasmuch as that's the way the world is (e.g. where there are wings there are beaks or bills, where there are predators there are prey, where there are fruit-eating birds there are fruit-bearing trees, etc.); and (2) dedicated mental modules selectively target these inputs for processing by domain-specific inferential structures (e.g. to produce natural taxonomies). In this way, the mind is able to take fragmentary instances of a person's experience (relative to the richness and complexity of the whole data-set) and spontaneously predict (project, generalize) the extension of those scattered cases to an indefinitely large class of intricately related cases (of larger relevance to our species and cultures). Thus, many different people, observing many different exemplars of dog under varying conditions of exposure to those exemplars, all still generate more or less the same general concept of *dog*. Within this evolutionary landscape of medium-sized objects that are snapshot in a single lifespan of geological time, biologically-poised mental structures channel cognitive development but do not determine it. Cultural life, including science, can selectively target and modify parts of this landscape but cannot simply ignore it or completely replace it.

4 The roots of scientific reasoning: infancy, modularity and the art of tracking

Peter Carruthers

This chapter examines the extent to which there are continuities between the cognitive processes and epistemic practices engaged in by human hunter–gatherers, on the one hand, and those which are distinctive of science, on the other. It deploys anthropological evidence against any form of 'no-continuity' view, drawing especially on the cognitive skills involved in the art of tracking. It also argues against the 'child-as-scientist' accounts put forward by some developmental psychologists, which imply that scientific thinking is present in early infancy and universal among humans who have sufficient time and resources to devote to it. In contrast, a modularist kind of 'continuity' account is proposed, according to which the innately channelled architecture of human cognition provides all the materials necessary for basic forms of scientific reasoning in older children and adults, needing only the appropriate sorts of external support, social context and background beliefs and skills in order for science to begin its advance.

1 Introduction

It needs no emphasis that there has been a staggering and explosive increase in scientific knowledge, together with associated technological ability, over the last five centuries. But to what extent has this depended upon extrinsic cultural–economic factors, and to what extent upon intrinsic cognitive ones? Undoubtedly changes of both kinds have taken place, and have played a significant role. The invention of the printing press, and the existence of a class of moneyed gentlemen with time to devote to systematic scholarship and scientific enquiry were surely important; as were new inferential practices – both mathematical, and those distinctive of the experimental method. And without doubt changes of both kinds have continued to be important, too – had it not been for the development of new scientific instruments, and without the economic

I wish to thank Alex Barber, Nick Chater, Alison Gopnik, Michael Siegal and Stephen Laurence for their helpful comments on earlier drafts; with thanks also to all those who participated in the discussion of this material at the *Cognitive Basis of Science* conference held in Sheffield in June–July 2000.

growth necessary for significant resources to be committed to scientific research, we would certainly not be in the epistemic position we are in today; but the development of statistical methods of reasoning, for example, have also been crucially significant.

1.1 The questions

The questions I want to address in this chapter are these: just how *fundamental* were the cognitive changes necessary for science to develop (and continue)? And what is the balance between the respective extrinsic and intrinsic factors underlying science? Although these questions are vague, they are important. To the degree that extrinsic factors dominate, to that extent we will be able to say that there is a fundamental continuity between the cognitive processes of scientists and those living in pre-scientific cultures. (I shall refer to this as a form of 'continuity hypothesis'.) But to the extent that differences in cognitive processing are important, we may need to say that the cognition of pre-scientific peoples is radically distinct from our own. (Exaggerating a great deal, I shall call this the 'no-continuity hypothesis'.)

Just what sort of question is it that I am asking, however? Plainly, it is not intended to be a question about the causally necessary conditions for the development of science. For it may well be the case that *both* the printing press *and* beliefs about experimental method (say) were causally necessary factors in the scientific revolution. My question rather concerns the *nature* and *extent* of the cognitive changes necessary for science to begin. Continuity views maintain that any changes in mental functioning were relatively peripheral and minor; whereas no-continuity views hold that major cognitive re-structuring and a 're-programming' of the mind were necessary.

Put somewhat differently (and non-equivalently – see the following paragraph), my question concerns the *innate basis* of science and scientific reasoning.[1] According to a continuity account, most of the kinds of cognitive processing and reasoning which are necessary for science form part of the innate human cognitive endowment, needing only to be supplemented by some changes in belief or desire, and perhaps also by changes in external social–economic resources, in order for science to become possible. According to

[1] Exactly how should the concept of *innateness* be understood? This is a complex and difficult matter. As a first attempt, we might regard any cognitive feature as innate if (a) it makes its appearance in the course of normal development, and yet (b) it is not learned. Applying the concept 'innate' to difficult cases would then require us to know what learning is. I wish I did. A slightly more informative definition might be that a cognitive feature is innate if (a) its development is *channelled*, following the same developmental trajectory with the same developmental outcome across a wide variety of different circumstances, and yet (b) that developmental process doesn't admit of a complete description in cognitive terms. See Samuels (forthcoming) for discussion of some of these issues.

no-continuity accounts, in contrast, our innate cognitive endowment is by no means sufficient for scientific reasoning, and most of the necessary cognitive materials (beliefs and/or reasoning processes) need to be socially constructed and learned before science can start its advance.

Notice that these two different ways of raising our question – was major re-programming of the mind necessary for science to begin? How close does our innate endowment come to being sufficient for science? – are not strictly equivalent. For someone might claim that our innate endowment falls a long way short of what is needed for science, and yet claim that there is a form of culturally transmitted, but universal and trans-historical, cognition which *is* close to being sufficient. This would then be a form of continuity thesis without any commitment to the innateness of our scientific abilities. But actually the thesis which interests me, and which I shall defend in this chapter, is that our innate endowment *does* come pretty close to being sufficient for science, and it is because of this that only relatively minor cognitive changes were necessary for science to originate.

Everyone should allow, of course, that the innate cognitive basis of the mind (whatever it is) is held in common between scientists and hunter–gatherers; for we are all now members of the same contemporary species. (And it is, in any case, very implausible to suppose that substantive genetic change might have taken place – and in parallel, on a global scale – in the mere 10,000 years or so since the beginning of the agricultural revolution.) So it is obvious that the differences between scientists and hunter–gatherers are not innate ones. As a result, my question certainly isn't whether some change in the human genotype explains the genesis and rise of science in recent centuries. That would be absurd. Rather (to repeat), I am asking just how *close* our innate cognitive endowment comes to being *sufficient* for scientific reasoning. My sort of continuity account maintains that it comes pretty close – relatively few changes of belief and/or social circumstances were necessary for science to begin. No-continuity accounts, in contrast, claim that our innate endowment falls a long way short of being sufficient for science, and that they needed to be radically extended or re-programmed.

Equally, everyone should allow that there are immense cognitive differences between a contemporary scientist (or even a scientist of the sixteenth century) and a hunter–gatherer. The differences in their belief-systems will be vast; and no doubt some of the sorts of reasoning process in which they engage will differ also. (After all, it requires *training* to become a scientist, and not all of this training is of a purely practical, visuomotor, variety – some is mathematical, and some is methodological.) But it still might be possible to claim that their cognitive processes are essentially similar, if the differences are (almost) all differences of belief, and if (almost) all of the basic (innate) sorts of general cognitive processing in which they engage are both shared and sufficient (or almost sufficient) for science. My sort of continuity theory maintains that this is the case, whereas no-continuity accounts deny it.

1.2 The four options

On one view, the innate basis of the mind is mostly domain-general in nature, having to do with general capacities for learning and/or reasoning, though perhaps containing some initial domain-specific information and/or attention-biases (Elman *et al.*, 1996; Gopnik and Meltzoff, 1997). On a contrasting view, much of the innate structure of the mind is domain-specific, embodying information about evolutionarily significant domains, and/or containing learning-principles specific to particular domains (Barkow, Cosmides and Tooby, 1992; Pinker, 1997).

The domain-general account of the innate basis of cognition is one or another version of the *general-purpose computer model* of the mind. In some versions (e.g. Dennett, 1991, 1995) what is given are a suite of massively parallel and distributed processors which nevertheless have the power to support a serial, language-involving, digital processor running on linguistic structures. (Dennett dubs this the 'Joycean machine' after the stream-of-consciousness writing of James Joyce's *Ulysses*.) This latter system is almost entirely programmed by enculturated language-use, acquiring both its contents and general patterns of processing through the acquisition of both information and habits of thought from other people, via linguistic communication and language-based instruction and imitation. On this view, the basic cognitive differences between ourselves and hunter–gatherers will be very large; and there will be a great deal of cognitive–linguistic programming required before the human mind becomes capable of anything remotely resembling scientific reasoning.

Quite a different sort of domain-general view is entailed by 'theorizing theory' accounts of the nature of human development through infancy and childhood (e.g. Gopnik and Meltzoff, 1997; Gopnik and Glymour, chapter 6 in this volume).[2] On this view, all human children are already little scientists, in advance of any exposure to scientific cultures – gathering data, framing hypotheses and altering their theories in the light of recalcitrant data, in essentially the same sort of way that scientists do. So on this view, the cognitive continuities between scientific and pre-scientific cultures will be very great, and almost all the

[2] Gopnik and Meltzoff themselves refer to their account as the 'theory theory' of child development. I prefer to reserve this term to designate the wider range of theories (including most forms of modularism) which hold that many of the cognitive abilities achieved by children are structured into the form of a *theory* of the domains they concern. The term 'theory theory' is best reserved for these synchronic accounts of our abilities such as folk-physics and mind-reading, rather than for Gopnik and Meltzoff's diachronic version, according to which the theories in question are *arrived at by a process of theorizing* (as opposed, for example, to emerging through the maturation of a module).

emphasis in an explanation of the rise of science over the last five hundred years will have to be on extrinsic factors. On Gopnik and Meltzoff's account, most adult humans (including hunter–gatherers) are scientists who have ceased to exercise their capacity for science, largely through lack of time and attention. But this account can at the same time emphasize the need for extrinsic support for scientific cognition (particularly that provided by written language, especially after the invention of the printing press) once theories achieve a certain level of complexity in relation to the data.

Domain-specific, more-or-less modular, views of cognition also admit of a similar divide between no-continuity and continuity accounts of science. On one approach, the modular structure of our cognition which those of us in scientific societies share with hunter–gatherers is by no means sufficient to underpin science, even when supported by the appropriate extrinsic factors. Rather, that structure needs to be heavily supplemented by culturally developed and culturally transmitted beliefs and reasoning practices. In effect, this account can share with Dennett the view that a great deal of the organization inherent in the scientific mind is culturally acquired – differing only in the amount of innate underlying modular structure which is postulated, and in its answer to the question whether intra-modular processing is connectionist in nature (as Dennett, 1991, seems to believe), or whether it rather involves classical transformations of sentence-like structures (as most modularists and evolutionary psychologists think: see, e.g. Fodor, 1983, 2000; Tooby and Cosmides, 1992; Pinker, 1997).[3]

Alternatively, a modularist may emphasize the cognitive continuities between scientific and pre-scientific cultures, sharing with the theorizing theory the view that the basic ingredients of scientific reasoning are present as part of the normal human cognitive endowment. However, such an account will also differ significantly from theorizing theory by postulating a domain-specific cognitive architecture, and in claiming that capacities for scientific reasoning are not a cognitive *given* at the start of development, but rather emerge at some point along its normal course. This is the view which I shall ultimately be proposing to defend.

The chapter will proceed as follows. In section 2 I shall argue against either form of no-continuity account of science. That is to say, I shall argue against the claim that the human mind needs to be radically re-programmed through immersion in a suitable linguistic-cultural environment in order for science to become possible, whether this programming takes place within a broadly

[3] So notice that Fodor, too, can endorse some version of Dennett's *Joycean machine* hypothesis (except that the sentences will be sentences of Mentalese rather than of natural language). Since Fodor believes that central cognition is entirely *holistic* in nature and without much in the way of innate architecture, he can believe that whatever substantive structure it has is dependent on language-involving enculturation, just as Dennett does. He, too, can believe that all of the *interesting* central-cognitive processes are culturally constructed and acquired.

modularist, or rather in a general-learning, framework. Then in section 3 I shall argue against the sort of domain-general continuity account espoused by those who adopt a theorizing-theory view of normal human cognitive development. Finally, in section 4 I shall sketch my favoured alternative, which is a modularist form of continuity account. But in the limited space available I shall be able do little more than comment on the main components of the story.

2 Against no-continuity views

According to no-continuity accounts, the human mind needs to be radically re-programmed by immersion in an appropriate language-community and culture – acquiring a whole suite of new cognitive processes – in order for anything resembling science to become possible (Dennett, 1991). But against this view it can be argued that hunter–gatherers actually engage in extended processes of reasoning – both public and private – which look a lot like science. Developing this point will require us to say something about the cognitive processes which are actually characteristic of science, however.

2.1 Scientific reasoning

On one view, the goal of science is to discover the causal *laws* which govern the natural world; and the essential activity of scientists consists in the postulation and testing of *theories*, and then applying those theories to the phenomena in question (Nagel, 1961; Hempel, 1966). On a contrasting view, science constructs and elaborates a set of *models* of a range of phenomena in the natural world, and then attempts to develop and *apply* those models with increasing accuracy (Cartwright, 1983; Giere, 1992). But either way science generates principles which are *nomic*, in the sense of characterizing how things *have to* happen, and in supporting subjunctives and counter-factuals about what would happen, or would not have happened, if certain other things were to happen, or hadn't happened.

Crucial to the activity of science, then, is the provision of theories and/or models to explain the events, processes and regularities observed in nature. Often these explanations are couched in terms of underlying mechanisms which have not been observed and may be difficult to observe; and sometimes they are given in terms of mechanisms which are unobservable. More generally, a scientific explanation will usually postulate entities and/or properties which are not manifest in the data being explained, and which may be unfamiliar – where perhaps the only reason for believing in those things is that if they did exist, then they would explain what needs explaining.

Science also employs a set of tacit principles for choosing *between* competing theories or models – that is, for making an inference to the *best* explanation of the data to be explained. The most plausible way of picturing this is that contained within the principles employed for *good* explanation are enough constraints to allow one to rank more than one explanation in terms of goodness. While no one any longer thinks that it is possible to codify these principles, it is generally agreed that the good-making features of a theory include such features as: *accuracy* (predicting all or most of the data to be explained, and explaining away the rest); *simplicity* (being expressible as economically as possible, with the fewest commitments to distinct kinds of fact and process); *consistency* (internal to the theory or model); *coherence* (with surrounding beliefs and theories, meshing together with those surroundings, or at least being consistent with them); *fruitfulness* (making new predictions and suggesting new lines of enquiry); and *explanatory scope* (unifying together a diverse range of data).

2.2 *Hunter–gatherer science*

Is there any evidence that hunter–gatherer communities engage in activities which resemble science? It is a now familiar and well-established fact that hunter–gatherers have an immense and sophisticated (if relatively superficial) understanding of the natural world around them. They have extensive knowledge of the plant and animal species in their environments – their kinds, life-cycles and characteristic behaviours – which goes well beyond what is necessary for survival (Atran, 1990; Mithen, 1990, 1996). But it might be claimed that the cognitive basis for acquiring this sort of knowledge is mere inductive generalization from observed facts. It might appear that hunter–gatherers don't really have to engage in anything like genuine theorizing or model-building in order to gain such knowledge. Nor (except in their magic) do they seem, on the face of it, to rely on inferences concerning the unobserved. On the contrary, it can easily appear as if mere careful observation of the environment, combined with enumerative induction, is sufficient to explain everything that they know.[4]

In fact this appearance is deceptive, and at least some of the knowledge possessed by hunter–gatherers concerns facts which they have not directly observed, but which they know by means of inference to the best explanation of signs which they can see and interpret (Mithen, chapter 2 in this volume). For example, the !Xõ hunter–gatherers of the Kalahari are able to understand some of the nocturnal calls of jackals as a result of studying their spoor the next day and deducing their likely activities; and they have extensive knowledge of the lives of nocturnal animals derived from study of their tracks, some of which has only

[4] On an old-fashioned *inductivist* conception of science, then, it is plain that hunter–gatherers will already count as scientists. But I am setting the criterion for science and scientific reasoning considerably higher, to include inference to the best (unobserved–unobservable) explanation of the data (whether that data is itself observed or arrived at through enumerative induction).

recently been confirmed by orthodox science (Liebenberg, 1990). But it is the reasoning in which hunters will engage when tracking an animal which displays the clearest parallels with reasoning in science, as Liebenberg (1990) argues at length in his wonderful but little-noticed study in anthropology and philosophy of science. (Anyone who was ever tempted to think that hunter–gatherers must be cognitively less sophisticated than ourselves should read this book.)

2.3 *The art of tracking*

It is true, but by no means obvious at first glance, that tracking will always have played a vital role in most human hunter–gatherer communities. This is not especially because tracking is necessary to locate a quarry. For while this is important in many contexts and for many types of game, it is not nearly so significant when hunting herd animals such as wildebeest. It is rather because, until the invention of the modern rifle, it would always have been rare for a hunter to bring an animal down immediately with an arrow or a spear. (And while hunting in large groups might have made it more likely that the target animal could be brought down under a volley of missiles, it would have made it much *less* likely that the hunters would ever have got close enough to launch them in the first place.)

In consequence, much of the skill involved in hunting consists in tracking a wounded animal, sometimes for a period of days. (Even the very simplest form of hunting – namely, running an animal down – requires tracking.[5] For almost all kinds of prey animal can swiftly sprint out of easy sight, except in the most open country, and need to be tracked rapidly by a runner before they have the opportunity to rest.) For example, the !Xõ will generally hunt in groups of between two and four, using barbed arrows which have been treated with a poison obtained from the larvae of a particular species of beetle. An initial shot will rarely prove immediately fatal, and the poison can take between 6 and 48 hours to take effect, depending on the nature of the wound and the size of the animal. So a wounded animal may need to be tracked for considerable periods of time before it can be killed.

As Liebenberg (1990) remarks, it is difficult for a city-dweller to appreciate the subtlety of the signs which can be seen and interpreted by an experienced tracker. Except in ideal conditions (e.g. firm sand or a thin layer of soft snow) a mere capacity to recognize and follow an animal's spoor will be by no means sufficient to find it. Rather, a tracker will need to draw inferences from the

[5] Many different species of animal – including even cheetahs! – can be hunted in this way given the right sort of open country. Under the heat of an African midday sun our combined special adaptations of hairlessness and sweat-glands give us a decisive advantage in any long-distance race. Provided that the animal can be kept on the move, with no opportunity to rest, it eventually becomes so exhausted that it can be approached close enough to be clubbed or speared to death.

precise manner in which a pebble has been disturbed, say, or from the way a blade of grass has been bent or broken; and in doing so he will have to utilize his knowledge of the anatomy and detailed behaviours and patterns of movement of a wide variety of animals.[6] Moreover, in particularly difficult and stony conditions (or in order to save time during a pursuit) trackers will need to draw on all their background knowledge of the circumstances, the geography of the area and the normal behaviour and likely needs of the animal in question to make educated guesses concerning its likely path of travel.

Most strikingly for our purposes, successful hunters will often need to develop speculative hypotheses concerning the likely causes of the few signs available to them, and concerning the likely future behaviour of the animal; and these hypotheses are subjected to extensive debate and further empirical testing by the hunters concerned. When examined in detail these activities look a great deal like science, as Liebenberg (1990) argues. First, there is the invention of one or more hypotheses (often requiring considerable imagination) concerning the unobserved (and now unobservable) causes of the observed signs, and the circumstances in which they may have been made. These hypotheses are then examined and discussed for their accuracy, coherence with background knowledge, and explanatory and predictive power.[7] One of them may emerge out of this debate as the most plausible, and this can then be acted upon by the hunters, while at the same time searching for further signs which might confirm or count against it. In the course of a single hunt one can see the birth, development and death of a number of different 'research programmes' in a manner which is at least partly reminiscent of theory-change in science (Lakatos, 1970).

2.4 *Tracking: art or science?*

How powerful are these analogies between the cognitive processes involved in tracking, on the one hand, and those which underlie science, on the other? First of all we should note one very significant *dis*analogy between the two. This is that the primary *goal* of tracking is not an understanding of some general set of processes or mechanisms in nature, but rather the killing and eating of a particular animal. And although knowledge and understanding may be sought in pursuit of this goal, it is knowledge of the past and future movements of a particular prey animal, and an understanding of the causal mechanisms which produced a particular set of natural signs which is sought, in the first instance.

[6] I use the masculine pronoun for hunters throughout. This is because hunting has always been an almost exclusively male activity, at least until very recently.

[7] I haven't been able to find from my reading any direct evidence that trackers will also place weight upon the relative *simplicity* and internal *consistency* of the competing hypotheses. But I would be prepared to bet a very great deal that they do. For these are, arguably, epistemic values which govern a great deal of cognition in addition to hypothesis selection and testing (Chater, 1999).

(Of course the hunters may also hope to obtain knowledge which will be of relevance to future hunts.)

This disanalogy is sufficient, in my view, to undermine any claim to the effect that tracking *is* a science. Since it doesn't share the same universal and epistemic aims of science, it shouldn't be classed as such. But although it isn't a science, it is perfectly possible that the cognitive processes which are involved in tracking and in science are broadly speaking the same. So we can still claim that the basic cognitive processes involved in each are roughly identical, albeit deployed in the service of different kinds of end. Indeed, it is just such a claim which is supported by the anthropological data.

First of all, it is plain that tracking, like science, frequently involves inferences from the observed to the unobserved, and often from the observed to the unobservable as well. Thus a tracker may draw inferences concerning the effects which a certain sort of movement by a particular kind of animal would have on a certain kind of terrain (which may never actually have been observed previously, and may be practically unobservable to hunter–gatherers if the animal in question is nocturnal). Or a tracker may draw inferences concerning the movements of a particular animal (namely the one he had previously shot and is now tracking) which are now unobservable even in principle, since those movements are in the past. Compare the way in which a scientist may draw inferences concerning the nature of a previously unobserved and practically unobservable entity (e.g. the structure of DNA molecules before the invention of the electron microscope), or concerning the nature of particles which are too small to be observable, by means of an inference to the best explanation from a set of broadly observational data.

Second, it is plausible that the patterns of inference engaged in within the two domains of tracking and science are isomorphic with one another. In each case inferences to the best explanation of the observed data will be made, where the investigators are looking for explanations which will be simple, consistent, explanatory of the observational data, coherent with background beliefs, maximal in explanatory scope (relevant to the aims of the enquiry, at least), as well as fruitful in guiding future patterns of investigation. And in each case, too, imaginative–creative thinking has a significant – nay, crucial – role to play in the generation of novel hypotheses.

Let me say something more about the role of creative thinking in tracking and in science. It is now widely accepted that inductivist methodologies have only a limited part to play in science. Noticing and generalizing from observed patterns cannot carry you very far in the understanding of nature. Rather, scientists need to propose *hypotheses* (whether involving theories or models) concerning the underlying processes which produce those patterns. And generating such hypotheses cannot be routinized. Rather, it will involve the imaginative postulation of a possible mechanism – guided and constrained by background knowledge, perhaps, but not determined by it. Similarly, a hunter may face a range of novel observational data which need interpreting. He has to propose a hypothesis concerning the likely causes of that data – where again, the hypothesis will be guided and constrained by knowledge of the circumstances, the season, the behaviour of different species of animal and so on, but is not by any means determined by such knowledge. Rather, generating such hypotheses requires creative imagination, just as in science.

These various continuities between tracking and science seem to me sufficient to warrant the following claim: that anyone having a capacity for sophisticated tracking will also have the basic cognitive wherewithal to engage in science. The differences will be merely these – differences in overall *aim* (to understand the world in general, as opposed to the movements of a given animal in particular); differences of *belief* (including methodological beliefs about appropriate experimental methods, say); as well as some relatively trivial differences in inferential practices (such as some of the dispositions involved in doing long-division sums, or in solving differential equations). In which case some version of the continuity hypothesis can be regarded as established. We can assert that the cognitive processes of hunter–gatherers and modern scientists are broadly continuous with one another, and that what was required for the initiation of the scientific revolution were mostly extrinsic changes, or changes relating to peripheral (non-basic) aspects of cognition, such as mere changes of belief.

It remains to enquire, though, roughly *when* and *how* the capacity for scientific reasoning emerges in the course of human development, and how it relates to other aspects of our cognition. In the next section I shall consider and criticize theorizing-theory accounts of development, according to which scientific abilities are innate and put to work very shortly after birth. Then in section 4 I shall sketch a view which also allows that scientific abilities are innately channelled, but which sees their development as dependent upon language – perhaps emerging only in later childhood or early adolescence (I shall remain open as to the exact timing).

3 Against the 'theorizing theory' continuity view

The orthodox position in cognitive science over the last two decades has been that many of our cognitive capacities – including the capacity to attribute mental states to oneself and others (i.e. the capacity to 'mind-read'), and the capacity to explain and predict many of the behaviours of middle-sized physical objects – are subserved by bodies of belief or knowledge, constituting so-called 'folk theories' of the domains in question. But a number of developmental psychologists have gone further in suggesting that these theories are developed in childhood *through a process of theorizing* analogous to scientific theorizing (e.g. Carey, 1985; Wellman, 1990). This theorizing-theory approach has been developed and defended most explicitly by Gopnik and Meltzoff (1997), and it is on their account that I shall focus my critique.

3.1 The absence of external supports

In my view, the main objection to a theorizing-theory account is that it ignores the extent to which scientific activity needs to be supported by external resources.[8] Scientists do not, and never have worked alone, but constantly engage in discussion, co-operation and mutual criticism with peers. If there is one thing which we have learned over the last thirty years of historically oriented studies of science, it is that the positivist–empiricist image of the lone investigator, gathering all data and constructing and testing hypotheses by himself or herself, is a highly misleading abstraction.

Scientists such as Galileo and Newton engaged in extensive correspondence and discussion with other investigators at the time when they were developing their theories; and scientists in the twentieth century, of course, have generally worked as members of research teams. Moreover, scientists cannot operate without the external prop of the written word (including written records of data, annotated diagrams and graphs, written calculations, written accounts of reasoning, and so on). Why should it be so different in childhood, if the cognitive processes involved are essentially the same?

I should emphasize that this point doesn't at all depend upon any sort of 'social constructivist' account of science, of the sort that Gopnik and Meltzoff find so rebarbative (and rightly so, in my view). It is highly controversial that scientific change is to be explained, to any significant extent, by the operation of wider social and political forces, in the way that the social constructivists claim (Bloor, 1976; Rorty, 1979; Latour and Woolgar, 1986; Shapin, 1994). But it is, now, utterly truistic that science is a social process in at least the minimal sense that it progresses through the varied social interactions of scientists themselves – co-operating, communicating, criticizing – and through their reliance on a variety of external socially provided props and aids, such as books, paper, writing instruments, a variety of different kinds of scientific apparatus, and (now) calculators, computers and the internet (Giere, chapter 15 in this volume). And this truism is sufficient to cause a real problem for the 'child as scientist' account of development.

One sort of response would be to claim that children do not *need* external props, because they have vastly better memories than adult scientists. But this is simply not credible, of course; for it is not true that children's event-memories are better than those of adults.[9] (Nor do young children have any memory for their own previous mental processes. And in particular, lacking any concept of belief as a representational state, children under four cannot remember their

[8] See Stich and Nichols (1998), Harris (chapter 17 in this volume) and Foucher *et al.* (chapter 18 in this volume) for development of a range of alternative criticisms.

[9] Admittedly the evidence here is confined to verbally expressible memories; so it could in principle be claimed that children have vastly better *non-conscious* memories than adults do. But there is no uncontentious evidence for this.

own previous false beliefs – Gopnik, 1993.)[10] And in any case it isn't true that science depends upon external factors *only* because of limitations of memory. On the contrary, individual limitations of rationality, insight and creativity all play an equally important part. For scientific discussion is often needed to point out the fallacies in an individual scientist's thinking; to show how well-known data can be explained by a familiar theory in ways that the originators hadn't realized; and to generate new theoretical ideas and proposals.[11]

Nor can these differences between children and adults plausibly be explained in terms of differences of attention, motivation, or time, in the way that Gopnik and Meltzoff try to argue. For adult scientists certainly attend very closely to the relevant phenomena, they may be highly motivated to succeed in developing a successful theory, and they may be able to devote themselves full-time to doing so. But still they cannot manage without a whole variety of external resources, both social and non-social. And still *radical* (conceptually-innovative) theory change in science (of the sort which Gopnik and Meltzoff acknowledge occurs a number of times over within the first few years of a child's life) is generally spread out over a very lengthy time-scale (often as much as a hundred years or more; Nersessian, 1992a and chapter 7 in this volume).

3.2 The extent of the data

According to Gopnik and Meltzoff (1997), the main difference between childhood theorizers and adult scientists lies in the extent and ease of availability of relevant data. In scientific enquiry, the relevant data are often hard to come by, and elaborate and expensive experiments and other information-gathering exercises may be needed to get it, perforce making scientific enquiry essentially social. But in childhood there are ample quantities of data easily available. Young children have plenty of opportunities to experiment with physical substances and their properties – knocking objects together, dropping or throwing them, pouring liquids, and mixing materials – when developing their naïve physics; and they have plenty of opportunity to observe and probe other agents

[10] The fact that younger children are incapable of thinking about their own previous mental states, as such, raises yet further problems for theorizing-theory. For while it may be *possible* to engage in science without any capacity for higher-order thinking – when revising a theory T one can think, '*These* events show that not-T', for example – in actual fact scientific theorizing is shot through with higher-order thoughts. Scientists will think *about* their current or previous theories as such, wonder whether the data are sufficient to show them false, or true, and so on. It therefore looks as if a theorizing-theorist will have to claim that these higher-order thoughts are actually mere epiphenomena in relation to the real (first-order) cognitive processes underlying science. But this is hardly very plausible.

[11] And note that tracking, in contrast, *is* generally social and externally supported in just this sense – for as we have seen, hunters will characteristically work collaboratively in small groups, pooling their knowledge and memories, and engaging in extensive discussion and mutual criticism in the course of interpreting spoor and other natural signs.

when developing their naïve psychology as well. Moreover, since infants come into the world with a set of innate domain-specific theories, on Gopnik and Meltzoff's account, they already possess theoretical frameworks which constrain possible hypotheses, determine relevant evidence, and so on.

This point, although valid as far as it goes, does not begin to address the real issue. If anything, the extent of the data available to the child is a further *problem* for the 'child as scientist' view, given the lack of external aids to memory, and the lack of any public process for sorting through and discussing the significance of the evidence. That plenty of data is *available* to an enquirer is irrelevant, unless that data can be recalled, organized and surveyed at the moment of need – namely, during theory testing or theory development.

Gopnik and Meltzoff also make the point that much relevant data is actually *presented to* children by adults, in the form of linguistic utterances of one sort or another. Now, their idea is not that adults *teach* the theories in question to children, thus putting the latter into the position of little science *students* rather than little scientists. For such a claim would be highly implausible – there is no real evidence that any such teaching actually takes place (and quite a bit of evidence that it does not). Their point is rather that adults make a range of new sorts of *evidence* available to the child in the form of their linguistic utterances, since those utterances are embedded in semantic frameworks which contain the target theories. Adult utterances may then provide crucial data for children, as they simultaneously elaborate their theories and struggle to learn the language of their parents.

This proposal does at least have the virtue of providing a social dimension to childhood development, hence in one sense narrowing the gap between children and adult scientists. But actually this social process is quite unlike any that an adult scientist will normally engage in, since scientists (as opposed to science students) are rarely in the position of hearing or engaging in discussions with those who have already mastered the theories which they themselves are still trying to develop. And in any case the proposal does nothing to address the fundamental problems of insufficient memory, bounded rationality and limited creativity which children and adults both face, and which the social and technological dimensions of science are largely designed to overcome.

3.3 *Simple theories?*

Might it be that the scientific problems facing adults are very much more complex than those facing young children, and that this is the reason why children, but not adult scientists, can operate without much in the way of external support? This suggestion is hardly very plausible, either. Consider folk psychology, for example, of the sort attained by most normal four- or five-year-old children.

This has a deep structure rivalling that of many scientific theories, involving the postulation of a range of different kinds of causally effective internal states, together with a set of nomic principles (or 'laws') concerning the complex patterns of causation in which those states figure. There is no reason at all to think that this theory should be easily arrived at. Indeed, it is a theory which many adult scientific psychologists have *denied* (especially around the middle part of the twentieth century) – viz. those who were behaviourists. It took a variety of sophisticated arguments, and the provision of a range of different kinds of data, to convince most people that behaviourism should be rejected in favour of cognitivism.

So why is it that *no* children (excepting perhaps those who are autistic) ever pass through a stage in which they endorse some form of behaviourism? If cognitivism were really such an easy theory to frame and establish, then the puzzle would be that adult scientific psychologists had such difficulty in converging on it. And given that it is *not* so easy, the puzzle is that all normal children *do* converge on it in the first four or five years of development (at least if they are supposed to get there by means of processes structurally similar to those which operate in science).[12] The natural conclusion to draw, at this point, is that folk psychology is both theoretically difficult and *not* arrived at by a process of theorizing, but rather through some sort of modular and innately channelled development.

I conclude this section, then, by claiming that the child-as-scientist account is highly implausible. Although this account (*contra* the no-continuity views discussed in section 2) at least has the virtue of emphasizing the continuities between childhood development, hunter–gatherer thinking and reasoning and adult scientific cognition, it suffers from a number of fatal weaknesses. Chief among these is that it ignores the extent to which scientific enquiry is a social and externally supported process.

4 A modularist continuity view

The anthropological data reviewed in section 2 give us reason to think that some version of the continuity thesis is correct – that is, for thinking that the basic cognitive processes of ourselves and hunter–gatherers are fundamentally the same. But now, in section 3, we have seen that there is good reason to reject the domain-general form of continuity thesis defended by theorizing-theorists such as Gopnik and Meltzoff. So it looks as if a domain-specific form

[12] Worse still, it is not just normal children who succeed in acquiring folk psychology. So do children with Down's Syndrome and Williams' Syndrome, who in many other respects can be severely cognitively impaired, and whose general learning and theorizing abilities can be well below normal.

of continuity view must be established by default. For this is the only alternative now remaining. Plainly, however, we cannot let matters rest here. We need to show how a modularist form of continuity account can be true, and how it can explain the data (both anthropological and developmental). All I can really hope to do here, however, is assemble the various materials which should go into a properly developed form of modularist continuity-view of science. I cannot at this stage pretend to be offering such an account in any detail.

4.1 Developing modules

The form of continuity view which I favour helps itself to a version of what might be called 'central-process modularism' (as distinct from Fodor's 'peripheral-process modularism', 1983). On this account, besides a variety of input and output modules (including early vision, face-recognition, and language, for example), the mind also contains a number of innately channelled conceptual modules, designed to process conceptual information concerning particular domains. While these systems might not be modular in Fodor's classic sense – they would not have proprietary inputs, and might not be fully encapsulated, for example – they would conform to at least some of the main elements of Fodorian modularity. They would be innate (in some sense or other) and subject to characteristic patterns of breakdown; their operations might be mandatory and relatively fast; and they would process information relating to their distinctive domains according to their own specific algorithms.

Plausible candidates for such conceptual modules might include a naïve physics system (Leslie, 1994b; Spelke, 1994; Baillargeon, 1995; Spelke, Vishton and von Hofsten, 1995), a naïve psychology or 'mind-reading' system (Carey, 1985; Leslie, 1994b; Baron-Cohen, 1995), a folk-biology system (Atran, 1990, 1998 and chapter 3 in this volume), an intuitive number system (Wynn, 1990, 1995; Gallistel and Gelman, 1992; Dehaerne, 1997), and a system for processing and keeping track of social contracts (Cosmides and Tooby, 1992). And evidence supporting the existence of at least the first two of these systems (folk-physics and folk-psychology) is now pretty robust. Very young infants already have a set of expectations concerning the behaviours and movements of physical objects, and their understanding of this form of causality develops very rapidly over the first year or two of life. And folk-psychological concepts and expectations also develop very early, and follow a characteristic developmental profile. Indeed, recent evidence from the study of twins suggests that three-quarters of the variance in mind-reading abilities among three-year-olds is both genetic in origin and largely independent of the genes responsible for verbal intelligence, with only one-quarter being contributed by the environment (Hughes and Plomin, 2000).

Some of the disputes which are internal to central-process modularism – in particular, concerning which conceptual modules are or are not developmentally primitive – are highly relevant to our present concerns. Thus Carey (1985) thinks that children's folk-biology is built from an initial given folk-psychology by a process of theorizing, whereas Atran (1990 and chapter 3 in this volume) conceives of our folk-biology system as a distinct module in its own right. If Carey is right, then plainly quasi-scientific theorizing abilities will need to be in place through the early school years, since children's grasp of the main principles of biological life-cycles, for example, is well established by about the age of seven. Whereas if folk-biology emerges at about this time through a process of maturation rather than of theorizing (albeit maturation in the context of a set of normal as well as environment-specific social inputs), then scientific abilities might not emerge until much later. I propose simply to set this dispute to one side for present purposes, on the simple but compelling grounds that I have nothing to contribute to it.

One important point to notice in this context is that many of these conceptual modules provide us with deep *theories* of the domains which they concern. Thus the folk-physics system involves a commitment to a number of unobservable forces or principles, such as the *impetus* of moving objects, or the *solidity* of impenetrable ones. And the folk-psychology module involves commitment to a variety of inner psychological states, together with the causal principles governing their production, interaction and effects on behaviour. These modules might therefore very plausibly be thought to provide us with a set of initial theoretical contents, on which our explanatory and abductive abilities can later be honed.

4.2 Cross-modular thinking and language

There is some tentative reason to think that language may have an important role to play in the genesis of scientific reasoning. For notice that the reasoning processes involved in tracking an animal will often require bringing together information from a variety of different intuitive domains, including folk-physics, folk-biology and folk-psychology. For example, in interpreting a particular scuff-mark in the dust a hunter may have to bring information concerning the anatomy and behaviours of a number of different animal species (folk-biology) together with inferences concerning the physical effects of certain kinds of impact between hoof and ground, say (folk-physics). And in predicting what an animal will do in given circumstances a hunter will rely, in part, on his folk-psychology – reasoning out what someone with a given set of needs and attitudes would be likely to do in those circumstances (Liebenberg, 1990).

On one sort of view of the role of natural language in cognition, then – the view, namely, that language serves as the link between a number of distinct cognitive systems or modules (Mithen, 1996; Carruthers, 1996a, 1998; Hermer-Vazquez, Spelke and Katsnelson, 1999) – speculative intelligent tracking would have depended on the evolution of the language-faculty. So it may be that hunting-by-tracking fully entered the repertoire of the hominid lineage only some 100,000 years ago, with the first appearance of anatomically modern, language-using, humans in Southern Africa, as Liebenberg (1990) suggests. And it may then be that the cognitive adaptations necessary to support scientific thinking and reasoning were selected for precisely because of their important role in hunting.[13] But it might also be that these abilities were selected for on other grounds, only later finding application in tracking and hunting. It would go well beyond the scope I have set myself in this chapter to try to resolve this issue here. For present purposes the point is just that sophisticated cross-modular abductive reasoning may crucially implicate the language faculty (as also might reasoning which extends or corrects our naïve modular theories, as frequently happens in science).

4.3 Developing imagination

In addition, a capacity for imaginative or creative thinking is vital in both tracking and science, as we have seen. Novel explanatory hypotheses need to be generated in both domains – hypotheses which may be unprecedented in previous experience, and may go well beyond the observational data. There is no doubt that such a capacity is a distinctive element in the normal human cognitive endowment. And on one natural view, it is the developmental function of the sort of pretend play which is such a striking and ubiquitous part of human childhood to enhance its growth, through *practice* in suppositional or pretend thinking (Carruthers, 1996b, 1998, 2002).

What cognitive changes were necessary for imaginative thinking and/or pretend play to make its appearance in the hominid lineage? Nichols and Stich (2000) argue that it required the creation of a whole new type of propositional attitude, which they represent as a separate 'box' in the standard flow-chart of cognition involving boxes for belief, desire, practical reasoning, theoretical inference, and so on – namely a 'possible worlds box'. And this might be taken to suggest that it would have required some powerful selectional pressure in order for imaginative thinking to become possible. But actually there is some reason to think that imagination comes in at least two forms, each of which

[13] Liebenberg (1990) speculates that it may have been the *lack* of any sophisticated tracking abilities which led to the near-extinction of the Neanderthals in Europe some 40,000 years ago. At this time the dramatic rise in world temperatures would have meant that they could no longer survive by their traditional method of hunting by simple tracking-in-snow through the long winter months, and then surviving through the brief summers on dried meats, scavenging, and gathering.

would be provided 'for free' by the evolution of other faculties. Let me briefly elaborate.

First, there is *experiential* imagination – namely, the capacity to form and to manipulate images relating to a given sense modality (visual, auditory, etc.). There is some reason to think that a basic capacity for this sort of imagination is a by-product of the conceptualizing processes inherent in the various perceptual input-systems (Kosslyn, 1994). There are extensive feed-back neural pathways in the visual system, for example, which are used in object-recognition when 'asking questions of' ambiguous or degraded input. And these very pathways are then deployed in visual imagination so as to generate quasi-perceptual inputs to the visual system. Evidence from cognitive archaeology (concerning the imposition of sophisticated symmetries on stone tools, which would have required a capacity to visualize and manipulate an image of the desired product) suggests that this capacity would have been present at about 400,000 years before the present (Wynn, 2000) – i.e. considerably before the evolution of full-blown language, if the latter only appeared some 100,000 years ago, as many believe.

Second, there is *propositional* imagination – the capacity to form and consider a propositional representation without commitment to its truth or desirability. There is some reason to think that this capacity comes to us 'for free' with language. For a productive language system will involve a capacity to construct new sentences, whose contents are as yet neither believed nor desired, which can then serve as objects of reflective consideration of various sorts. In which case a capacity for propositional imagination is likely to have formed a part of the normal human cognitive endowment for about the last 100,000 years.[14]

Of course a mere capacity for creative generation of new sentences (or images) will not be sufficient for imaginative thinking as we normally understand it. For there is nothing especially imaginative about generating any-old new sentence or image. Rather, we think imagination consists in the generation of *relevant* and/or *fruitful and interesting* new ideas. And such a capacity will not come to us 'for free' with anything. But then this may be precisely the developmental function of pretend play, if through frequent practice such play helps to build a consistent disposition to generate novel suppositions which *will* be both relevant and interesting (Carruthers, forthcoming).

Both forms of imagination play a role in science (Giere, 1992, 1996a). And both forms are important in tracking as well. Thus by visualizing the distinctive

[14] This is not to say that such a capacity would always have been frequently *employed* throughout this time. On the contrary, what may well have happened between the first emergence of anatomically modern humans and the 'cultural–creative explosion' which occurred worldwide only some 40,000 years ago was selection for a disposition to engage in pretend play in childhood – selection which may have been driven by the evident usefulness of imaginative thinking, not only in tracking but in many other forms of activity. See Carruthers (1998 and forthcoming).

gait of a wildebeest a tracker may come upon the hypothesis that some faint marks in the dust on a windy day were produced by that animal. And by recalling that a zebra was seen in the vicinity the previous afternoon, he may frame the alternative hypothesis that the marks were produced by that animal instead.

4.4 Developing abductive principles

How are the principles involved in inference to the best explanation (or 'abductive inference') acquired? Are they in some sense innate (emerging without learning in normal humans in the course of development in normal environments)? Or are they learned? Plainly, the mere fact that such principles are employed by adult humans in both hunter–gatherer and scientific societies is not sufficient to establish their innateness. For it may be that broadly scientific patterns of thinking and reasoning were an early cultural invention, passed on through the generations by imitation and overt teaching, and surviving in almost all extant human societies because of their evident utility.

In fact, however, it is hard to see how the principles of inference to the best explanation could be other than substantially innate (Carruthers, 1992). For they are certainly not explicitly taught, at least in hunter–gatherer societies.[15] While nascent trackers may acquire much of their background knowledge of animals and animal behaviour by hearsay from adults and peers, very little overt teaching of tracking itself takes place. Rather, young boys will practise their observational and reasoning skills for themselves, first by following and interpreting the tracks of insects, lizards, small rodents and birds around the vicinity of the camp-site, and then in tracking and catching small animals for the pot (Liebenberg, 1990). They will, it is true, have many opportunities to listen to accounts of hunts undertaken by the adult members of the group, since these are often reported in detail around the camp fire. So there are, in principle, opportunities for learning by imitation. But in fact, without at least a reasonably secure grasp of the principles of inference to the best explanation, it is hard to see how such stories could even be so much as *understood*. For of course those principles are never explicitly articulated; yet they will be needed to make sense of the decisions reported in the stories; and any attempt to uncover them by inference would need to rely upon the very principles of abductive inference which are in question.

The question of how a set of abductive principles came to be part of the normal human cognitive endowment is entirely moot, at this point. One possibility

[15] Nor are they taught to younger school-age children in our own society. Yet experimental tests suggest that children's reasoning and problem-solving is almost fully in accord with those principles, at least once the tests are conducted within an appropriate scientific–realist framework (Koslowski, 1996, Koslowski and Thompson, chapter 9 in this volume). This is in striking contrast with many other areas of cognition, where naïve performance is at variance with our best normative principles (see Evans and Over, 1996 and Stein, 1996, for reviews).

is that they were directly selected for because of their crucial role in one or more fitness-enhancing activities, such as hunting and tracking. It may be, for example, that successful hunting (and hence tracking) assumed an increasing importance for our ancestors, either during the transition to anatomically modern humans some 100,000 years ago or shortly thereafter. Note, however, that evidence from contemporary hunter–gatherers suggests that the selective force at work would have been sexual rather than 'natural'. For almost all hunter–gatherer tribes which have been studied by anthropologists are highly egalitarian in organization, with meat from successful hunts being shared equally among all family groups. In which case successful hunting would not have improved the survival-chances of the hunter or his off-spring directly. Nevertheless, successful hunters are highly respected; and there is evidence, both that they father significantly more children than do other men, and (somewhat more puzzling) that their children have a greater chance of surviving to reproductive age (Boyd and Silk, 1997).[16]

Another possibility (although not one which I really understand, I confess) is that the principles of abductive inference originally evolved to govern the intra-modular processing of our naïve physics or naïve psychology systems (or both), and that they somehow became available to domain-general cognition through some sort of 'representational re-description' (Karmiloff-Smith, 1992). For notice that the contents of these modules are at least partially accessible to consciousness (they are not fully encapsulated), since we can articulate some of the principles which they employ. In which case it is possible that whatever developmental and/or evolutionary processes resulted in such accessibility might also have operated in such a way as to make accessible the abductive inference-principles which they use. (I suppose it is some support for this proposal that mechanics and folk psychology continue to form our main paradigms of causal explanation, to which we are inclined to assimilate all others.)

The evolutionary explanation of abductive inference-principles falls outside the remit of this chapter, fortunately. For present purposes what matters is that there is good reason to think that they *are* a part of our natural cognitive endowment.

4.5 *How to build a tracker or a scientist*

I am suggesting, then, that a number of distinct cognitive elements had to come together in just the right way – by accident or natural selection – in order for

[16] Does this proposal predict that men should be much better than women at abductive inference? Not necessarily, since sexually selected traits are normally initially possessed by both sexes, as a result of the genetic correlation between the sexes (Miller, 2000). Then provided that abductive inference soon came to be useful to women as well, it would thereafter have been *retained* by both sexes. And certainly the reproductive success of hunter–gatherer women today is determined much more by their own resourcefulness than by fertility (Hrdy, 1999).

both sophisticated tracking and scientific reasoning to become possible. The ingredients are as follows:

(1) A variety of innately channelled conceptual modules, including folk physics, folk psychology, folk biology, and perhaps some sort of 'number sense'.
(2) An innately channelled language-system which can take inputs from any of these modular systems, as well as providing outputs to them in turn. This system thus has the power to link together and combine the outputs of the others.
(3) An innate capacity for imagination, enabling the generation of new ideas and new hypotheses.
(4) An innately channelled set of constraints on theory choice, amounting to a distinct non-modular faculty of inference to the best explanation.

With all these ingredients in place, humans had the capacity to become trackers or scientists, depending on their motives, interests and the use to which those ingredients were put.

4.6 And is that all?

If this sketch is on the right lines, then the changes which had to take place in facilitating the scientific revolution were mostly extrinsic ones, or involved only minor cognitive changes. People needed to begin taking as an explicit goal the systematic investigation and general understanding of the natural world. (It may be that the pursuit of knowledge, as such, was already an innately given goal – see Papineau, 2000.) They needed to have the leisure to pursue that goal. They needed a variety of external props and aids – the printing press, scientific instruments and a culture of free debate and exchange of information and ideas – in order to help them to do it. And on the cognitive side, they needed the crucial belief that in evaluating theories it is a good idea to seek for experimental tests. But they did not need any fundamentally new cognitive re-programming.

Of course, much depends on what one counts as 'fundamental', here. In particular, a powerful case can be made for the significance of mathematical knowledge and mathematical skills and practices in making the scientific revolution possible. And it might be claimed that a mind which reasons in accordance with such knowledge is fundamentally different from one which does not. In which case the claim that our innate cognitive endowment is *nearly* sufficient for science would turn out to be a considerable exaggeration.

Here I am perfectly prepared to be concessive. For in the end it really doesn't matter whether or not the label 'continuity thesis' is applicable. What matters is the detailed account of the cognitive capacities and social practices underpinning science, and the facts concerning which of these are part of our innate endowment, and which are culturally acquired. If the views I have been sketching and defending in this chapter are correct, then we can conclude that the former (innately endowed) set is much more extensive than is generally recognized, and in a way which is quite distinct from that proposed by developmental theorizing-theorists.

5 Conclusion

I have argued that the cognitive processes of hunter–gatherers and scientists are broadly continuous with one another, and that it did not require any extensive cognitive re-programming of the human mind to make the scientific revolution possible. But I have also argued that it is unlikely that a capacity for scientific reasoning is present in normal human infants soon after birth, and is then put to work in developing a series of naïve theories. Rather, the capacity for scientific reasoning will emerge at some later point in childhood, through the linking together of a number of innately-given cognitive faculties, both modular and non-modular.

Part two

Science and cognition

5 Science without grammar: scientific reasoning in severe agrammatic aphasia

Rosemary Varley

The issue explored in this chapter is whether the forms of reasoning that are utilized by scientists necessarily require the resources of the language faculty. The types of reasoning investigated include the processes involved in hypothesis generation, testing and revision, as well as various forms of causal understanding. The introspective evidence from human problem-solvers suggests that performances on cognitively demanding tasks are often accompanied by inner speech, and that the stages in the solution of the problem, or possible strategies at each stage, are laid out in fully explicit linguistic propositions. But although such introspective reports are pervasive, the question remains as to whether the natural language propositions of inner speech are an optional gizmo that serve to support and scaffold other cognitive performances, or whether such propositions are fundamental to sophisticated cognition – so that without access to such propositions, forms of thinking such as those that underpin science would be impossible.

This issue was explored through the examination of the performances of two men with severe aphasia (SA and MR) on a series of reasoning tasks that involved components of scientific thinking. Both patients display severe grammatical impairments that result in an inability to access language propositions in any modality of use (spoken or written). Despite profound disruption of propositional language, SA showed preserved ability on all tasks. By contrast, MR showed satisfactory performance on the causal reasoning tasks, but severely impaired ability on the hypothesis generation, testing and revision task. The dissociation between propositional language and scientific thinking evidenced by SA suggests that scientific reasoning is not dependent upon access to explicit natural language propositions.

1 Introduction

The human species is uniquely endowed with the capability for language, science, art and religion. In the case of language, it is the capacity for grammar

My thanks to SA and MR for their enthusiastic participation in the experimental tasks; to Peter Carruthers, Caroline Denby, Harriet Read and Michael Siegal for their help in developing the ideas behind the experiments; and to Denis Hilton for his suggestions for further explorations of verb comprehension. The basket-lifter machine was designed and built by R.M. and R.W. Parry.

and the construction of propositions that appears to be the singular possession of humans. Other species of primate and monkey appear capable of primitive lexical abilities, occurring either naturally or as a result of systematic language training. These lexical abilities are in no way comparable either quantitatively or qualitatively with the human lexicon. The lexicons of non-human species are subject to capacity constraints and probably do not show the cross-domain linkage of information evident in human lexical systems. The co-occurrence of language, and grammar in particular, with other distinctively human achievements (e.g. science, sophisticated social cognition, art and religion) can lead to the assumption that there is a causal relationship between such capacities. Specifically, the claim is advanced that propositional language has extended the human behavioural repertoire to beyond the capacity for complex inter-personal communicative exchange, and enabled new forms of thinking.

There are two variants of the claim that language enables cognitive advances in other domains. The first claim is phylogenetic and ontogenetic – that language re-engineers the human mind and it has done so both in the evolution of our species and during the development of each contemporary human (Dennett, 1991; Bickerton, 1995; Mithen, 1996). The second variant of claim regarding the role of language in cognition has both strong and weak forms. They both address the role of language in occurrent (here-and-now) reasoning. In the strong form, the claim would be that language is the vehicle for thinking and reasoning in a range of cognitive domains outside of the home domain of language (Carruthers, 1996a; Segal, 1996; de Villiers and de Villiers, 1999). The weaker variant of the claim is that language, in the form of inner or public speech, can be used as a tool in support of thinking, particularly at times of high cognitive demand (Berk, 1994; Clark, 1998).

Testing hypotheses regarding the role of language in various forms of thinking is a difficult business. The phylogenetic and ontogenetic claims can be addressed only through cases of anomalous development, as in cases of specific language impairment (SLI) or of profoundly deaf children raised in a non-signing environments. SLI might appear to be an important test ground, since the developmental disruption is specific to language and the child will display a relatively normal cognitive profile in other domains. This contrasts with the severe forms of language impairment found in children raised in conditions of extreme deprivation, such as Genie (Curtiss, 1977), or those with all-pervasive cognitive impairments. However, children with SLI rarely display the complete absence of linguistic knowledge or radical loss in a particular sub-domain of language such as grammar. Van der Lely, Rosen and McClelland (1998) report the case of AZ, a boy with a marked and persisting grammatical deficit, who displayed an impairment at the level of morphology particularly. Despite these linguistic difficulties, AZ showed a normal profile of scores on non-linguistic cognitive tasks. However, it is clear that AZ was able to construct simple language sentences (e.g. 'My dad go to work') and thus had sufficient linguistic ability to conjoin arguments to form propositions. In view of these characteristics of SLI, investigations with profoundly deaf individuals raised in non-signing environments may provide the best test ground for hypotheses regarding the phylogenetic and ontogenetic claims for the role of language in configuring the human mind (Schaller 1995; Goldin-Meadow and Zheng, 1998).

With regard to the testing of claims of the role of language in occurrent thinking, the evidence from aphasia is important. Aphasia is an acquired disorder of language that results from damage to the peri-sylvian language zones and associated sub-cortical structures in the language-dominant hemisphere (Gazzaniga, Ivry and Mangun, 1998). However, just as in the case of SLI, aphasia is a heterogeneous disorder and most individuals with aphasia have extensive residual grammatical and lexical knowledge. But there are important differences between aphasia and SLI. First, aphasia disrupts a mature and highly modularized mental architecture, and second, there is less plasticity in the adult brain, which acts to limit the reorganization of function following a lesion. As a result of these factors, there is the possibility of very profound loss of linguistic knowledge within the spectrum of aphasic disorders. This might include substantial loss of both lexical and grammatical processing ability as in global aphasia, or the excision of grammatical abilities in severe agrammatic aphasia (Varley, 1998; Varley and Siegal, 2000). People with aphasia usually have normal language processing abilities prior to the brain insult, and language will have been in place and played its postulated role in configuring the human mind. The evidence from aphasia is therefore unable to address any evolutionary or ontogenetic claims regarding the role of language in wider cognitive domains.

In this chapter, I will examine the role of language, and specifically grammar, in occurrent scientific reasoning. A plausible hypothesis, motivated by the unique co-incidence of propositional language and scientific thinking in humans, is that grammatical language in some way enables elements of this form of reasoning. This represents an association or correlation between two phenomena – language and science. However, one of the early elements in formal scientific training is the awareness that correlation does not imply causation. In the case of language and its co-occurrence with other unique human capabilities, there can be no simple assumption that such an association justifies causal claims of the type that propositional language determines the second capacity. A correlation between X and Y may occur because of a 'hidden' third variable that is critical to both. For example, the ability to sustain a series of representations in working memory might lead both to a pressure for grammar and the ability to understand cause–effect associations between things remote in space and time. The ability to hold a series of representations in an internal cognitive workspace might permit the combination of elements into propositions and the need for grammar to codify the relationships between those elements. Similarly, the capacity to combine and manipulate a series of representations drawn from on-going perception or from long-term memory might facilitate the development of causal associations between entities (e.g. seeds seen in the autumn at a particular location, and a plant seen in the summer at the same location). In this way, grammar and scientific thinking may invariably co-occur, but only because they are both co-products of the third capacity. One way to investigate a correlation in order to establish if there is any causal link between two phenomena is to seek evidence of co-variation – if one phenomenon alters, does one observe a stepwise alteration in the supposedly linked phenomenon? If the assumed direction of the relationship is that grammar enables scientific thinking, then as grammatical competence is diminished, one would expect a stepwise decrease in the capacity for scientific cognition.

This chapter reports the performance of two people with severe aphasic disorders on tasks that involve components of scientific reasoning. Both patients displayed severe impairments in the comprehension and production of propositional language. The purpose of this investigation was to examine the degree to which scientific thinking requires explicit natural language propositions, and specifically grammar, to support it. Are language propositions the substance in which scientific thinking takes place (the strong claim for a role for language in occurrent thinking), or do they act as scaffolding to scientific thinking (the weaker version of the claim)? In either case, we might expect to see impairments in the performances of the two participants on reasoning tasks. Alternatively, evidence that these patients perform well on the tasks might suggest either that such thinking takes place in a propositional code that is not natural language (Fodor, 1981; Pinker, 1994), or that this form of thinking does not involve propositional forms at all.

Neither of the patients reported here were professional scientists, or had received any formal scientific education beyond the age of 16. The concerns of this study are therefore with naïve or folk scientific thinking. It is assumed, however, that professional science is a refinement of reasoning abilities that are present in all humans (Dunbar, 1995, chapter 8 in this volume; Carruthers, chapter 4 in this volume). The components of reasoning that are central to scientific thinking and which are tested here are (1) developing hypotheses about possible causal mechanisms underpinning a phenomenon or event; and (2) the processes involved in hypothesis generation, the subsequent testing of hypotheses against evidence and the modification of hypotheses in the light of evidence. These forms of reasoning were examined through performances on three tasks. The Wisconsin Card Sorting Test (Heaton *et al.*, 1993) was used to examine ability to generate, test and modify hypotheses. Two tasks were developed to examine reasoning about causes of events. The first of these involved generating possible causes for both usual and unusual events; while the second represented a more practical reasoning task involving identifying the cause of a series of faults in a complex machine.

2 Methodology

2.1 Participants

The participants in this study were SA and MR. Both patients were pre-morbidly right-handed and experienced severe aphasic disorders following damage to the left cerebral hemisphere. These disorders include profound difficulties in understanding and producing language propositions in both spoken and written modalities. If one adopts the position that inner speech is impaired in similar ways to public speech in aphasic disorders, neither participant would be able to use either public or private propositional language to support reasoning. Previous investigations have shown that both SA and MR are capable of counter-factual 'theory of mind' reasoning, and elements of causal reasoning as measured by the WAIS Story Arrangement Test (Wechsler, 1981) and a test of simple causal association (Varley, 1998; Varley and Siegal, 2000; Varley, Siegal and Want, 2001). The scores obtained by SA and MR on the WAIS Story Arrangement Test placed them on the 84th percentile and 63rd percentile of an aged-matched population, respectively.

SA is a 50 year-old male who is a retired police sergeant. Eight years prior to the investigations reported here, he had a sub-dural empyema (a focal bacterial infection) in the left sylvian fissure with accompanying meningitis. An MRI scan revealed an extensive lesion of the left hemisphere peri-sylvian zones. This brain insult resulted a severe agrammatic aphasia and a profound apraxia of speech that rendered his speech virtually unintelligible. Following this lesion, SA communicated with great effectiveness through a combination of single-word writing, drawing, gesture and occasional single-word spoken utterances. Despite his profound linguistic and phonetic problems, SA is alert and extremely able in interpreting the world around him. He is able to drive, he plays chess against both human and machine opponents, and is responsible for various household chores, as well as taking responsibility for financial planning for his family.

The results obtained by SA on tests of lexical and grammatical language processing are presented in table 1. Tests of input and output lexical processing in both spoken and written modalities suggested that, although impaired, SA still had substantial lexical knowledge available to him. Word–picture matching tasks (testing understanding of nouns) revealed performances significantly above chance (PALPA 47 and 48, Kay, Lesser and Coltheart, 1992). Picture naming was used to test the ability to retrieve noun vocabulary. Results are reported only for written naming, as spoken naming responses were difficult to

Table 1. *Performance of SA and MR on measures of language processing*

Test	SA score	MR score	Chance score
PALPA Test 47 Spoken word–picture matching	35/40*	33/40*	8/40
PALPA Test 48 Written word–picture matching	37/40*	31/40*	8/40
PALPA Test 54 Spoken picture naming	–	39/60	–
PALPA Test 54 Written picture naming	24/60	1/20	–
Written Grammaticality Judgements	26/40	34/40	20/40
PALPA Test 55 Spoken sentence–picture matching	23/60	36/60	20/60
PALPA Test 56 Written sentence–picture matching	15/30	10/30	10/30
Spoken reversible sentence–picture matching	50/100	55/100	50/100
PALPA Test 57 Auditory comprehension of verbs and adjectives	Verbs 18/26 Adjectives 14/15	Verbs 13/27 Adjectives 9/15	13/26 7.5/15
PALPA Test 13 Digit span	3 items	3 items	–

Notes: *Scores significantly above chance at the $p < 0.001$ level.

All tests were taken from the PALPA battery (Kay, Lesser and Coltheart, 1992) with the exception of grammaticality judgements and the spoken reversible sentence–picture matching test. Word–picture matching tests involve matching a word (spoken or written) to a picture, in the presence of four semantic and/or visual foils (e.g. to match a picture of a candle in response to the word 'candle' from pictures of a candle, match, lamp, lipstick and glove). Spoken and written picture naming tests require the subject to say or write down the name of a pictured object. Test stimuli are drawn from three word-frequency bands. The grammaticality judgement test requires the participant to categorize written sentences as grammatical or ungrammatical. The PALPA sentence–picture matching tasks involve matching a sentence to a picture in the presence of two foils. The spoken reversible sentence–picture matching test was developed for the purposes of this study. It was designed to overcome some of the difficulties with the PALPA interpret because of intelligibility problems. The score indicated that SA could retrieve a third of targets (PALPA 54). In contrast with his performance in nominal processing, SA showed severely impaired performance in all elements of grammatical processing. On making judgements as to whether a sentence was grammatical, SA's score was not significantly above chance. Similarly, on tests involving matching spoken or written sentences to pictures, SA's scores were at chance in both modalities (PALPA 55, 56, and the spoken reversible sentence comprehension test).

Many psycholinguistic theories suggests that verbs are central in the process of both understanding and producing sentences (e.g. Garrett, 1982), and that the processing of a verb results in automatic access to a bundle of information, including the argument structure and thematic roles of that verb. Verbs were absent from SA's spoken and written production, and he showed parallel difficulties in verb comprehension. The testing of verb comprehension is complex. The usual task of word–picture matching used in assessing nominal lexical processing can produce ambiguous results when testing the comprehension of verbs. Many verbs have related nominal forms (e.g. walk, yawn, laugh) and, if the patient has residual noun processing ability, it is then not clear whether a word(verb)–picture matching test has measured what it was intended to measure. On a task involving indicating whether a definition of a word was correct or incorrect, SA performed at chance level on verb definitions. On a parallel set of adjective definitions, SA's performance was above chance (PALPA 57). This suggests that there was a specific difficulty in comprehending verbs, matching the parallel deficit of verb production.

In a further attempt to probe SA's understanding of verbs, a task based on a study by Brown and Fish (1983) was developed. Brown and Fish categorized a series of verbs as to whether causal attribution was given to the Agent in action verbs (e.g. '*John* helps Mary' (because John is a helpful person)), or the Stimulus in mental state verbs (e.g. 'John likes *Mary*' (because Mary is a likeable sort of person)). A set of sixteen verbs that differed in causal attribution was selected, with eight verbs of action–Agent attribution pattern and eight

Note to Table 1 (*cont.*) sentence tests, where the majority of stimulus sentences are in the active voice. A patient who is using a order-of-mention strategy in determining which noun is playing which role ('assume first mentioned noun is the Subject/Agent') is thus able to achieve an above-chance score on reversible sentence comprehension. The reversible sentence comprehension test contained equal numbers of active and passive sentences (e.g. the man killed the lion/the man was killed by the lion), and each noun appeared in both Agent and Patient position (e.g. the man killed the lion/the lion killed the man). The auditory comprehension of verbs and adjectives test requires the subject to indicate whether a definition for a verb or adjective (e.g. 'washing') is correct or incorrect. Digit span (PALPA 13) was tested in a recognition paradigm, vs. the usual verbal recall of digits, in order to minimize lexical retrieval and speech production factors in performance.

of state–Stimulus attribution type. SA was presented with a Subject–Verb–Object sentence in both spoken and written form (e.g. 'Bill hated Mary'). He was then asked 'Why did Bill hate Mary?', with the question supported with a prompt card depicting a questioning, puzzled person and the written word 'Why?'. The purpose of the task was to examine if SA was able to access the direction of causal attribution that was embedded within the semantics of each verb and, in the above example, if he would provide a response that indicated reasons why Mary might be perceived as hateful. All sixteen verbs were presented twice, and testing took place across two sessions separated by an interval of seven weeks. The orders of the noun arguments were reversed across the two trials. Responses to the thirty-two trials consisted of fifteen (47%) in which an unambiguous causal attribution was made to one of the arguments of the proposition, and eleven (34%) responses followed the pattern of causal attribution given by Brown and Fish's normative subjects. Given that previous investigations with SA indicated that he was able to understand notions of causality (Varley, 1998; Varley and Siegal, 2000), this low score suggests that he had considerable difficulty in accessing the semantics of the verbs that were tested.

SA's grammatical difficulties could not be attributed to a deficit in phonological working memory (Basso *et al.*, 1982; Martin and Feher, 1990; Martin, 1993). His digit span was measured at three items which, although below normal capacity, was sufficient to complete the tasks in the language assessment. In addition, the lack of any dissociation in ability between auditory processing of grammar and written processing (where information was permanent and available for repeated reference) suggested that limited working memory capacity was not a determining factor in his grammatical difficulties.

In addition to difficulties in the understanding of grammatical information, SA showed a parallel lack of grammatical structure in both spoken and written output. Spoken output consisted of single-element utterances, usually nouns. SA's severe motor speech disorder might account for this restricted spoken output, however a parallel lack of grammatical structure was evident in his writing. No clause structures occurred in writing. Some phrasal constructions could be found, but these were limited to article–noun, adjective–noun and quantifier–noun combinations. When pressed to write in sentences, SA either produced lists of nouns, or appropriate nouns randomly interspersed with grammatical function words.

The second patient in this investigation is MR, a 60 year-old right-handed male who was tested three years post-onset of a left middle cerebral artery stroke. A CT scan revealed a large peri-sylvian infarct affecting temporal, anterior parietal, posterior frontal zones and extending medially to striate cortex. Following this lesion, MR had a dense right hemiplegia and was classified as profoundly aphasic, with severe impairment of performance in all modalities

1. A little bit but ok now
2. Yes its ok
3. Yes that's ok now
4. But that
5. Is the
6. Its ok that way and now that way its...
7. Oh I can't explain it
8. Its ok now
9. Its the
10. Its ok
11. Its ok now
12. Its just ..
13. Oh I can't explain
14. Its ok though
15. Oh you know X
16. You know ...
17. Charles
18. You know
19. Oh dear
20. Its awful
21. But cancer
22. Its bad
23. Oh its bad
24. Well I don't know
25. I think so
26. Its awful
27. Well I don't know
28. Its awful though
29. Anyway
30. Its ok
31. That's ok now
32. And you know
33. Is there
34. (Neologism)
35. Also he's ...
36. Its ehr long ago
37. Far away
38. Its ok there
39. Its on and on and on and on
40. Its
41. I don't know
42. In there
43. Anyway
44. Anyway
45. Bikes
46. All the time
47. Its ok though
48. Anyway
49. Its ok
50. Dentist as well
51. Its ehr just
52. But oh dear me no
53. That is
54. Anyway that's ok
55. I'm afraid so
56. I think so
57. Oh look at that
58. Yes and I don't know
59. On and on and on and on and on
60. That's nice though
61. Now then I can't explain
62. I can't explain now
63. Right anyway
64. Oh dear bloody hell
65. Its ok
66. Its ... anyway
67. Now you that's ok but
68. Oh dear
69. Yes but its ok I think
70. I think
71. Well I really don't know
72. Its ok though
73. You see

Fig. 1 Consecutive utterances in spontaneous speech (excluding 'yes' and 'no') from speaker MR

of language use. Comprehension problems were evident and his speech output was restricted to stereotyped forms (see figure 1). Reading and writing were severely impaired. Prior to his stroke, MR worked as a lecturer in further education, in a non-science subject area. Pre-morbidly he was a proficient amateur musician, and much of his time following his stroke was spent in listening to music, often in conjunction with the orchestral score. A series of experiments on MR's musical processing ability revealed a strong dissociation between his symbolic abilities in music and in language (Read and Varley, in preparation).

The performance of MR on tests of lexical and grammatical processing is displayed in table 1. In lexical processing, MR's performance was similar to that of SA. The two patients showed similar nominal lexical comprehension scores in both in auditory and written modalities (PALPA 47 and 48). In lexical production, MR had a severe impairment of writing. He was tested on written picture naming of 20 high-frequency nouns, and as he was able to retrieve only a single noun within this set (nouns from mid- and low-frequency ranges were not tested). In contrast, MR's spoken noun retrieval score was surprisingly good in comparison to the level of nominal lexical retrieval he displayed in spontaneous speech (PALPA 54).

Although SA and MR were well matched on lexical scores, MR showed some preservation of grammatical knowledge, as measured by the ability to make grammaticality judgements. Despite his ability to recognize grammatical violations, MR had difficulties in decoding the meaning of sentences. This is evident in particular on the spoken reversible sentence comprehension test, where his score was at a chance level. The reading of sentences was severely impaired and performance on written sentence–picture matching was at a chance level (PALPA 56). MR had a digit span of three items. Examples of spoken output are presented in figure 1. There is little evidence of generative grammatical capacity, or the ability to utilize his nominal lexical knowledge in spontaneous output. His speech is largely restricted to stereotypic forms (which were often produced with different intonational contours in order to signal different attitudinal states).

In order to test their remaining folk-scientific abilities, SA and MR completed a series of tasks. Each task was administered in a separate session.

2.2 *Procedure 1: Wisconsin Card Sorting Test (WCST)*

This is a standardized clinical test of 'executive function' (Heaton *et al.*, 1993). The subject is presented with four key cards (depicting a single red triangle, two green stars, three yellow crosses and four blue circles). The subject is then given a stack of 128 cards to sort. The cards depict each shape in all possible combinations of colour and number. The subject is required to sort each card to one of the key cards (by shape, colour, or number), and the experimenter gives feedback on the accuracy of each categorization. The different possible categorization variables are not suggested to the subject and neither is the subject instructed on which variable the categorization should be based. In this way, the subject has to discover, on the basis of the feedback given by the experimenter, both the possible sorting variables and the accuracy of the variable that has been selected for matching. After the subject has completed ten consecutively correct matches by the first criterion (colour), the experimenter, without any indication that the sorting variable has changed, switches the categorization rule to shape. The subject has again to discover the new sorting principle on the basis of the feedback received from the experimenter. The test continues until six categories have been completed.

The WCST, as a test of executive function, is claimed to test 'abstract reasoning ability and the ability to shift cognitive strategies in response to changing environmental contingencies' (Heaton *et al.*, 1993, p. 1). In this study, it was used to model the processes involved in empirical science. These include hypothesis generation (albeit from a fixed number of possible hypotheses), hypothesis testing against data (assessing the validity of the hypothesis against the feedback data obtained from the experimenter) and hypothesis change (modifying the hypothesis when the feedback indicates that it is wrong). The standard WCST procedure was used to score the test. This includes the number of categories completed, the number of correct responses, the number of perseveratory responses (Persisting with a previous classificatory principle despite a change to a new sorting criterion) and the number of errors.

2.3 Procedure 2: causes of possible and improbable events

The purpose of the test was to evaluate the participant's ability to generate possible causes of a particular event. Two categories of event were used: possible events (e.g. house-fires, shipwrecks); and improbable events (e.g. a baby who knocked over a wall, a tennis player who knocked down buildings). These two categories were developed in order to establish not only whether the participant had access to learned causal associations but could also engage in novel or generative reasoning about causes of the improbable events. In terms of a model of scientific thinking, this task aimed to evaluate reasoning about causal mechanisms behind empirical observations. A series of coloured pictures was developed for each trial. These depicted an initial state (e.g. a baby crawling by a wall), and then the changed state (the baby amid the rubble of the wall). The participant was asked to sort the events into possible and improbable categories. For those events that were correctly categorized, the participant was subsequently presented with a response card depicting a puzzled figure, and they were asked to generate three possible causes of the event. Any form of communicative response was accepted, including speech, writing, drawing, or gesture. Causes for possible events were elicited first. Responses were scored as correct if they were both valid (e.g. faulty wiring, arson and lit cigarettes are all valid causes of a house-fire) and generative. Hence if a cause was given more than once (even across different events and even if a valid cause in both situations), it was given only credit the first time it was used. SA, for example, used 'dream' as a possible cause of two improbable events, and this was credited to his score only on the first occasion of use. In addition, if a response was a variant of a cause that already been given (e.g. SA gave two responses addressing the properties of a wall that a baby could knock down – made of Lego, made of cardboard), then only one response was credited.

2.4 *Procedure 3: the basket-lifter machine*

In this task, both participants were shown a complex machine – the output of which was to raise or lower a basket. The machine was constructed from *Capsela* components, many of which are made of transparent plastic and thus allow the internal operation of the machine to be observed. The machine was complex in the sense that it had multiple component systems. The basket-lifting function was achieved from a battery power source, driving a motor, which then turned a drive-shaft mechanism. The drive-shaft turned a series of cogs and a belt, which in turn drove a cog. This cog turned the winding arm of a pulley system responsible for raising and lowering the basket. The machine also contained multiple decoy systems. For example, the battery leads could be switched to a second motor that powered a sail rather than the basket lifter. The drive shaft could be switched at two points with the result that either there was no output from the machine, or a second sail system was activated.

Both participants observed the machine in operation and, in particular, their attention was directed to the function of the machine (raising and lowering the basket). The participant was then blind-folded, and the machine was sabotaged. On the first two trials, a single fault was introduced into the machine, while in the subsequent three trials, multiple lesions of the machine were made (with a total of ten faults across trials). Whereas some faults were clearly visible (e.g. switching of the battery wires from one motor to the second), the majority of the sources of faults were less apparent (e.g. connecting and disconnecting the cog systems on the drive shaft). After the machine had been damaged, the blindfold was then removed from the participant, and the experimenter demonstrated the post-lesion dysfunction of the machine. The participant was required to indicate where the machine was damaged. There was no requirement to fix the machine as, owing to motor and sensory problems affecting the right hand, both patients had marked difficulties with fine manual movement.

3 Results

3.1 *Wisconsin Card Sorting Test*

SA was able to complete the full six sets of categorizations. He did this in 78 trials, producing only 10 errors within those trials. This error score placed him on the 91st percentile of an age- and education-matched normative sample (a score consistent with that he obtained on the WAIS Story Arrangement Test). In contrast, MR found the task difficult. He failed to complete a single category in 128 trials. He produced 60 correct responses, but his high error score placed him on the 5th percentile of an age- and education-matched normative sample, and within the 'severely impaired' range of performance. The first sorting principle on the WCST is colour, and MR consistently failed to use this principle. Throughout the test he attempted to sort by form and number, with a strong tendency to perseverate on number. An Ishihara test of colour vision was subsequently performed on MR and this indicated some mild problems with colour processing (Ishihara, 1983). In order to exclude the possibility that a colour perception problem had influenced performance on the task, a revised WCST was devised, in which colour was replaced by a 'fill-in' variable (blue/solid; yellow/blank; green/stripes; red/diamonds).

MR was re-tested on the revised WCST eight weeks after the first attempt. Despite removal of the colour variable, MR's performance remained severely impaired. He completed a single category, but then was unable in 40 subsequent trials to establish the next sorting criterion, and the revised task was abandoned. MR's pattern of responding again showed perseveration to the number variable (despite the consistent feed-back that this was the incorrect sorting principle). He occasionally switched to other sorting principles but, as in the first attempt at the task, his performance suggested that he was unable to determine that there were three potential sorting principles embedded within the task and to systematically test each one. MR's score on the WCST stands in contrast to his performance on the WAIS Story Arrangement Test, where he performed at the 63rd percentile.

3.2 Causes of possible and improbable events

SA correctly categorized the five possible events as possible and went on to generate thirteen causes of those events (the maximum score was fifteen, as the participant was required to produce three causes for each event). He categorized four of five improbable events as improbable, and gave nine out of a possible twelve causes for the four correctly-categorized improbable events. Examples of his responses are presented in table 2.

MR correctly categorized all possible and improbable events. He generated ten of fifteen causes for possible events, and five of fifteen causes for improbable events. His errors were qualitatively different from those of SA (who was predominantly penalized for 'repeat' errors). Errors on possible events were largely describing the effect of the event rather than the causes; while on improbable events, errors were predominantly no responses, descriptions, or consequences of the events.

Table 2. *Examples of causes of possible and improbable events given by SA and MR*

Event	SA response	MR response
Possible event – house fire	1 Wiring fault 2 Arson 3 Cigarette	1 Cigarette 2 Arson 3 Burglar alarm rings (error)
Possible event – parachute from plane	1 fire 2 computer error 3 no fuel	1 sky-diver 2 crash 3 –
Improbable event – tennis player knocking down tower	1 dreaming 2 playing on the moon 3 very strong player	1 atmosphere allowed ball to go up high and fall down onto the tower 2 ball hit tower (error)
Improbable event – bullet bouncing off buffalo	1 blank bullet 2 toy gun 3 wooden buffalo	1 tail movement deflects bullet 2 pop gun 3 buffalo tells him to clear off (error)

3.3 Basket-lifter machine

Both SA and MR were able to identify all ten breakage points on the machine task. These included trials on which a single fault had been introduced into the machine, and trials in which multiple faults had to be detected. SA was faster than MR in detecting the faults. He also introduced spontaneous improvements into the design of the machine at the end of the experiment (e.g. altering the position of the sail towers in order to reduce slippage of the drive belt). MR identified all the faults introduced into the machine, but was slow on the first trial to switch attention from the basket lifter component of the machine to other systems (the fault in the machine involved the disconnection of a wire from the battery switch). MR also introduced a single irrelevant modification to the machine (switching one of the wheels on the drive belt) which introduced noise into the machine, although it did not disrupt function.

4 Discussion

In this neuropsychological study of components of scientific reasoning, one patient (SA) showed well preserved performances on all tasks despite a severe language impairment. A second patient (MR) was able to identify faults in a complex machine and to suggest causes for both possible and impossible events. However, although MR was able to demonstrate some capacity for sophisticated causal cognition, he showed severely impaired performance on a hypothesis generation and testing task. On the Wisconsin Card Sorting Test he had profound difficulty in identifying all the potentially relevant variables of the categorization task and in using the feedback from the experimenter to influence his subsequent categorization decisions. Before any conclusion is drawn that his performance reflects a significant dissociation of the cognitive processes involved in causal cognition and those implicated in hypothesis testing, an alternative view should be considered.

MR showed subtle difficulties in both causal cognition tasks. In producing causes to both possible and improbable events, he displayed less generativity than SA and also produced some non-causal responses. Similarly, in the machine task MR introduced an irrelevant modification. He also had difficulty at the beginning of the task in switching attention from the most salient component of the machine – the basket-lifting mechanism. This difficulty in shifting attention was similar to the perseveration on number categorization that was observed on the WCST. The WCST involves abstract reasoning and the ability to detach from elements of the stimulus. The apparent dissociation between tasks displayed by MR might then result from the differences between tasks in demand for abstraction and detachment from a salient stimulus. The WCST is high in demand on this dimension and hence MR's performance was at its most impaired on this task. However, it is clear that MR – a patient arguably with more language than SA – showed a greater degree of impairment on tasks of scientific reasoning than did SA.

Of the two patients, SA is the most critical to a test of the relationship between propositional language and scientific thinking. On tests operating at an explicit level of grammatical processing, SA displays severe impairments in making grammaticality judgements, in comprehending spoken and written sentences and has difficulty in accessing the semantic representation of verbs. In production, SA is unable to retrieve verbs and is unable to construct natural language propositions in either speech or writing. Despite these substantial disruptions to grammatical processing, SA displays sophisticated cognition both in activities of daily living and in the laboratory tasks reported here and in previous studies (Varley and Siegal, 2000). Without recourse to explicit language propositions, SA was able to demonstrate clear knowledge of causal relationships. His ability in the domain of causal thinking was not restricted to the retrieval of learned associations from memory, but extended to the ability to generate possible causes of novel and abstract phenomena. In addition, SA showed superior ability in relation to a normal population in the cognitive processes involved in testing and revising hypotheses. In these ways SA was an extremely competent folk scientist, and hence the forms of cognition used in occurrent naïve science appear to be independent of explicit grammatical knowledge.

On the basis of the evidence from SA's performance, there needs to be clear shift in the focus of claims that are made for the role of language in cognition. The view that the propositions of inner speech are the substance of certain forms of thought cannot be maintained. Without the ability to fully comprehend a simple sentence, or the ability to construct a simple clause, SA demonstrates theory of mind reasoning (which, if substantiated by language propositions, might involve the embedding of one clause within another), causal understanding (involving the co-ordination of two clauses) and conditional reasoning involved in hypothesis testing. If the evidence from SA could not be replicated, it would be unwise to generalize from the single case to the population as a whole. However, the data from MR provides a replication of the findings both in the domain of causal thinking, and in theory of mind understanding (Varley, Siegal and Want, 2001).

A rather different claim for the role of language in cognition is that it permits permeability of information across cognitive domains (Dennett, 1991). Mithen (1996) develops this proposal and suggests that the origins of art, religion and science can be found at the point when the contents of macro-domains such as 'social intelligence', 'natural history intelligence' and 'technical intelligence' become available to other domains. The mechanism of this cognitive permeability and fluidity is claimed to be language – manifesting an intra-personal communicative function that allowed the integration of the contents of previously rigidly encapsulated systems. Hence if natural history intelligence (knowledge of the environment and the plants and animals found within it) can be combined with the resources of technical intelligence, then the products of animals, such as bone and antlers, can be used to develop a whole new technology of tools. Mithen sees the fluidity of intelligence engineered by language as critical in the cognitive and cultural development of the human species. Mithen's claims generally address issues of phylogeny and ontogeny, and he makes no specific claims that could be tested against the evidence from aphasia for the role of language in occurrent cognition. If one examines the tasks used in the present study, moreover, they do not represent an adequate test of Mithen's views when applied to occurrent thinking. The tasks do not involve inter-domain fluidity. For example, the basket lifter task might, in Mithen's terms, be classed as operating solely within technical intelligence. But there may be considerable difficulty in developing adequate empirical tests of the hypothesis of cognitive fluidity in the terms given by Mithen.

The intelligences described by Mithen represent macro-domains and do not represent 'domains' in neurobiological terms. For example, the knowledge represented by 'natural history intelligence' would be instantiated by numerous neurocognitive systems. Thus knowledge of dogs simply at a sensori-perceptual level includes a distributed network of information including what they smell like, look like, sound like, feel like – and, in some cultures, taste like. These different strands of knowledge are represented in different areas of cortex (Martin, Ungerleider and Haxby, 2000). In addition to knowledge gained from immediate sensory experience, there is additional information built up from repeated direct and vicarious experiences of dogs. Because it is then difficult to reconcile 'intelligences' with neurobiological systems, it is then equally difficult to describe an adequate empirical test of the hypothesis in the terms of Mithen's macro-domains.

The notion that language creates cognitive permeability between otherwise encapsulated systems has, however, been explored in the occurrent reasoning of normal subjects by Hermer-Vasquez, Spelke and Katsnelson (1999). In this study, the domains investigated have substantial neurobiological reality. Hermer-Vasquez and her colleagues report a series of experiments on spatial memory involving a task which required subjects to combine geometric and landmark information. These two forms of visual information are believed to be mediated by separate brain systems – specifically the dorsal 'where' system, responsible for visual–spatial processing, and the ventral 'what' system, responsible for visual–object processing (Ungerleider and Mishkin, 1982). In these experiments, subjects were blocked from using the resources of the language faculty by a verbal shadowing task (repeating back a heard story at the same time as completing the memory task). Whereas adults who were not engaged in verbal shadowing had no difficulty in combining spatial and landmark–object information, those subjects who were unable to utilize language to support the memory task were unable to conjoin these two forms of visual information. The interference of memory in the verbal shadowing condition could not be attributed to the general loading of information processing capacity by dual task performance. In a control condition, subjects who were engaged in non-linguistic auditory shadowing showed no disruption of visuo-spatial memory. Replicating these findings with people with aphasia is an important future direction for research, particularly as it is then possible to tease apart the issue of whether language propositions are necessary to the conjoining of information, or whether sub-components of the language faculty such as the lexicon are critical.

In the study reported here, the main focus has been on the role of propositional language and grammar in scientific reasoning. Both of the patients reported here had substantial lexical knowledge still available to them. In addition to extending the empirical work to address issues of the integration of information across domains, the role of the lexicon deserves some address. Bickerton (1995) suggests that the lexicon and specifically the availability of a linguistic label is important in creating cohesive conceptual structures. As in the argument for the distributed representation for 'dog' above, Bickerton claims that the lexicon may allow conceptual cohesion. The word may act as a mental glue between the distributed bundle of perceptual elements and binds them together to form a coherent concept. Given this view, the degree of impairment shown by people with global aphasia on a variety of within-domain and between-domain reasoning tasks should be substantial.

The current evidence from aphasia suggests that explicit grammatical knowledge is not necessary to support occurrent within-domain cognition, and specifically the forms of thinking that are utilized in science. There are many riders within this conclusion (what about implicit grammatical knowledge?, what about the lexicon?, what about inter-domain cognition?), and the need for additional empirical work is therefore clear.

6 Causal maps and Bayes nets: a cognitive and computational account of theory-formation

Alison Gopnik and Clark Glymour

In this chapter, we outline a more precise cognitive and computational account of the 'theory theory' of cognitive development. Theories and theory-formation processes are cognitive systems that allow us to recover an accurate 'causal map' of the world: an abstract, coherent representation of the causal relations among events. This kind of knowledge can be perspicuously represented by the formalism of directed graphical causal models, or 'Bayes nets'. Human theory formation may involve similar computations.

1 The theory theory

Cognitive psychologists have argued that much of our adult knowledge, particularly our knowledge of the physical, biological and psychological world, consists of 'intuitive' or 'naïve' or 'folk' theories (Murphy and Medin, 1985; Rips, 1989). Similarly, cognitive developmentalists argue that children formulate and revise a succession of such intuitive theories (Carey, 1985; Gopnik, 1988; Keil, 1989; Wellman, 1990; Gopnik and Meltzoff, 1997; Wellman and Gelman, 1997). This idea, which we have called the 'theory theory', rests on an analogy between everyday knowledge and scientific theories. Advocates of the theory theory have drawn up lists of features that are shared by these two kinds of knowledge. These include static features of theories, such as their abstract, coherent, causal, counter-factual-supporting character; functional features of theories such as their ability to provide predictions, interpretations and explanations; and dynamic features such as theory changes in the light of new evidence (see Gopnik and Wellman, 1994; Gopnik and Meltzoff, 1997). The assumption behind this work is that there are common cognitive structures

Versions of this chapter were presented at the International Congress on Logic, Methodology and Philosophy of Science, Cracow, Poland (August 1999); and at seminars at the University of Chicago, the California Institute of Technology, and the Cognitive Science program and Department of Statistics at Berkeley. We are grateful to those who commented. Conversation with Steve Palmer, Lucy Jacobs and Andrew Meltzoff played a major role in shaping these ideas. John Campbell and Peter Godfrey-Smith also read drafts of the chapter and made very helpful suggestions.

and processes, common representations and rules, that underlie both everyday knowledge and scientific knowledge.

If this is true it should be possible to flesh out the nature of those cognitive structures and processes in more detail. Formulating the analogy between science and development has been an important first step but, like all analogies, it is only a first step. Moreover, as with all analogies, there is a risk that we will end up in endless disputes about whether specific features of the two types of cognition are similar or different. Light waves aren't wet, after all, and natural selection takes place in an entirely different time-scale than artificial selection. Scientific theory change is different from child development in many ways – it is a social process, it involves a division of labour, and so on (see Faucher *et al.*, chapter 18 in this volume). Rather than wrangling over whether these differences invalidate the analogy it would be more productive to describe in some detail the representations and rules that could underpin both these types of knowledge. Ideally, such an account should include ideas about the computational character of these representations and rules, and eventually even their neurological instantiation. This is the project that has been so successful in other areas of cognitive science, particularly vision science.

In this chapter we will outline a more developed cognitive and computational account of the theory theory. In particular, we will argue that many everyday theories and everyday theory changes involve a type of representation we will call a 'causal map'. A causal map is an abstract representation of the causal relationships among kinds of objects and events in the world. Such relationships are not, for the most part, directly observable, but they can often be accurately inferred from observations. This includes both observations of patterns of contingency and correlation among events as well as observations of the effects of experimental interventions. We can think of everyday theories and theory-formation processes as cognitive systems that allow us to recover an accurate causal map of the world.

We will argue that this kind of knowledge can be perspicuously represented by the formalism of directed graphical causal models, more commonly known as Bayes nets (Pearl, 1988, 2000; Spirtes, Glymour and Scheines, 1993). Interestingly, this formalism has its roots in work in the philosophy of science. It is the outcome, at least in part, of an attempt to characterize the inductive inferences of science in a rigorous way. The formalism provides a natural way of representing causal relations, it allows for their use in prediction and experimental intervention, and, most significantly, it provides powerful tools for reliably inferring causal structures from patterns of evidence. In recent work in artificial intelligence, systems using Bayes nets can infer accurate, if often incomplete, accounts of causal structure from suitable correlational data. We will suggest that human causal inference and theory formation may involve more heuristic versions of similar computations.

2 Theory-formation and the causal inverse problem

The most successful theories in cognitive science have come from studies of perception. The visual system, whether human or robotic, has to recover and reconstruct three-dimensional information from the retinal (or fore-optic) image. One aspect of vision science is about how that reconstruction can be done computationally, and about how it is done in humans. Although accounts are very different in detail they share some general assumptions, in particular these: (1) Visual systems, whether human or automated, have an objective problem to solve: they need to discover how three-dimensional moving objects are located in space. (2) The data available are limited in particular ways. Organisms have no direct access to the external world. Rather, the external world causes a flow of information at the senses that is only indirectly related to the properties of the world itself. For example, the information at the retina is two-dimensional, while the world is three-dimensional. (3) Solutions must make implicit assumptions about the ways that objects in the world produce particular patterns – and successions of patterns – on the retina. The system can use those assumptions to recover spatial structure from the data. In normal conditions, those assumptions lead to veridical representations of the external world. But these assumptions are also contingent; if the assumptions are violated then the system will generate incorrect representations of the world (as in perceptual illusions). (See Palmer, 1999.)

We propose an analogous problem about discovering the causal structure of the environment. (1) There are causal facts, as objective as facts about objects, locations and states of relative motion, used and evidenced in prediction, intervention and control, and partially revealed in correlations. (2) The data available are limited in particular ways. Children and adults may observe associations they cannot control or manipulate; they may observe features they can control or manipulate only indirectly, through other objects or features; the associations they observe, with or without their own interventions, may involve an enormous number of features, only some of which are causally related. (3) Human beings have a theory-formation system, like the visual system, that recovers causal facts by making implicit assumptions about the causal structure of the environment. Those assumptions are contingent; where they are false, causal inference, whether in learning new causal relationships or in deploying old ones, may fail to get things right.

3 Causal maps

What kinds of representations might be used to solve the causal inverse problem? The visual system seems to use many very different types of representations and rules to solve the spatial problem. In some cases, like the case of translating two-dimensional retinal information to three-dimensional representations, the kinds of representations and the rules that generate them may be relatively fixed and 'hard-wired'. However, other, more flexible, kinds of spatial representations are also used

In particular, since Tolman (1932), cognitive scientists have suggested that organisms solve the spatial inverse problem by constructing 'cognitive maps' of the spatial environment (O'Keefe and Nadel, 1978; Gallistel, 1990). These cognitive maps provide animals with representations of the spatial relations among objects. Different species of animals, even closely related species, may use different types of cognitive map. There is some evidence suggesting the sorts of computations that animals use to construct spatial maps, and there is even some evidence about the neurological mechanisms that underpin those computations. In particular, O'Keefe and Nadel (1978) proposed that these mechanisms were located in the rat hippocampus.

There are several distinctive features of cognitive maps. First, such maps provide non-egocentric representations. Animals might navigate through space, and sometimes do, egocentrically, by keeping track of the changing spatial relations between their bodies and objects as they move through the spatial environment. In fact, however, cognitive maps are not egocentric in this way. They allow animals to represent geometric relationships among objects in space, independently of their own relation to those objects. A cognitive map allows an animal who has explored a maze by one route, to navigate through the maze even if it is placed in a different position initially. This aspect of cognitive maps differentiates them from the kinds of cognitive structures proposed by the behaviourists, which depend on associations between external stimuli and the animal's own responses. This, of course, made Tolman one of the precursors of the cognitive revolution.

Second, cognitive maps are coherent. Rather than just having particular representations of particular spatial relations, cognitive maps allow an animal to represent many different possible spatial relations, in a generative way. An animal who knows the spatial layout of a maze can use that information to make many new inferences about objects in the maze. For example, the animal can conclude that if A is north of B, and B is north of C, then A will be north of C. The coherence of cognitive maps gives them their predictive power. An animal with a spatial map can make a much wider variety of predictions about where an object will be located than can an animal restricted to egocentric spatial navigation. It also gives cognitive maps a kind of interpretive power; an animal with a spatial map can use the map to resolve ambiguous spatial information.

Third, cognitive maps are learned. Animals with the ability to construct cognitive maps can represent an extremely wide range of new spatial environments, not just one particular environment. A rat moving towards a bait in a familiar maze is in a very different position than, say, a moth moving towards a lamp. The moth appears to be hard-wired to respond in set ways to particular stimuli in the environment. The rat in contrast moves in accordance with a learned representation of the maze. This also means that spatial cognitive maps may be defeasible. As an animal explores its environment and gains more information about it, it will alter and update its map of that environment. In fact, it is interesting that the hippocampus, which, in rats, seems to be particularly involved in spatial map-making, also seems particularly adapted for learning and memory. Of course, this general learning ability depends on innate learning mechanisms that are specialized and probably quite specific to the spatial domain such as, for example, 'dead-reckoning' mechanisms (see Gallistel, 1990).

Our hypothesis is that human beings construct similar representations that capture the causal character of their environment. This capacity plays a crucial role in the human solution to the causal inverse problem. These causal maps are what we refer to when we talk about everyday theories. Everyday theories are non-egocentric, abstract, coherent, learned representations of causal relations among events, and kinds of events, that allow causal predictions, interpretations and interventions.

Note that we are not proposing that we actually use spatial maps for the purpose of representing or acquiring causal knowledge, or that we somehow extend spatial representations through processes of metaphor or analogy. Rather we want to propose that there is a separate cognitive system with other procedures devoted to uncovering causal structure, and that this system has some of the same abstract structure as the system of spatial map making with which it must in many cases interact. We also do not mean that knowledge of causal relations is developed entirely independently of knowledge of spatial facts, but that there are special problems about learning causal relationships, and special types of representations designed to solve those problems.

Just as cognitive maps may be differentiated from other kinds of spatial cognition, causal maps may be differentiated from other kinds of causal cognition. Given the adaptive importance of causal knowledge, we might expect that a wide range of organisms would have a wide range of devices for recovering causal structure. Animals, including human beings, may have some hard-wired representations which automatically specify that particular types of events lead to other events. For example, animals may always conclude that when one object collides with another the second object will move on a particular trajectory. These sorts of specific hard-wired representations could capture particular important parts of the causal structure of the environment. This is precisely the proposal that Heider (1958) and Michotte (1962) made regarding the 'perception' of both physical and psychological causality. There is evidence for such representations even in young infants (Leslie and Keeble, 1987; Oakes and Cohen,

1990). Animals might also be hard-wired to detect specific kinds of causal relations that involve especially important events, such as the presence of food or pain. Such capacities appear to be involved in phenomena like classical conditioning or the Garcia effect, in which animals avoid food that leads to poisoning (Palmerino, Rusiniak and Garcia, 1980).

Animals could also use a kind of egocentric causal navigation, they might calculate the causal consequences of their own actions on the world and use that information to guide further action. Operant conditioning is precisely a form of such egocentric causal navigation. Operant conditioning allows an animal to calculate the novel causal effects of its own actions on the world, and to take this information into account in future actions. More generally, trial-and-error learning seems to involve similar abilities for egocentric causal navigation.

Causal maps, however, would confer the same sort of advantages as spatial maps (Campbell, 1995). With a non-egocentric causal representation of the environment, an animal could predict the causal consequences of an action without actually having to perform it. The animal could produce a new action that would bring about a particular causal consequence, in the same way that an animal with a spatial map can produce a new route to reach a particular location. Similarly, an animal with a causal map could update the information in that map simply by observing causal interactions in the world, for example, by observing the causal consequences of another animal's actions, or by observing causal phenomena in the environment. The animal could then use that information to guide its own goal-directed actions. The coherence of causal maps allows a wide range of predictions. Just as an animal with a spatial map could make transitive spatial inferences (if A is north of B, and B is north of C, then A will be north of C) animals with causal maps could make transitive causal inferences (if A causes B, and B causes C, then A will cause C). A causal map would allow for a wide range of causal predictions and also allow a way of interpreting causally ambiguous data.

Since causal maps are learned they should give animals an opportunity to master new causal relations, not just whatever limited set might be 'hard-wired' perceptually. We would expect that animals would perpetually extend, change and update their causal maps just as they update their spatial maps.

It may well be that human beings are, in fact, the only animals that construct causal maps. In particular, there is striking recent evidence that chimpanzees, our closest primate relatives, rely much more heavily on egocentric trial-and-error causal learning. While chimpanzees are extremely adept at learning how to influence their environments, quite possibly more adept than human beings are, they have a great deal of difficulty appreciating causal relations among objects that are independent of their own actions (Tomasello and Call, 1997; Povinelli, 2000). Chimpanzees are, at best, severely restricted in their ability to learn by observing the interventions of others, or by observing causal relations among objects and events.

4 Theories as causal maps

'Everyday' or 'folk' theories seem to have much of the character of causal maps. Such everyday theories represent causal relations among a wide range of objects and events in the world independently of the relation of the observer to those actions. They postulate coherent relations among such objects and events which support a wide range of predictions, interpretations and interventions. Because of their causal character, they support counter-factual reasoning. Moreover, theories, like causal maps, are learned through our experience of and interaction with the world around us. Because of this, the theory theory has been especially prominent as a theory of cognitive development.

These are also features that unite everyday theories and scientific theories. While not all scientific theories are causal, causal claims and inferences do play a central role in most scientific theories (see Salmon, 1984; Cartwright, 1989). Scientific theories also involve learned, coherent, non-egocentric networks of causal claims which support prediction, interpretation, counter-factual reasoning and explanation. Moreover, when scientific theories are less concerned with causal structure it tends to be because these theories involve formal mathematical structure instead. However, this is also one way in which scientific theories appear to be unlike everyday theories. Scientific theories seem to be a-causal chiefly only in so far as they are formulated in explicit formal and mathematical terms. Everyday theories are rarely formulated in such terms.

Moreover, the idea of causal maps seems to capture the scope of 'theory theories' very well. The theory theory has been very successfully applied to our everyday knowledge of the physical, biological and psychological worlds. However, the theory theory does not seem to be as naturally applicable to other types of knowledge, for example, purely spatial knowledge, syntactic or phonological knowledge, musical knowledge, or mathematical knowledge. Nor does it apply to the much more loosely organized knowledge involved in empirical generalizations, scripts or associations (Gopnik and Meltzoff, 1997). But these types of knowledge also do not appear to involve causal claims in the same way. Conversely some kinds of knowledge that do involve causal information, like the kinds of knowledge involved in operant or classical conditioning, do not seem to have the abstract, coherent, non-egocentric character of causal maps, and we would not want to say that this sort of knowledge was theoretical.

Some earlier accounts have proposed that theories, both scientific and everyday, should be cognitively represented as connectionist nets (Churchland, 1990) or as very generalized schemas (Giere, 1992). The difficulty with these proposals is that they seem to be too broad to capture what makes theories special; practically any kind of knowledge can be represented as nets or schemas. On the other hand, more modular accounts, such as that of Keil (1995) or Atran (chapter 3 in this volume) have proposed that there are only a few specific explanatory schemas, roughly corresponding to the domains of physics, biology and psychology, that are used in everyday theories. These proposals do not seem to capture the wide range of explanations that develop in everyday life, nor the way that in cognitive development and in science we move back and forth among these domains. The idea of causal maps seems to capture both what is general and what is specific about everyday theories.

5 Bayes nets as causal maps

We propose, then, that children and adults construct causal maps: non-egocentric, abstract, coherent representations of causal relations among objects. An adequate representation of how such maps work must do three things: (1) It must show how they can be used to enable an agent to infer the presence of some features from other features of a system – it must allow for accurate predictions. (2) It must show how an agent is able to infer the consequences of her own or others' actions – it must allow for appropriate interventions. And (3) it must show how causal knowledge can be learned from the agent's observations and actions.

There has recently been a great deal of computational work investigating such representations and mechanisms. The representations commonly called Bayes nets can model complex causal structures and generate appropriate predictions and interventions. Moreover, we can use Bayes nets to infer causal structure from patterns of associations, whether from passive observation or from action. A wide range of normatively accurate causal inferences can be made, and, in many circumstances, they can be made in a computationally tractable way. The Bayes net representation and inference algorithms allow one sometimes to uncover hidden unobserved causes, to disentangle complex interactions among causes, to make inferences about probabilistic causal relations and to generate counter-factuals (see Pearl, 1988, 2000; Spirtes, Glymour and Scheines, 1993, 2000; Jordan, 1998).

This work has largely taken place in computer science, statistics and philosophy of science, and has typically been applied in one-shot 'data-mining' in a range of subjects, including space physics, mineralogy, economics, biology, epidemiology and chemistry. In these applications there is typically a large amount of data about many variables that might be related in a number of complex ways. Bayes net systems can automatically determine which underlying causal structures are compatible with the data, and which are ruled out. But these computational theories might also provide important suggestions about how human beings, and particularly young children, recover and represent causal information. Work in computer vision has provided important clues about the nature of the visual system, and this work might provide similar clues about the nature of the theory formation system. Causal maps might be a kind of Bayes net.

6 'Screening off' in causal inference

Bayes net systems are elaborations of a much simpler kind of causal inference, long discussed in the philosophy of science literature. Causal relations in the world lead to certain characteristic patterns of events. In particular, if A causes B, then the occurrence of A will change the probability that B will occur. We might think that this could provide us with a way of solving the causal inverse problem. When we see that A is correlated with B – that is, that the probability if A is consistently related to the probability of B – we can conclude that A caused B (or vice versa).

But there is a problem. The problem is that other events might also be causally related to B. For example, some other event C might be a common cause of both A and B. A doesn't cause B but whenever C occurs both A and B will occur together. Clearly, what we need in these cases is to have some way of considering the probability of A and B relative to the probability of C. Many years ago the philosopher of science Hans Reichenbach proposed one natural way of doing this, which he called 'screening off' (Reichenbach, 1956). We can represent this sort of reasoning formally as follows. If A, B and C are the only variables and A is only correlated with B conditional on C, C 'screens off' A as a cause of B – C rather than A is the cause. If A is correlated with B independent of C, then C does not screen off A and A causes B. This sort of 'screening off' reasoning is ubiquitous in science. In experimental design we control for events that we think might be confounding causes. In observational studies we use techniques like partial correlation to control for confounding causes.

The trouble with the reasoning we've described so far is that it's limited to these rather simple cases. But, of course, in real life events may involve causal interactions among dozens or even hundreds of variables rather than just three. And the relations among variables may be much more complicated than either of the simple structures we described above. The causal relations among variables may also have a variety of different structures – A might be linearly related to B, or there might be other more complicated functions relating A and B, or A might inhibit B rather than facilitating it. The causal relations might involve Boolean combinations of causes. Or A and B together might cause C, though either event by itself would be insufficient. And there might be other unobserved hidden variables, ones we don't know about, that are responsible for patterns of correlation. Moreover, in the real world, even when the causal relations we are uncovering are deterministic, the evidence for them is almost invariably noisy and probabilistic. Is there a way to generalize the 'screening off' reasoning we use in the simple cases to these more complicated ones? Would a similar method explain how more complicated causal relations might be learned? The Bayes net formalism provides such a method.

7 Bayes nets and their uses

Bayes nets are directed graphs, like the one shown in figure 1. The nodes or vertices of the graph represent variables, whose values are features or properties of the system, or collections of systems to which the net applies. 'Colour', for example, might be a variable with many possible values; 'weight' might be a variable with two values, heavy and light, or with a continuum of values. When Bayes nets are given a causal interpretation, a directed edge from one node or variable to another – X to Y, for example – it says that an intervention that varies the value of X but otherwise does not alter the causal relations among the variables will change the value of Y. In short, changing X will cause Y to change.

For each value of each variable, a probability is assigned, subject to a fundamental rule, the Markov assumption. The Markov assumption is a generalization of the 'screening off' property we just described. It says that if the edges of the graphs represent causal relations, then there will only be some patterns of probabilities of the variables, and not others. The Markov assumption constrains the probabilities that can be associated with a network. It says that the various possible values of any variable, X, are independent of the values of any set of variables in the network that does not contain an effect (a descendant of X), conditional on the values of the parents of X. So, for example, applied to the directed graph in figure 1, the Markov assumption says that X is independent of {R, Z} conditional on any values of variables in the set {S, Y}.

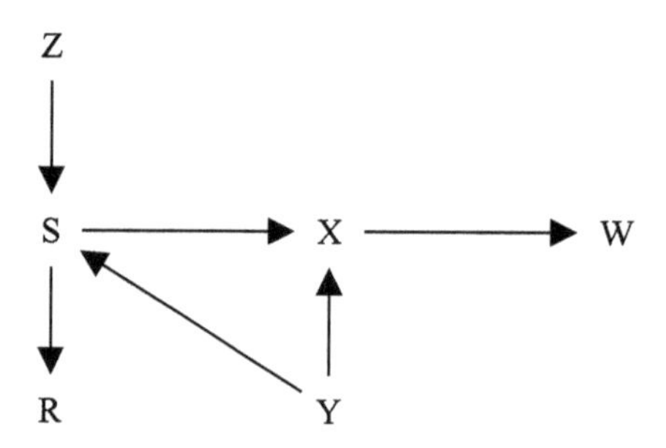

Fig. 1 A causal graph

Bayes nets allow causal predictions. Information that a system has some property or properties often changes the probabilities of other features of the system. The information that something moves spontaneously, for example, may change the probability that it is animate. Such changes are represented in Bayes nets by the conditional probability of values of a variable given values for another variable or variables. Bayes net representations simplify such calculations in many cases. In the network above, the probability of a value of X conditional on a value of R may be calculated from the values of p(R | S), p(S) and p(X | S). (See Pearl, 1988.) This allows us to predict the value of X if we know the value of R. Bayes nets also provide a natural way of assessing and representing counter-factual claims (Pearl, 2000).

In planning we specifically predict the outcome of an action. The probabilities for various outcomes of an action that directly alters a feature are not necessarily the same as the probabilities of those outcomes conditional on that altered feature. Suppose R in figure 1 has two values, say red and pink. Because the value of S influences R, the conditional probabilities of values of S given that R = red will be different from the conditional probabilities of values of S given that R = pink. Because S influences X, the probabilities of values of X will also be different on the two values of R. *Observing* the value of R gives information about the value of X. But R has no influence on S or X, either direct or indirect, so if the causal relations are as depicted, acting from outside the causal relations represented in the diagram to change the value of R will do nothing to change the value of S or X. It is possible to compute over any Bayes network which variables will be indirectly altered by an action or intervention that directly changes the value of another variable. It is also possible to compute the probabilities that indirectly result from the intervention. These calculations are sometimes possible even when the Bayes net is an incomplete representation of the causal relations. (See Pearl and Verma, 1991; Spirtes, Glymour and Scheines, 1993; Glymour and Cooper, 1999.)

Bayes nets thus have two of the features that are needed for applying causal maps: they permit prediction from observations, and they permit prediction of the effects of actions. With an accurate causal map – that is, the correct Bayes net representation – we can accurately predict that y will happen when x happens, or that a particular change in x will lead to a particular change in y, even when the causal relations we are considering are quite complex. Similarly, we can accurately predict that if we intervene to change x then we will bring about a change in y. Bayes nets have another feature critical to cognition: they can be learned.

8 Learning Bayes nets

In 'data-mining' applications Bayes nets have to be inferred from uncontrolled observations of variables. To do this, the Markov assumption is usually supplemented by further assumptions. The additional assumptions required depend on the learning procedure. (A detailed survey of several learning algorithms is given in the contributions in Glymour and Cooper, 1999.) One family of algorithms uses Bayes Theorem to learn Bayes nets. Another class of algorithms learns the graphical structure of Bayes nets entirely from independence and conditional independence relations among variables in the data, and requires a single additional assumption. We will describe some features of the latter family of algorithms.

The additional assumption required is *faithfulness*: the independence and conditional independence relations among the variables whose causal relations are described by a Bayes net must all be consequences of the Markov assumption applied to that network. For example, in figure 1, it is possible to assign probabilities so that S and X are independent, although the Markov assumption implies no such independence. (We can arrange the probabilities so that the association of S and X due to the influence of S on X is exactly cancelled by the association of S and X due to the influence of Y on both of them.) The faithfulness assumption rules out such probability arrangements.

The faithfulness assumption is essentially a simplicity assumption. It is at least logically possible that the contingencies among various causes could be randomly arranged in a way that would 'fool' a system that used the causal Markov condition. The faithfulness condition assumes that in the real world such sinister coincidences will not take place.

The learning algorithms for Bayes nets are designed to be used either with or without background knowledge of specific kinds. In addition to the Markov assumption and the faithfulness assumption we may add other assumptions about how causes are related to events. For example, an agent may, and a child typically will, know the time order in which events occurred, and may believe that some causal relations are impossible and others certain. Information of that sort is used by the algorithms. For example, suppose the child, or whatever agent, knows that events of kind A come before events of kind B which come before events of kind C. Suppose the true structure were as represented in figure 2. Given data in which A, B and C are all associated, a typical Bayes net learning algorithm such as the TETRAD II 'Build' procedure (Scheines *et al.*, 1994) would use the information that A precedes B and C to test only whether B and C are independent conditional

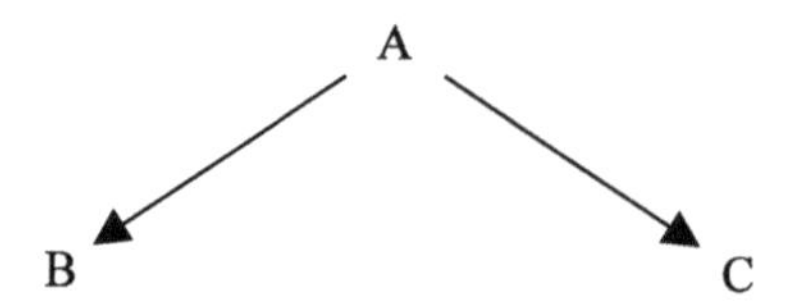

Fig. 2 The true structure of a causal graph

on A. Finding that conditional independence, the algorithm will conjecture the structure in figure 2. No other structure is consistent with the associations, the conditional independence, the time order, and the Markov and faithfulness assumptions.

Given the Markov and faithfulness assumptions, then, we can construct algorithms that will arrive at the correct causal structure if they are given information about the contingencies among events. These systems can learn about causal structure from observations and interventions.

9 Bayes nets and adults

Human adults seem to have causal maps that go beyond the causal representations of classical or operant conditioning. Is there any empirical evidence that these maps also involve Bayes net-like representations? In fact, there is some evidence that adults make causal judgements in a way that respects the assumptions of a Bayes net formalism. There is a long literature, going back to Kelley in social psychology, about the way that adults perform a kind of causal 'discounting' (Kelley, 1973). Adults seem to unconsciously consider the relationships among possible causes, that is, to consider alternative causal graphs, when they decide how much causal influence one event has on another event. In particular, Patricia Cheng's 'causal power' theory turns out to be equivalent to a particular common parameterization of causal graphs in Bayes net theories (Cheng, 1997). Cheng's theory, which was empirically motivated and developed independently of work on the Bayes net, makes the same assumptions about the relation between causal graphs and probabilities that are made in these AI models (Glymour, 2001). In effect, Cheng's work suggests that adults may use a Bayes net-like representation to make causal predictions.

While Bayes nets provide tools for prediction and intervention, they also admit algorithms for learning new causal relations from patterns of correlation. Interestingly, however, there is little work on how adults learn about new causal relations. This is probably because adults rely overwhelmingly on their prior causal knowledge of the world in making causal judgements. They already have rich, powerful, well-confirmed theoretical assumptions about what will cause what. Because of the enormous causal knowledge adults bring with them, experimentation on adult causal learning is virtually forced to imaginary or artificial scenarios to separate learning from the application of background knowledge. In everyday life, adults may rarely be motivated to revise their earlier causal knowledge or construct new knowledge of a general kind (of course, adults learn new causal particulars every day). The cognitive problem for adults is to apply that knowledge appropriately in particular situations.

The situation is very different for children. Interestingly it is also different in the special conditions in which adult human beings do science. By definition scientific inquiry is precisely about revising old causal knowledge and constructing new causal knowledge – science is quintessentially about learning. It is no coincidence that work on 'causal inference' and in particular the 'Bayes net formalism' has largely been done by philosophers of science, rather than cognitive psychologists. Human capacities for learning new causal facts about the world may be marginal for understanding much everyday adult cognition, but they are central for understanding scientific cognition.

10 Bayes nets and learning

We propose that the best place to look for powerful and generalized causal learning mechanisms, learning of the sort that might be supported by Bayes net algorithms, is in human children. Unlike adults, children cannot just rely on prior knowledge about causal relations. Prior knowledge isn't prior until after you've acquired it. And empirically, we have evidence that massive amounts of learning, particularly causal learning, take place in childhood. Indeed, in some respects the cognitive agenda for children is the reverse of the agenda for adults. Children are largely protected from the exigencies of acting swiftly and efficiently on prior knowledge, adults take those actions for them. But children do have to learn a remarkable amount of new information, in a relatively short time, with limited but abundant evidence.

Moreover, unlike non-human animals, children's learning must extend well beyond the limited set of causal relations that involve adaptively important mechanisms or involve the effects of one's own actions. Both human adults and children themselves have a large store of information about causal relations that do not involve positive or negative reinforcement and are not the result of the actions (this, of course, was one of the lessons of the cognitive revolution).

We have *prima facie* evidence that children do, in fact, learn an almost incredible amount about the causal structure of the world around them. That is the evidence that supports the theory theory in general. Similarly, we have *prima facie* evidence that, in the special circumstances of science, adult human minds can uncover new causal structure in the world. In the case of science, philosophers for many years simply abandoned the hope of uncovering a logic for this sort of causal induction. In cognitive psychology, the computational models of learning that have been proposed have been either the highly constrained 'parameter setting' models of modularity theories (see, e.g., Pinker, 1984), or the highly unconstrained and domain-general regularity detection of connectionist modelling (see, e.g. Elman *et al.*, 1997). Neither of these alternatives has been satisfactory as a way of explaining children's learning of everyday theories, or scientific change. Bayes net representations and computations provide a promising alternative. Such representations might play an important role in the acquisition of coherent causal knowledge.

In several recent studies, we have begun to show that children as young as two years old, in fact, do swiftly and accurately learn new causal relations – they create new causal maps. They do so even when they have not themselves intervened to bring about an effect, and when they could not have known about the relation through an evolutionarily determined module or through prior knowledge.We present children with a machine, 'the blicket detector', that lights up and plays music when some objects but not others are placed upon it. Children observe the contingencies between the objects and the effects and have to infer which objects are 'blickets'. That is, they have to discover which objects have this new causal power. Children as young as two years old swiftly and accurately make these inferences. They identify which objects are blickets and understand their causal powers (Gopnik and Sobel, 2000; Nazzi and Gopnik, 2000).

Moreover, they do so in ways that respect the assumptions of the Bayes net formalism. In particular, even very young children use a form of 'screening-off' reasoning to solve these problems (Gopnik *et al.*, 2001).

There are some important caveats here. Whenever we apply computational work to psychological phenomena we have no guarantee that the human mind will behave in the same way as a computer. We even have important reasons to think that the two will be different. Clearly, the computations we propose would be performed unconsciously both by children and adults. (Since the three-year-olds we are considering are still unable to consciously add two plus two it is rather unlikely that they would be consciously computing exact conditional probabilities.) Moreover, it is likely – indeed, almost certain – that human children rely more heavily on prior knowledge and on various heuristics than the current Bayes net 'data-mining' systems do.

Nevertheless we would again draw an analogy to our understanding of vision and spatial cognition. In this area there has been a thoroughgoing and extremely productive two-way interaction between computational and psychological work. While computational vision systems clearly have different strengths and weaknesses than human vision, they have proved to be surprisingly informative. Moreover, and rather surprisingly, the human visual system often turns out to use close to optimal procedures for solving the spatial inverse problem.

The program we propose is therefore not to theorize that children or scientists are optimal data-miners, but rather to investigate in general how human minds learn causal maps, and how much (and, possibly, how little) their learning processes accord with Bayes net assumptions and heuristics.

11 Conclusion

In a book published as recently as 1997 the first author of this chapter expressed pessimistic sentiments about the prospect of a computational account of everyday theory-formation and change.

> Far too often in the past psychologists have been willing to abandon their own autonomous theorising because of some infatuation with the current account of computation and neurology. We wake up one morning and discover that the account that looked so promising and scientific – S–R connections, gestalt field theory, Hebbian cell assemblies – has vanished and we have spent another couple of decades trying to accommodate our psychological theories to it. We should summon up our self-esteem and be more stand-offish in future. (Gopnik and Meltzoff, 1997)

We would not entirely eschew that advice. Pessimism may, of course, still turn out to be justified – what we have presented here is a hypothesis and a research program rather than a detailed and well-confirmed theory. Moreover, we would emphasize that, as in the case of computer vision, we think the computational accounts have as much to learn from the psychological findings as vice versa. Nevertheless, sometimes it is worth living dangerously. We hope that this set of ideas will eventually lead, not to another infatuation, but to a mutually rewarding relationship between cognition and computation.

7 The cognitive basis of model-based reasoning in science

Nancy J. Nersessian

The issue of the nature of the processes or 'mechanisms' that underlie scientific cognition is a fundamental problem for cognitive science, as is how these facilitate and constrain scientific practices for science studies. A full theory of human cognition requires understanding the nature of what is one of its most explicitly constructed, abstract and creative forms of knowing. A rich and nuanced understanding of scientific knowledge and practice must take into account how human cognitive abilities and limitations afford and constrain the practices and products of the scientific enterprise. Here I want to focus on the issue of the cognitive basis of certain model-based reasoning practices – namely, those employed in creative reasoning leading to representational change across the sciences. Investigating this issue provides insights into a central problem of creativity in science: how are genuinely novel scientific representations created, given that their construction must begin with existing representations? I will start by considering methodological issues in studying scientific cognition; then address briefly the nature of specific model-based reasoning practices employed in science; and finally provide outlines of an account of their cognitive basis, and of how they are generative of representational change.

1 How to study scientific cognition?

The project of understanding scientific cognition is inherently inter-disciplinary and collaborative. It requires a detailed knowledge of the nature of the actual cognitive practices employed by scientists; knowledge of a wide extent of existing cognitive science research pertinent to explaining those practices, such as on problem-solving, conceptual change and imagery; and employment of the customary range of cognitive science methods used in investigations of specific aspects of scientific cognition. In its approach to studying science, cognitive science has been working under the assumption made by Herbert Simon at its outset: that scientific problem-solving is just an extension of ordinary problem-solving – study the latter and you will understand the former.

This research was supported by National Science Foundation Scholar's Award SES 9810913.

Cognitive practices that take place within the context of doing scientific work have not received much scrutiny by cognitive scientists – in part because of the complexity of scientific work, and in part because the methodological practices of earlier cognitive science did not afford study of scientific cognition in the contexts in which it occurs. Computational analyses of 'scientific discovery' have tended to focus on a small range of computationally tractable reasoning practices gleaned from selective historical cases (Langley *et al.*, 1987). Psychological studies of scientific reasoning are customarily carried out in the context of studies of expert–novice problem-solving (Chi, Feltovich and Glaser, 1981) and protocol analysis of scientists or students solving science-like problems posed to them by cognitive researchers (Clement, 1989; Klahr, 1999). It is only recently, as some cognitive scientists have begun to examine scientific cognition in its own right and in the contexts within which it occurs, that they have begun to see that although it may be an extension, there are features of it that afford insight into cognition not provided by studies of mundane cognition, and that studying scientific cognition could lead to revising how we view mundane problem solving.

Study of scientific cognition has been facilitated by an important methodological shift in the field of cognitive science towards more observational studies conducted in naturalistic settings – that is, the settings in which the cognition under study naturally takes place. For example, Kevin Dunbar (1995, chapter 8 in this volume) has proposed two methods for the cognitive study of science, *in vivo* and *in vitro* studies. *In vivo* studies are those of scientific practices in 'naturalistic' – or real-world – settings, such as research laboratories. These studies employ standard protocol analysis and ethnographic methods. *In vitro* studies, on his account, employ the traditional methods of experimental psychology to investigate how subjects in studies solve authentic discovery problems in the traditional experimental settings. I will extend *in vitro* studies to encompass also 'toy' science problems given to either expert or novice subjects, and computational modelling of scientific discovery processes. Although both *in vivo* and *in vitro* studies provide valuable windows into scientific cognition, they can supply only a partial view of the nature of scientific practice. To obtain a more complete view, findings from another mode of analysis, which (following Dunbar's mode of expression) could be called *sub specie historiae* studies need to be integrated into the analysis of scientific cognition.

Sub specie historiae studies provide the perspective of how scientific practices develop and are used over periods that can extend lifetimes rather than hours and days. These practices can be examined at the level of individuals and at the level of communities. They are set in the context of training, earlier research, the knowledge base, community, collaborators, competitors and various material and socio-cultural resources. The findings derive from examining a multiplicity of sources, including notebooks, publications, correspondence and instruments.

They often involve extensive meta-cognitive reflections of scientists as they have evaluated, refined and extended representational, reasoning and communicative practices.

These studies of past science employ a *cognitive–historical* method (Nersessian, 1992a, 1995). Cognitive–historical analysis creates accounts of the nature and development of science that are informed by studies of historical and contemporary scientific practices, and cognitive science investigations of aspects of human cognition pertinent to these practices. The 'historical' dimension of the method is used to uncover the practices scientists employ, and for examining these over extended periods of time and as embedded within local communities and wider cultural contexts. The 'cognitive' dimension factors into the analysis how human cognitive capacities and limitations could produce and constrain the practices of scientists. Thus the practices uncovered are examined through a cognitive 'lens', i.e. in light of cognitive science investigations of similar practices in both ordinary and in scientific circumstances. The objectives of this line of research are to identify various cognitive practices employed in scientific cognition; to develop explanatory accounts of the generativity of the practices; and to consider, reflexively, the implications of what is learned for understanding basic cognitive processes generally. For example, my own research on conceptual change has centred on using historical case studies to identify candidate generative 'mechanisms' leading to conceptual change in science, to develop an explanatory account of how the reasoning processes employed are generative, and to use this account reflexively in addressing issues pertaining to mundane cognition, such as the nature of visual analogy.

Cognitive–historical analyses make use of the customary range of historical records for gaining access to practices and draw on and conduct cognitive science investigations into how humans reason, represent and learn. These records include notebooks, diaries, correspondence, drafts, publications and artefacts, such as instruments and physical models. The cognitive science research pertinent to the practices spans most cognitive science fields. What research is utilized and conducted depends on the issues that arise in the specific investigation. Dimensions of scientific change amenable to cognitive–historical analysis include, but are not limited to: designing and executing experiments (real-world and thought), concept formation and change, using and inventing mathematical tools, using and developing modelling tools and instruments, constructing arguments, devising ways of communicating and training practitioners.

Underlying the cognitive–historical method is a 'continuum hypothesis': the cognitive practices of scientists are extensions of the kinds of practices humans employ in coping with their physical and social environments and in problem-solving of a more ordinary kind. Scientists extend and refine basic cognitive strategies in explicit and critically reflective attempts to devise methods for understanding nature. That there is a continuum, however, does not rule out the possibility that there are salient differences between scientific and ordinary cognition. Most of the research in cognitive science has been conducted on mundane cognition in artificial contexts and on specific cognitive processes considered largely in isolation from other processes. Further, the point as argued from the perspective of situated cognition about mundane cognition (Greeno, 1998) clearly applies even more strongly to scientific cognition. Scientific 'cognition refers not only to universal patterns of information transformation that transpire inside individuals but also to transformations, the forms and functions of which are shared among individuals, social institutions and historically accumulated artefacts (tools and concepts)' (Resnick, Levine and Teasley, 1991, p. 413). To fathom scientific cognition we must examine it in a contextualized fashion.

The complex nature of scientific cognition forces integration and unification of cognitive phenomena normally treated in separate research domains such as analogy, imagery, conceptual change and decision making. In so doing, investigating scientific cognition opens the possibility that aspects of cognition previously not observed or considered will emerge, and may require enriching or even altering significantly current cognitive science understandings. Thus the cognitive–historical method needs to be reflexive in application. Cognitive theories and methods are drawn upon insofar as they help interpret the historical and contemporary practices, while at the same time cognitive theories are evaluated as to the extent to which they can be applied to scientific practices. The assumptions, methods and results from both sides are subjected to critical evaluation, with corrective insights moving in both directions. Practices uncovered in cognitive–historical investigations can provide a focal point for observational studies and for designing experiments. The point is that all three kinds of investigation are needed to develop an understanding of this complex phenomenon.

2 Model-based reasoning in conceptual change

One aspect of scientific cognition that has received significant attention in the cognitive–historical literature is conceptual change. This form of representational change has also been the focus of much research in history and philosophy of science. This research has established that conceptual innovations in 'scientific revolutions' are often the result of multiple, inter-connected, problem-solving episodes extending over long periods and even generations of scientists. The nature of the specific conceptual, analytical and material resources and constraints provided by the socio-cultural environments, both within and external to the scientific communities in which various episodes have taken place, have been examined for many episodes and sciences. What stands out from this research is that in numerous instances of 'revolutionary' conceptual change across the sciences the practices of analogy, visual representation and thought experimenting are employed. My own historical investigations centre on practices employed in physics (Nersessian, 1984a, 1984b, 1985, 1988, 1992a, 1992b, 1995, 2001a, 2001b), but studies of other sciences by philosophers, historians, and cognitive scientists establish that these practices are employed across the sciences (Rudwick, 1976; Darden, 1980, 1991; Holmes, 1981, 1985; Latour, 1986, 1987; Tweney, 1987, 1992; Giere, 1988, 1992, 1994; Griesemer and Wimsatt, 1989; Gooding, 1990; Lynch and Woolgar, 1990; Griesemer, 1991a, 1991b; Thagard, 1992; Shelley, 1996; Gentner *et al.*, 1997; Trumpler, 1997).

In historical cases, constructing new representations in science often starts with modelling, followed by the quantitative formulations found in the laws and axioms of theories. The same modelling practices often are used in communicating novel results and 'instructing' peers within the community in the new representations. They have been shown to be employed in conceptual change in science in *in vivo* (Dunbar, 1995, 1999a) and *in vitro* studies (Klahr 1999; Clement, 1989), and also in computational studies (Thagard, 1992; Gentner, 1997; Griffith, Nersessian, and Goel 1996, 2001; Griffith, 1999). Although these practices are ubiquitous and significant they are, of course, not exhaustive of the practices that generate new representational structures.

The practices of analogical modelling, visual modelling, and thought experimenting (simulative modelling) are frequently used together in a problem-solving episode. For example, figure 1 is a drawing constructed by James Clerk Maxwell in his derivation of the mathematical representation of the electromagnetic field concept, that provides a visual representation of an analogical model that is accompanied by verbal instructions for simulating it correctly in thought. Such co-occurrences underscore the significant relationships among these practices and have led me to attempt a unified account of them as forms of 'model-based reasoning'. In this chapter I take it as a given that model-based reasoning is generative of representational change in science. The project of the chapter is to determine the cognitive capacities that underlie it, and to provide an explanation of how it is generative.

Within philosophy, where the identification of reasoning with argument and logic is deeply ingrained, these practices have been looked upon quite unfavourably. Traditional accounts of scientific reasoning have restricted the notion of reasoning primarily to deductive and inductive arguments. Some philosophical accounts have proposed abduction as a form of creative reasoning, but the nature of the processes underlying abductive inference and hypothesis generation have largely been left unspecified. Conceptual change has customarily been portrayed as something inherent in conceptual systems rather than as a reasoned process, with the philosophical focus on *choice* between competing systems rather than *construction* of the alternatives. The main problem with embracing modelling practices as 'methods' of conceptual change in science is that it requires expanding philosophical notions of scientific reasoning to encompass forms of creative reasoning, many of which cannot be reduced to an

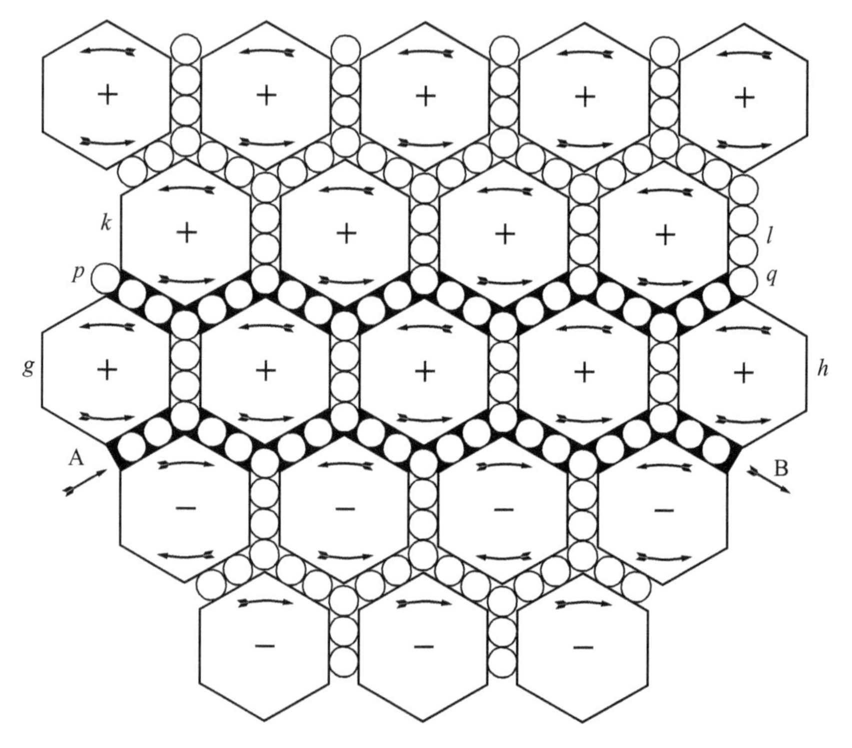

Fig. 1 Maxwell's drawing of the vortex-idle wheel model
Source: Maxwell (1890, 1, plate VII).

algorithm in application and are not always productive of solutions, and where good usage can lead to incorrect solutions. The cognitive–historical approach challenges the *a priori* philosophical conception of reasoning both with historical case studies that serve as counter-examples and as data for a richer account of scientific reasoning, and with cognitive science research that leads to a more expansive notion of reasoning.

3 The cognitive basis of model-based reasoning

Although it is not possible to go into the details in depth within the confines of this chapter, the account of model-based reasoning derives from extensive historical and cognitive research. The historical research includes my own studies – mainly of, but not limited to, nineteenth- and early twentieth-century field physicists – and pertinent research by historians and philosophers of science into other scientific domains and periods, such as noted above. As stated earlier, the nature of these scientific practices is determined by historical research and *in vivo* investigations. These provide the focal points for examining cognitive science research in search of findings that help to explain the cognitive underpinnings of the scientific practices, to formulate hypotheses about why these practices are effective and to discern ways in which the cognitive research might be challenged by the findings from examinations of scientific cognition. The cognitive science research pertinent to model-based reasoning is drawn, primarily, from the literatures on analogy, mental modelling, mental simulation, mental imagery, imagistic and diagrammatic reasoning, expert–novice problem-solving and conceptual change. In this section, a cognitive basis for model-based reasoning in science will be established by considering the representational and reasoning processes underpinning modelling practices. I will first locate my analysis of model-based reasoning within the mental modelling framework in cognitive science. I will then discuss the roles of analogy, visual representation and thought experimenting in constructing new conceptual structures.

3.1 The mental modelling framework

Akin to the traditional philosophical view, the traditional psychological view holds that the mental operations underlying reasoning consist of applying a mental logic to proposition-like representations. The work by Jean Piaget and Barbel Inhelder provides an exemplar of this position in its explicit identification of reasoning with the propositional calculus (Inhelder and Piaget, 1958). For some time critics of this view have contended that a purely syntactical account of reasoning can account neither for significant effects of semantic information exhibited in experimental studies of reasoning, nor for either the logical competence or the systematic errors displayed by people with no training in logic (Wason, 1960, 1968; Johnson-Laird, 1982, 1983; Mani and Johnson-Laird, 1982; McNamara and Sternberg, 1983; Perrig and Kintsch, 1985; Oakhill and Garnham, 1996). Instead, they propose adopting a hypothesis, first put forth by Kenneth Craik (1943), that in many instances people reason by carrying out thought experiments on internal models. In its development within contemporary cognitive science, the hypothesis of reasoning via 'mental modelling' serves as a framework for a vast body of research that examines understanding and reasoning in various domains including: reasoning about causality in physical systems (DeKleer and Brown, 1983); the role of representations of domain knowledge in reasoning (Gentner and Gentner, 1983); logical reasoning (Johnson-Laird, 1983); discourse and narrative comprehension (Johnson-Laird, 1983; Perrig and Kintsch 1985); and induction (Holland *et al.*, 1986). Additionally, there is considerable experimental protocol evidence collected by cognitive psychologists that supports claims of mental modelling as significant in the problem-solving practices of contemporary scientists (Chi, Feltovich and Glaser, 1981; Clement, 1989; Dunbar, 1995, 1999a).

Advocates of mental modelling argue that the original capacity developed as a means of simulating possible ways of manoeuvring within the physical environment. It would be highly adaptive to possess the ability to anticipate the environment and possible outcomes of actions, so it is likely that many organisms have the capacity for mental modelling from perception. Given human linguistic abilities, it should be possible to create mental models from both perception and description. This hypothesis receives support from research in narrative and discourse comprehension. It is also plausible that, as human brains developed, this ability extended to wider understanding and reasoning contexts, including science. Additionally, the differences in novice and expert reasoning skill in solving scientific problems (Chi, Feltovich and Glaser, 1981) lend support to the possibility that skill in mental modelling develops with learning (Ippolito and Tweney, 1995; Nersessian, 1995). That is, the nature and richness of models one can construct and one's ability to reason develops with learning domain-specific content and techniques. Thus facility with mental modelling is a combination of an individual's biology and learning.

The notion of understanding and reasoning via 'mental modelling' is best considered as providing an explanatory 'framework' for studying cognitive phenomena. There is not a single unitary hypothesis about the specific nature of the format of the representation of a mental model. Further, little is known about the nature of the generative processes underlying the construction and use of mental models. Given the constraints of this chapter it will not be possible to go into the details of various forms of the hypothesis invoked in explanations. Rather we will briefly consider hypotheses about the format of a mental model and discuss the reasoning processes associated with these formats.

In the first place, a mental model is a form of knowledge organization. There are two main usages of the term 'mental model' that tend to get conflated in the literature: (1) a structure in long-term memory (LTM), and (2) a temporary structure created in working memory (WM) during comprehension and reasoning processes. The first usage focuses on how the mental representation of knowledge in a domain is organized in LTM and the role it plays in supporting understanding and reasoning. Numerous studies have led to the claim that the LTM structures representing knowledge in a domain are not organized as abstract, formal structures with rules for application. Rather, it is proposed that knowledge is organized by means of qualitative models capturing salient aspects of objects, situations and processes such as the structure and causal behaviours of various systems in a domain. Mental models are schematic in that they contain selective representation of aspects of the objects, situations and processes and are thus able to be applied flexibly in many reasoning and comprehension tasks. These models are hypothesized to be generative in reasoning processes because specific inferences can be traced directly to a model, such as inferences about electrical phenomena based on a model of electricity as particulate ('teeming crowds') or as a flowing substance ('flowing water') (Gentner and Gentner, 1983), or of the operation of a thermostat based on either a valve or threshold model of its operation (Kempton, 1986). Much of the research in this area has focused on specifying the content of the mental models in a domain, with issues about the format of the mental model usually not addressed.

The second usage focuses on the nature of the structure employed in WM in a specific comprehension or reasoning task. This literature maintains that mental models are created and manipulated during narrative and discourse comprehension, deductive and inductive logical reasoning and other inferential processes such as in learning and creative reasoning. The LTM knowledge such reasoning processes draw upon need not be represented in a model – e.g. Holland *et al.* (1986) hold that LTM knowledge is organized in proposition-like schemas. Although Philip Johnson-Laird's own research focus has been on mental modelling in deductive and inductive reasoning tasks and not mental modelling in other domains, his 1983 book provides a general account of mental models as temporary reasoning structures that has had a wide influence. He holds that a mental model is a structural analogue of a real-world or imaginary situation, event, or process that the mind constructs in reasoning. What it means for a mental model to be a structural analogue is that it embodies a representation of the salient spatial and temporal relations among, and the causal structures connecting, the events and entities depicted, and whatever other information is relevant to the problem-solving task. This characterization needs expansion to include functional analogues as well.

The mental model is an analogue in that it preserves constraints inherent in what is represented. Mental models are not mental images, although in some instances an accompanying image might be employed. The representation is intended to be isomorphic to dimensions of the real-world system salient to the reasoning process. Thus, for example, in reasoning about a spring the mental model need not capture the three-dimensionality of a spring if that is not taken to be relevant to the specific problem-solving task. The nature of the representation is such as to enable simulative behaviour in which the models behave in accord with constraints that need not be stated explicitly. For example, for those tasks that are dynamic in nature, if the model captures the causal coherence of a system it should, in principle, be possible to simulate the behaviours of the system. Thus, the claim that the inferential process is one of direct manipulation of the model is central to the WM usage. The specific nature of the model-manipulation process is linked to the nature of the format of the representation.

The format issue is significant because different kinds of representations – linguistic, formulaic, imagistic, analogue – enable different kinds of operations. Operations on linguistic and formulaic representations, for example, include the familiar operations of logic and mathematics. These representations are interpreted as referring to physical objects, structures, processes, or events descriptively. Customarily, the relationship between this kind of representation and what it refers to is 'truth', and thus the representation is evaluated as being true or false. Operations on such representations are rule-based and truth-preserving if the symbols are interpreted in a consistent manner and the properties they refer to are stable in that environment. Additional operations can be defined in limited domains provided they are consistent with the constraints that hold in that domain. Manipulation, in this case, would require explicit representation of salient parameters including constraints and transition states. I will call representations with these characteristics 'propositional', following the usual philosophical usage as a language-like encoding possessing a vocabulary, grammar, and semantics (Fodor, 1975) rather than the broader usage sometimes employed in cognitive science which is co-extensive with 'symbolic'.

On the other hand, analogue models, diagrams and imagistic (perceptual) representations are interpreted as representing demonstratively. The relationship between this kind of representation and what it represents – that, following Peirce, I will call 'iconic' – is 'similarity' or 'goodness of fit'. Iconic representations are similar in degrees and aspects to what they represent, and are thus evaluated as accurate or inaccurate. Operations on iconic representations involve transformations of the representations that change their properties and relations in ways consistent with the constraints of the domain. Significantly, transformational constraints represented in iconic representations can be implicit, e.g. a person can do simple reasoning about what happens when a rod is bent without having an explicit rule, such as 'given the same force a longer rod will bend farther'. The form of representation is such as to enable simulations in which the model behaves in accord with constraints that need not be stated explicitly during this process. Mathematical expressions present an interesting case in that it's conceivable they can be represented either propositionally or iconically (Kurz and Tweney, 1998).

Dispersed throughout the cognitive science literature is another distinction pertinent to the format of mental models, concerning the nature of the symbols that constitute propositional and iconic representations – that between 'amodal' and 'modal' symbols (Barsalou, 1999). Modal symbols are analogues of the perceptual states from which they are extracted. Amodal symbols are arbitrary transductions from perceptual states, such as those associated with language. A modal symbol representing a cat would retain perceptual aspects of cats; an amodal symbol would be the strings of letters 'cat' or 'chat' or 'Katze'. Propositional representations, in the sense discussed above, are composed of amodal symbols. Iconic representations can be composed of either. For example, a representation of the situation 'the circle is to the left of the square which is to the left of the triangle' could be composed of either the perceptual correlates of the tokens, such as •–■–Δ, or amodal tokens standing for these entities, such as C–S–T. One can find all possible flavours in the mental modelling literature: propositional, amodal iconic and modal iconic mental models.

Among the WM accounts of mental modelling, Holland *et al.* (1986) maintain that reasoning with a mental model is a process of applying condition–action rules to propositional representations of the specific situation, such as making inferences about a feminist bank-teller on the basis of a model constructed from knowledge of feminists and bank-tellers. Johnson-Laird's mental models are not propositional in nature, rather they are amodal iconic representations. Making a logical inference such as *modus ponens* occurs by manipulating amodal tokens in a specific array that captures the salient structural dimensions of the problem and then searching for counter-examples to the model transformation. 'Depictive mental models' (Schwartz and Black, 1996) provide an example of modal iconic mental models. In this case, manipulation is by using tacit knowledge embedded in constraints to simulate possible behaviours, such as in an analogue model of a set-up of machine gears. In both instances of iconic models operations on a mental model transform it in ways consistent with the constraints of the system it represents.

Although the issues of the nature of the LTM representations and the WM format and processes involved in reasoning with mental models need eventually to be resolved in mental models theory, these do not have to be settled before it is possible to make progress on an account of model-based reasoning in science. My analysis of model-based reasoning adopts only a 'minimalist' hypothesis: that in certain problem-solving tasks humans reason by constructing a mental model of the situations, events and processes in WM that in dynamic cases can be manipulated through simulation. The WM model is held to be iconic but leaves open the questions of the nature of the representation in long-term memory, and whether the format of the WM representation employed in reasoning is amodal or modal.

3.2 Conceptual change and generic modelling

To explain how model-based reasoning could be generative of conceptual change in science requires a fundamental revision of the understandings of concepts, conceptual structures, conceptual change and reasoning customarily employed explicitly in philosophy and at least tacitly in the other science-studies fields. Only an outline of my account will be developed here. A basic ingredient of the revision is to view the representation of a concept as providing sets of constraints for generating members of classes of models. Concept formation and change is then a process of generating new, and modifying existing, constraints. A productive strategy for accomplishing this is through iteratively constructing models embodying specific constraints until a model of the *same type* with respect to the salient constraints of the phenomena under investigation – the 'target' phenomena – is achieved.

In the model-construction process, constraints drawn from both the target and source domains are domain-specific and need to be understood in the reasoning process at a sufficient level of abstraction for retrieval, transfer and integration to occur. I call this type of abstraction 'generic'. Although the instance of a model is specific, for a model to function as a representation that is of the same kind with respect to salient dimensions of the target phenomena inferences made with it in a reasoning process need to be understood as generic. In viewing a model generically, one takes it as representing features, such as structure and behaviours, common to members of a class of phenomena. The relation between the generic model and the specific instantiation is similar to the type–token distinction used in logic. Generality in representation is achieved by interpreting the components of the representation as referring to object, property, relation, or behaviour types rather than tokens of these.

One cannot draw or imagine a 'triangle in general' but only some specific instance of a triangle. However, in considering what it has in common with all triangles, humans have the ability to view the specific triangle as lacking specificity in its angles and sides. In considering the behaviour of a physical system such as a spring, again one often draws or imagines a specific representation. However, to consider what it has in common with all springs, one needs to reason as though it as lacked specificity in length and width and number of coils; to consider what it has in common with all simple harmonic oscillators, one needs to reason as though it lacked specificity in structure and aspects of behaviour. That is, the reasoning context demands that the interpretation of the specific spring be generic.

The kind of creative reasoning employed in conceptual innovation involves not only applying generic abstractions but creating and transforming them during the reasoning process. There are many significant examples of generic abstraction in conceptual change in science. In the domain of classical mechanics, for example, Newton can be interpreted as employing generic abstraction in reasoning about the commonalities among the motions of planets and of projectiles, which enabled him to formulate a unified mathematical representation of their motions. The models he employed, understood generically, represent what is common among the members of specific classes of physical systems, viewed with respect to a problem context. Newton's inverse-square law of gravitation abstracts what a projectile and a planet have in common in the context of determining motion; for example, that within the context of determining motion, planets and projectiles can both be represented as point masses. After Newton, the inverse-square-law model of gravitational force served as a generic model of action-at-a-distance forces for those who tried to bring all forces into the scope of Newtonian mechanics.

My hypothesis is that analogies, visual models and thought experiments are prevalent in periods of radical conceptual change because such model-based reasoning is a highly effective means of making evident and abstracting constraints of existing representational systems and, in light of constraints provided by the target problem, effective means of integrating constraints from multiple representations such that novel representational structures result. I will now provide brief encapsulations of how this occurs.

3.3 *Analogical modelling*

There is a vast cognitive science literature on analogy, and a vast number of historical cases that substantiate its employment in conceptual change. This literature provides theories of the processes of retrieval, mapping, transfer, elaboration and learning employed in analogy and the syntactic, semantic and pragmatic constraints operating on these processes (Gick and Holyoak, 1980, 1983; Gentner, 1983, 1989; Holyoak and Thagard, 1989, 1996; Thagard *et al.*, 1990; Gentner *et al.*, 1997). Most of the analogy research in cognitive science examines cases in which the source analogies are provided, either implicitly or explicitly. This had led to a focus on the nature of the reasoning processes involved with the objective of finding and making the correct mappings. There is widespread agreement on criteria for good analogical reasoning, drawn from psychological studies of productive and non-productive use of analogy and formulated by Gentner (Gentner 1983, 1989): (1) 'structural focus' – preserves relational systems; (2) 'structural consistency' – isomorphic mapping of objects and relations; and (3) 'systematicity' – maps systems of interconnected relationships, especially causal and mathematical relationships. Generic abstraction can been seen to be highly significant in analogical modelling, since objects and relations often need to be understood as lacking specificity along certain dimensions in order for retrieval and transfer to occur. This is especially evident in instances that involve recognizing potential similarities across disparate domains, and abstraction and integration of information from these. Gick and Holyoak's (1983) analysis of how knowledge gained in one analogical problem-solving task is transferred to another by creating a 'schema' common to both target and source domains provides an example of its use in mundane reasoning.

As employed in model-based reasoning, I propose that analogies serve as sources of constraints for constructing models. In this use of analogy the source domain(s) provide constraints that, in interaction with constraints drawn from the target domain, lead to the construction of initial as well as subsequent models. Model construction utilizes knowledge of the generative principles and constraints for models in a known 'source' domain, selected on the basis of target constraints. The constraints and principles can be represented in different informational formats and knowledge structures that act as explicit or tacit assumptions employed in constructing and transforming models during problem-solving. Evaluation of the analogical modelling process is in terms of how well the salient constraints of a model fit the salient constraints of a target problem, with key differences playing a significant role in further model generation (Griffith, Nersessian and Goel, 1996; Griffith, 1999). Unlike the situation typically studied for mundane cognition, in science the appropriate analogy or even analogical domain is often unknown. And it even happens that no direct analogy exists once a source domain is identified, and construction of the source analogy itself is required. In the case of Maxwell's (1890) construction of a mechanical model of the electromagnetic aether, the initial source domain was continuum mechanics and the target domain electromagnetism. No direct analogy existed within continuum mechanics, the initial source domain, and Maxwell integrated constraints first from electromagnetism and continuum mechanics to create an initial model and later used constraints from machine mechanics to modify the model, creating a hybrid model consistent with the constraints of electromagnetic induction (Nersessian, 1992a, 2001a, 2001b).

The cognitive literature agrees with the position that analogies employed in conceptual change are not 'merely' guides to reasoning but are generative in the reasoning processes in which they are employed. For example, in investigations of analogies used as mental models of a domain, it has been demonstrated that inferences made in problem-solving depend significantly upon the specific analogy in terms of which the domain has been represented. One example already mentioned is the study where subjects constructed a mental model of electricity in terms of either an analogy with flowing water or with swarming objects, and then specific inferences – sometimes erroneous – could be traced directly to the analogy (Gentner and Gentner, 1983). Here the inferential work in generating the problem solution was clearly done through the analogical models.

3.4 Visual modelling

A variety of perceptual resources can be employed in modelling. Here I focus on the use of the visual modality since it figures prominently in cases of conceptual change across the sciences. A possible reason why is that employing the visual modality may enable the reasoner to bypass specific constraints inherent in current linguistic and formulaic representations of conceptual structures. The hypothesis that the internal representations can be imagistic does not mean that they need to be picture-like in format. The claim is that they are modal in format and employ perceptual and possibly motor mechanisms in processing. They can be highly schematic in nature. Thus the fact that some scientists such as Bohr claim not to experience mental pictures in reasoning is not pertinent to the issue of whether this kind of perceptual modelling is playing a role in the reasoning.

There is a vast cognitive science literature on mental imagery that provides evidence that humans can perform simulative imaginative combinations and transformations that mimic perceptual spatial transformation (Kosslyn, 1980; Shepard and Cooper, 1982). These simulations are hypothesized to take place using internalized constraints assimilated during perception and motor activity (Kosslyn, 1994). Other research indicates that people use various kinds of knowledge of physical situations in imaginary simulations. For example, when objects are imagined as separated by a wall, the spatial transformations exhibit latency time-consistent with having simulated moving around the wall rather than through it. There are significant differences between spatial transformations and transformations requiring causal and other knowledge contained in scientific theories. Although the research on imagery in problem-solving is scant, cognitive scientists have recently undertaken several investigations examining the role of causal knowledge in mental simulation involving imagery – for example, experiments with problems employing gear rotation provide evidence of knowledge of causal constraints being utilized in imaginative reasoning (Hegarty, 1992; Hegarty and Just, 1994; Hegarty and Sims, 1994; Schwartz and Black, 1996).

As used in physics, for example, imagistic representations participate in modelling phenomena in several ways, including providing abstracted and idealized representations of aspects of phenomena and embodying aspects of theoretical models. Thus, early in Faraday's construction of an electromagnetic field concept, the imagistic model he constructed of the lines of force provided an idealized representation of the patterns of iron filings surrounding a magnet (see figure 2). However, cognitive–historical research substantiates the interpretation that later in his development of the field concept, the imagistic model functioned as the embodiment of a dynamical theoretical model of the transmission and inter-conversion of forces generally, through stresses and strains in, and various motions of, the lines (Gooding, 1981, 1990; Nersessian, 1984b, 1985; Tweney, 1985, 1992). But, as I have argued, the visual representation Maxwell presented of the idle wheel–vortex model was intended as an embodiment of an imaginary system, displaying a generic dynamical relational structure, and not as a representation of the theoretical model of electromagnetic field actions in the aether (figure 1).

External visual representations (including those made by gesturing and sketching) employed during a reasoning process are a significant dimension of cognitive activity in science and should be analysed as part of the cognitive system. These can be interpreted as providing support for the processes of constructing and reasoning with a mental model. In model-based reasoning

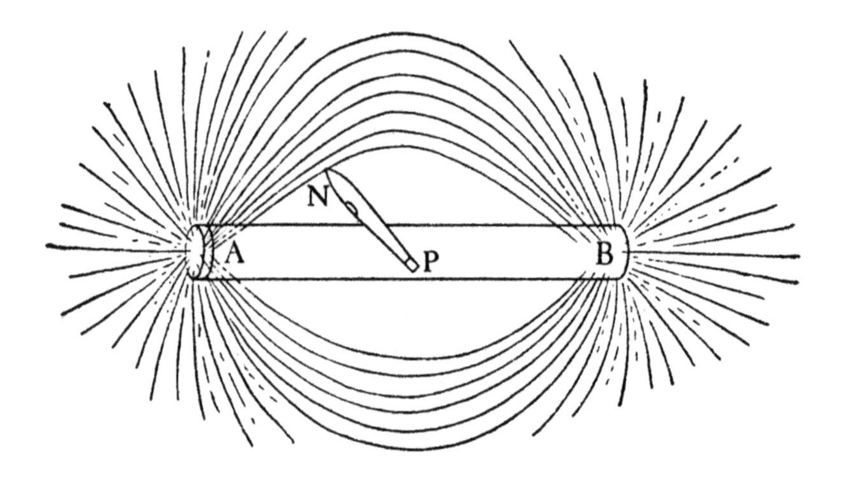

Fig. 2 Faraday's drawing of the lines of force surrounding a bar magnet. *Source*: Faraday (1839–55, vol. 1, plate 1). A, B mark the ends of the magnetic poles and P, N delineate a silver knife blade laid across the lines of force.

processes they function as much more than the external memory aids they are customarily considered to be in cognitive science. They aid significantly in organizing cognitive activity during reasoning, such as fixing attention on the salient aspects of a model, enabling retrieval and storage of salient information and exhibiting salient constraints, such as structural and causal constraints, in appropriate co-location. Further they facilitate construction of shared mental models within a community and transportation of scientific models out of the local milieu of their construction.

3.5 Simulative modelling

As a form of model-based reasoning, thought experimenting can be construed as a specific form of the simulative reasoning that can occur in conjunction with the other kinds of model-based reasoning. In simulative reasoning inferences are drawn by employing knowledge embedded in the constraints of a mental model to produce new states. Constructing a thought-experimental model requires understanding the salient constraints governing the kinds of entities or processes in the model and the possible causal, structural and functional relations among them. Conducting a simulation can employ either tacit or explicit understanding of the constraints governing how those kinds of things behave and interact and how the relations can change. A simulation creates new states of a system being modelled, which in turn creates or makes evident new constraints. Changing the conditions of a model enables inferences about differences in the way that a system can behave. Various kinds of knowledge of physical situations is employed in imaginary simulations. Because the simulation complies with the same constraints of the physical system it represents, performing a simulation with a mental model enables inferences about real-world phenomena. Note that understanding of the mathematical constraints governing a situation is one kind of knowledge that can be used in simulative reasoning by scientists.

In the case of scientific thought experiments implicated in conceptual change, the main historical traces are in the form of narrative reports, constructed after the problem-solving has taken place. These have often provided a significant means of effecting conceptual change within a scientific community. Accounting for the generative role of this form of model-based reasoning begins with examining how these thought-experimental narratives support modelling processes and then making the hypothesis that the original experiment involves a similar form of model-based reasoning. What needs to be determined is: (1) how a narrative facilitates the construction of a model of an experimental situation in thought, and (2) how one can reach conceptual and empirical conclusions by mentally simulating the experimental processes.

From a mental modelling perspective, the function of the narrative form of presentation of a thought experiment would be to guide the reader in constructing a mental model of the situation described by it and to make inferences through simulating the events and processes depicted in it. A thought-experimental model can be construed as a form of ‘discourse’ model studied by cognitive scientists, for which they argue that the operations and inferences are performed not on propositions but on the constructed model (Johnson-Laird, 1982, 1989; Perrig and Kintsch, 1985; Morrow, Bower and Greenspan, 1989).

Simulation is assisted in that the narrative delimits the specific transitions that govern what takes place. The thought-experimental simulation links the conceptual and the experiential dimensions of human cognitive processing (see also Gooding, 1992). Thus, the constructed situation inherits empirical force by being abstracted both from experiences and activities in the world and from knowledge, conceptualizations and assumptions of it. In this way, the data that derive from thought experimenting have empirical consequences and at the same time pinpoint the locus of the needed conceptual reform.

Unlike a fictional narrative, however, the context of the scientific thought experiment makes the intention clear to the reader that the inferences made pertain to potential real-world situations. The narrative has already made significant abstractions, which aid in focusing attention on the salient dimensions of the model and in recognizing the situation as prototypical (generic). Thus, the experimental consequences are seen to go beyond the specific situation of the thought experiment. The thought-experimental narrative is presented in a polished form that 'works', which should make it an effective means of generating comparable mental models among the members of a community of scientists.

The processes of constructing the thought-experimental model in the original experiment would be the same as those involved in constructing any mental model in a reasoning process. In conducting the original thought experiment a scientist would make use of inferencing mechanisms, existing representations and scientific and general world knowledge to make constrained transformations from one possible physical state to the next. Simulation competence should be a function of expertise. As with real-world experiments, some experimental revision and tweaking undoubtedly goes on in conducting the original experiment, as well as in the narrative construction, although accounts of this process are rarely presented by scientists.

Finally, in mundane cases the reasoning performed via simulative mental modelling is usually successful because the models and manipulative processes embody largely correct constraints governing everyday real-world events. Think, for example, of how people often reason about how to get an awkward piece of furniture through a door. The problem is usually solved by mentally simulating turning over a geometrical structure approximating the configuration of the piece of furniture through various rotations. The task employs often implicit knowledge of constraints on such rotations, and is often easier when the physical item is in front of the reasoner acting to support the structure in imagination. In the case of science where the situations are more removed from human sensory experience and the assumptions more imbued with theory, there is less assurance that a simulative reasoning process, even if carried out correctly, will yield success. Clearly scientists create erroneous models – revision and evaluation are crucial components of model-based reasoning. In the evaluation process, a major criterion is goodness of fit to the constraints of the target phenomena, but success can also include such factors as enabling the generation of a viable mathematical representation that can push the science along while other details of representing the phenomena are still to be worked out, as Newton did with the concept of gravitation, and Maxwell with the concept of electromagnetic field.

4 Reflexivity: cognitive hypotheses

There are several key ingredients common to the various forms of model-based reasoning practices under consideration. The problem-solving processes involve constructing models that are of the *same kind* with respect to salient dimensions of target phenomena. The models are intended as interpretations of target physical systems, processes, phenomena, or situations. The modelling practices make use of both highly specific domain knowledge and knowledge of abstract general principles. Further, they employ knowledge of how to make appropriate abstractions. Initial models are retrieved or constructed on the basis of potentially satisfying salient constraints of the target domain. Where the initial model does not produce a problem solution, modifications or new models are created to satisfy constraints drawn from an enhanced understanding of the target domain and from one or more source domains (same as target domain or different). These constraints can be supplied by means of linguistic, formulaic, and imagistic (all perceptual modalities) informational formats, including equations, texts, diagrams, pictures, maps, physical models and various kinaesthetic and auditory experiences. In the modelling process, various forms of abstraction, such as limiting case, idealization, generalization and generic modelling, are utilized, with generic modelling playing a highly significant role in the generation, abstraction and integration of constraints. Evaluation and adaptation take place in light of structural, causal, and/or functional constraint satisfaction and enhanced understanding of the target problem that has been obtained through the modelling process. Simulation can be used to produce new states and enable evaluation of behaviours, constraint satisfaction, and other factors. Figure 3 illustrates the interactive nature of these construction processes.

What cognitive account of representational and reasoning processes involved in model-based reasoning might the scientific practices support? In the preceding I have attempted to carry out the analysis by remaining as neutral as possible on some contentious issues within cognitive science, in order to show that progress can be made on understanding conceptual change in science with certain significant issues unresolved. In this section I conclude by briefly noting the cognitive hypotheses made in the analysis that would bear further investigation by cognitive science. The modelling practices exhibited by scientists utilize and engage internal modelling processes that are highly effective means

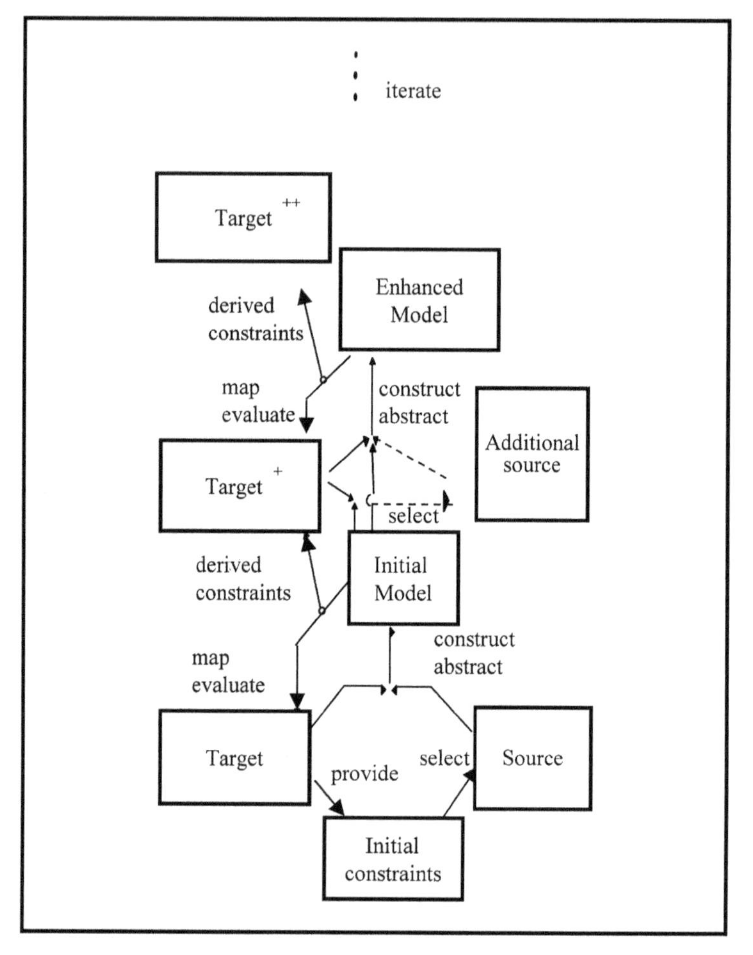

Fig. 3 Modelling processes

of generating representations and transmitting novel representations through a community. Model-based reasoning is not reducible to operations of applying mental logic to proposition-like representations. The representations are iconic in nature. The reasoning process is through model manipulation and involves processing mechanisms used in perceptual–motor activity. This implies that modal symbols are employed to some extent. Concept formation and change involve processes of generating new and modifying or replacing existing constraints. This assumes that the representations of scientific concepts provide sets of constraints for generating members of classes of models. Model-based reasoning is generative of conceptual change in science because analogical modelling, visual modelling and thought experimenting (simulative modelling) are effective means of abstracting and examining constraints of existing conceptual systems in light of constraints provided by the target problem; effective means of bypassing constraints inherent in linguistic and formulaic representations of conceptual systems; and effective means of generating and synthesizing constraints into new–revised conceptual systems. Generic abstraction enables the extraction and integration of constraints from diverse sources, leading to genuine novelty.

8 Understanding the role of cognition in science: the *Science as Category* framework

Kevin N. Dunbar

Scientists, philosophers, cognitive scientists, sociologists and journalists have often pondered over the nature of science and attempted to identify the cognitive strategies and procedures that scientists use in making discoveries. One thing that is clear from looking at science from this multiplicity of perspectives is that while many have argued for one simple cognitive procedure that characterizes science, this is fraught with problems. The goal of having a simple definition of science has now all but vanished from contemporary accounts of science. Following T. S. Kuhn's *The Structure of Scientific Revolutions* (1962) there has been a shift to a more contextually grounded account of science. In this chapter, I will show that recent research in cognitive science points to a new way of conceptualizing scientific thinking that shows what mental procedures scientists use and provides insights into the cognitive underpinnings of science and the scientific mind. I will first provide an overview of some of our recent findings on what it is that scientists do while they conduct their research. Then I will introduce the *Science as Category* framework, and discuss the implications of this framework for cognitive models of science and human development.

1 Understanding science and the scientific mind: how do we do it?

Over the past hundred years there have been many efforts to provide a unified account of science. Interestingly, many of these accounts have placed at their core a cognitive activity such as induction or deduction. The goal of the scholar has been to show that science is built upon a cognitive activity. Numerous theories have been proposed that have enriched our understanding of induction and deduction. However, while these accounts have definitely resulted in richer and more detailed accounts of logic and probability theory, most accounts rely on a partial historical record and attempt to account for some important discoveries in science using this incomplete historical record. While this continues to be a valuable and important approach to understanding science, there are large lacunae in our understandings of what it is that scientists do and how they think. More complete cognitive accounts of scientific thinking require that theorists have access to what it is that scientists *really* do when they are conducting their science, rather than relying on lab notes, biographies, or interviews. Gaining access to scientists' thinking and reasoning while they are conducting their research is imperative to understanding the nature of science and in evaluating many claims about science and the cognitive activities of scientists.

2 Science *In Vivo* as a window on the scientific mind

Capturing the types of thinking and reasoning strategies that scientists use in their work is not an easy process. Three standard methods in cognitive science have been to conduct interviews (Thagard, 1999), gather verbal protocols (Newell and Simon, 1972; Ericsson and Simon, 1984), or analyse historical documents and the history of a scientific concept (Tweney and Chitwood, 1995). All three of these methods provide much important information. However, interviews are subject to many problems (Nisbett and Wilson, 1977), and verbal protocols are difficult to obtain. Furthermore, all three approaches fail to capture the more detailed and momentary aspects of the cognitive components of science. Clearly, if we are to build cognitive models of science we need to be able to capture key aspects of the cognitive processes involved in scientists' thinking. By finding out what scientists really do while they conduct their research we can move beyond the stereotypes of science as a simple application of one process such as induction, causal reasoning, or deduction to more complex and realistic accounts of science and the scientific mind.

In this chapter I provide an overview of a new, but related, method of discovering the cognitive underpinnings of science. What we have done over the past decade is to observe and analyse scientists as they work in their laboratories (Dunbar, 1993, 1995, 1997, 1999a, 1999b, 1999c, 2000, 2001a, 2001b). In particular, we have used the weekly laboratory meeting as a way of investigating the ways that scientists think. At the lab meetings theories are proposed, experiments designed, data interpreted and arguments can happen. Thus the lab meeting provides a cross-section of the types of mental activity that scientists use. We record the meetings, transcribe every statement made and code the data along a variety of cognitive and social dimensions. I have termed this method '*In Vivo* cognition' as it is the investigation of science 'live' or '*in vivo*'. We have investigated scientists reasoning *In Vivo* at leading molecular biology and immunology labs in the USA, Canada and Italy. Furthermore, we have investigated these labs for extensive periods, ranging from four months to several years.

Using the *In Vivo* method we have been able to study labs at both a very detailed level and for long periods of time. The labs not only let us record the meetings, but allowed us to conduct interviews, have drafts of papers, grant proposals (and grant rejections), record planning sessions and attend social functions. Using this approach, we have captured discoveries as they unfold, experiments that failed and much of the cognitive life of a lab. While the lab meetings are our primary source of data, these other sources are used to supplement the primary data. (See Dunbar 1997, 2001 for an explicit description of the details of this methodology.) Clearly, we can capture only a fragment of what it is that scientists do and how they think using this method. The most important point is that this data allows us to investigate the way that scientists think and reason in a naturalistic context. This way of investigating the cognitive processes actually used by scientists allows us to bypass scientists' faulty memories and avoid being dependent on the already highly filtered cognitive processes that are apparent in scientists' lab books.

In the next section of this chapter I provide an overview of our basic results. Then I use the results to propose a new cognitive account of science: the *Science as Category* (SAC) framework. Finally, I use the SAC framework to address a number of important controversies that have raged around cognitive models of science. By investigating scientists as they work in their labs it is possible to go beyond many of the stereotypes of science and of scientists that have motivated much of the research on scientific thinking. By developing models of the way that working scientists think and reason we can then look at the genesis of scientific thinking and assess recent models of scientific thinking in human development.

3 Basic results from *In Vivo* investigations of scientists

By analysing scientists' thinking, reasoning and interactions at lab meetings we have observed four main classes of cognitive activity that scientists engage in: *Causal Reasoning* (which involves deductive and/or inductive reasoning), *Analogy*, *Categorization* and *Distributed Reasoning*. None of these mental activities are mutually exclusive (for example Analogy and Causal Reasoning have been thought to be forms of induction) and the borders between any one type of activity and the others are necessarily fuzzy. Each of these cognitive activities can result in minor and major changes in scientists' representations of their knowledge. A more detailed discussion of these results has been given elsewhere (Dunbar, 1993, 1995, 1997, 1999a, 2001). Here I will discuss these findings as a way of laying the groundwork for the *Science as Category* (SAC) framework.

3.1 Causal reasoning

Scientists engage in causal reasoning where they propose a cause, an effect and a mechanism that leads from a cause to an effect (Ahn and Bailenson, 1999; Cheng, 1999; Baker and Dunbar, 2000; Dunbar, 2001a). This type of causal reasoning occurs in virtually every scientific activity including interpreting data, designing experiments and theory building. What we have done over the past decade is to catalogue all the casual reasoning episodes that we find at a lab meeting. What we have found is that causal reasoning is one of the key types of cognitive activity that the scientists engage in – accounting for over 80% of the statements made at a meeting. We have defined causal reasoning episodes as sequences of statements in which scientists attempt to propose a mechanism that will lead from a cause to an effect.

Causal reasoning may be involved in a scientist designing an experiment in which they predict that a sequence of biological processes will occur, such as the steps involved in HIV cutting open host DNA and joining itself to the DNA. Here the scientists proposed a causal mechanism consisting of a dynamic sequence of steps leading from one state to another. What is interesting about the reasoning here is that while deduction is a component of the reasoning, a causal episode will consist of dozens of deductions, problem-solving techniques, analogies and inductions by generalizing over particular statements. Thus while it might be tempting to label this type of reasoning as one type of cognitive process, many types of cognitive processes are actually involved in a causal reasoning episode. What happens in causal reasoning is that the entire cognitive machinery is put together in the service of a goal. In designing an experiment, the goal is an experiment with results that can be interpreted both in terms of current hypotheses, but also in terms of the overall goal. Thus, the overall goal might be to propose a theory of retroviral integration, but a particular lab meeting might be composed of a series of causal reasoning episodes, each with a particular goal state of proposing or delineating a mechanism that takes the scientists from a cause to an effect. Causal reasoning in science is not a unitary cognitive process, as argued by Gopnik and Glymour (see chapter 6 in this volume), but a combination of very specific cognitive process that are co-ordinated to achieve a causal explanation such as an unexpected finding.

The ways that scientists react to unexpected findings makes apparent the components underlying causal reasoning. As I have shown elsewhere (Dunbar, 1995, 1997, 1999a, 2001), one of the goals of weekly lab meetings is to solve problems that crop up in the scientists' research. Many of these problems concern unexpected findings, which are findings that are not what were predicted. These unexpected findings are due to technical problems, discovery of new mechanisms and mechanisms that the scientists had not originally considered but were familiar with. We have found that 30–70 percent of results presented at a particular lab meeting are unexpected and that scientists have a clear set of cognitive strategies for dealing with these findings. First, they categorize their findings. This categorization consists of labelling their findings in terms of their current knowledge and predictions. For example, they might say that this is the 'ROP 1 gene' or 'this is P30'. Expected findings are not dwelt upon and often allow the scientists to move to the next step in their planned sequence of experiments. When the findings are unexpected the scientists still attempt to categorize the finding and usually blame the result on a methodological problem. Here, scientists often try to fix the experimental problem by drawing an analogy to a related experiment. That is, they provide methodological explanations using local analogies that suggest ways of changing their experiments to obtain the desired result. If the changes to experiments do not provide the desired results, then scientists switch from blaming the method to formulating hypotheses. This involves the use of more distant analogies and collaborative reasoning in which groups of scientists build models. Thus, only when scientists see a *series* of unexpected findings do they attempt to change their models and theories. One type of reasoning sequence is a combination of classification, blame the method, make an analogy and then propose a new experiment. Another reasoning sequence is classify the finding, generalize over a series of findings, propose a new model, make an analogy to a related domain and propose new research questions. The second sequence is much less frequent than the first.

Overall, rather than causal reasoning being the product of a unitary causal reasoning faculty, causal reasoning consists of a combination of cognitive components that are put together to achieve a particular goal such as fixing an experiment or determining the locus of an unexpected finding. Causal reasoning is made up of a number of basic cognitive process that can vary according to context. As the next section demonstrates, analogy is frequently a key component of this causal reasoning process.

3.2 Analogy

The mental operations specified in the previous section were at a very general level. We can now examine some of these operations in more detail. Here we will look at scientists' use of analogy. Contemporary cognitive scientists distinguish between two main components of an analogy – the source and the target. The target is the problem that a scientist is working on and the source is the known phenomenon that the scientist uses to understand the target. For example, scientists have used an analogy to dust devils in deserts to explain changes in the visual features of the planet Mars. When scientists use an analogy they map features of the source onto the target. By mapping the features of the source onto the target, scientist can understand the target better. (For an in-depth coverage of recent research on analogy, Gentner, Holyoak and Kokinov, 2001, provides an excellent compendium.)

For our initial investigations, we analysed the use of analogy at sixteen laboratory meetings in four laboratories (Dunbar, 1995, 1997, 1999a). We found over 99 analogies in total, and that scientists used anywhere from three to fifteen analogies in a one-hour laboratory meeting. What we also found is that the majority of analogies that scientists used were biological or within-domain analogies. Molecular biologists and immunologists used analogies from the biological or immunological domains and not from domains such as economics, astrophysics, or personal finances. However, the picture of analogy is slightly more complicated than this. The goals that the scientists have will determine the way in which they use analogy. What we found was that when the goal was to fix experimental problems, the sources and targets did indeed share superficial features such as using the same temperature. For example, a scientist might draw an analogy between two different experiments using a gene on the HIV virus. The gene, the proteins used and the temperature used might be identical, but the incubation time might be different. Thus superficial features would be very similar.

It is important to note that analogies based upon superficial features are very useful and help solve problems that the scientists frequently encounter. For example, if a scientist is obtaining uninterpretable data and another scientist makes an analogy to an experiment that worked in which the temperature was different, and suggests changing the temperature, this can potentially fix the problem. Many of the superficial analogies were of this type and solved the problem. These types of analogy are very frequent, but were rarely mentioned by the scientists in their interviews and never appeared in their lab books. When asked why they did not mention these analogies, the scientists thought that they were so obvious that they were not worth talking about. Despite the obviousness of these analogies, they are one of the most frequent cognitive workhorses of the scientific mind.

When scientists switched goals to formulating a hypothesis, the types of analogy used also changed. We found that the distance between the source and the target increased and that the superficial features in common between the source and target also decreased. Here, scientists used structural and relational features to make an analogy between a source and a target, rather than superficial features. Thus, while only 25% of all analogies that scientists used were based on structural rather than superficial features, over 80% of these structural analogies were used to formulate hypotheses. The scientists based their hypotheses on the similarity of the underlying genetic sequences of the source and the target. For example, if a gene in clams and a gene in the parasite plasmodium had a similar genetic sequence, the scientists might postulate that this genetic sequence has the same function in plasmodium as in clams. While there is no superficial similarity between clams and plasmodium, there is an underlying structural similarity. It is important to note that even though the scientists were basing their analogies on sources and targets within the domain of biology, the analogies were based upon underlying sets of structural relations rather than any superficial similarity between clams and plasmodium. These results indicate that while scientists use many analogies that share superficial features between the source and the target, they can and do produce analogies that are based on deeper structural features. Thus, analogy-use appears to be flexible, and to change with the goals of the analogizer.

These findings on analogy provide a very different perspective on analogy to that offered in most cognitive models of science. First, mundane analogies are the workhorse of the scientific mind rather than the strange analogies that many theorists such as Boden (1990) have focused upon. Second, analogies work as a type of cognitive scaffolding and are thrown away after they are used. Because of this, most analogies do not make it into scientists' lab books or biographies. Third, analogy is used with other cognitive processes such as deduction and is also used with very specific triggers such as explanations, hypothesis formulation and fixing experiments. Overall, this points to the fact that scientific thinking is composed of many inter-related sub-components that work in the context of particular goals.

3.3 Collaborative and distributed scientific thinking

One important problem for cognitive models of science has been to build models that participate in and are sensitive to the social and cultural contexts of science. Even a cursory visit to a scientific laboratory reveals that there is a vast social network of scientists who interact and share values and norms, and that this is an integral part of science. We have found that lab meetings have large amounts of social interaction and that during these interactions significant changes in scientists' representations of their knowledge occur. (See Dunbar and Dama, in preparation, for a detailed model of the relationship of social processes to cognitive factors; see also Giere, chapter 15 in this volume.) What I will focus on here is one component of this: the *Representational Change Cycle* or RCC.

What we have found is that certain types of social interaction are conducive to changes in scientists' representations of their knowledge and that there are complex interactions of status, knowledge and willingness to change representations. Furthermore, we have found that for many types of cognitive change, such as generalizing a new concept from a set of findings (induction via generalization), many people can be involved. One person might provide one result, another the next result, and the third will contribute the generalization. Thus, even basic inductive processes are frequently completed by groups. As mentioned above, the trigger for group participation in reasoning is often a series of unexpected findings. Here, different sources of knowledge that different members of the group possess can be applied to a problem. Overall, these results indicate that social and cognitive processes interact to produce significant changes in knowledge and that these changes are subject to both social and cognitive constraints. (See Dunbar and Dama, in preparation, Dama and Dunbar, 1996, for a detailed account of how this works.) This research on distributed reasoning in science again points to the fact that scientific thinking is not a closed cognitive process, but a process in which inputs can come from many agents. What our representational change cycle is allowing us to do is predict when and whether input from others will produce changes in representations.

4 What's special about science?: an investigation of politics and political thinking

The results of our analysis of scientists' thinking and reasoning in naturalistic settings reveals that *Analogy*, *Categorization*, *Causal Reasoning*, *Collaborative Reasoning* and *Problem-Solving* are frequently used. Furthermore these cognitive processes are frequently used together to subserve particular goals. Can we define science as the amalgamation of these different strategies? Even a superficial consideration of this question would lead to a negative answer. An examination of the types of reasoning of people engaged in a non-scientific activity should provide an important contrast for understanding science. We decided to investigate politicians' reasoning at political rallies. One reason for picking politics is that it has often been portrayed as the opposite of science (e.g. Popper, 1959). Other theorists have argued that science is inherently political and that science is merely a specialized form of political discourse (Latour and Woolgar, 1986). What I will do here is look at some of the reasoning that politicians engage in and ask whether it is similar to or different from that of scientists. One way that we have investigated this question is by looking at politicians' use of analogy.

Having investigated the use of analogy in one naturalistic context – science – we decided to investigate analogy use in another naturalistic context – politics (Blanchette and Dunbar, 2001). We investigated the use of analogy in a referendum on independence that took place in Quebec in 1995. The electorate was presented with the choice of voting to keep the province of Quebec within the country of Canada or voting to make Quebec a new country. We analysed politicians' and journalists' use of analogy in newspaper articles during the final week of this very important referendum campaign. What we did was to take every newspaper story that appeared in three Montreal newspapers during the last week of the referendum campaign and search for all uses of analogy in the paper. We found over 400 articles referring to the referendum and then searched for every analogy that was used. This analysis resulted in the discovery of over 200 analogies.

The analysis of the sources that were used revealed that only 24% of the sources were from politics. That is, over 75% of the analogies were not based on superficial features. Rather, the analogies were based on higher-order relations and structural features. The range of source categories was very diverse, ranging from agriculture, the family, sport, magic, to religion. Overall, these analyses revealed that the politicians and journalists frequently used sources from domains other than politics. We also looked at political meetings and rallies 'live'. The results we obtained were virtually identical to our analyses of analogies in newspapers. We found that over 70% of the analogies used were from non-political domains. Politicians were not basing their analogies on superficial features. Their analogies were based upon structural features. In both science and politics we have found that analogies based on higher-order relations are common.

Overall, scientists and politicians frequently use analogy as a component of their thinking. However, analogy-use in both science and politics appears somewhat different. Politicians rarely used analogies based upon superficial features, whereas scientists more frequently used analogies based upon superficial features. Furthermore, our analyses show that scientists tend not to use distant analogies at their lab meetings, whereas politicians use analogies from non-political domains. When we look at the types of goal that scientists and politicians have at their meetings, the reasons for the different uses of analogy become readily apparent. One of the primary goals of scientists at their lab meetings is to fix problems with their experiments. To achieve this goal, the scientists make analogies to superficially similar situations. In the political debates that we have analysed, the goal of the politicians is not to fix small problems, but engage in massive political change. In this context, superficial analogies were rare. Thus, the goals appear to be shaping the types of analogies used. Why do the scientists rarely use distant analogies? Again the answer appears to lie in the goals that the scientists have. Our analyses show that scientists will use distant analogies to explain entities to others. Thus, when understood in the light of the relevant goals, analogy-use and distance of the source from the target are virtually identical in both science and politics. Analogy is used in both domains and analogy-use operates under the constraint of the current goal.

The comparisons of the political and scientific uses of analogy highlight the fact that goals provide a large constraint on the ways that analogies are invoked. Rather than being a general-purpose reasoning mechanism that is used in the same way in all contexts, the goals of the analogizer will determine how near or how far the analogy will be from the current problem, and the types of features that will be used in mapping from a source to a target. Again, this points to a need to move from unitary abstract models of scientific thinking to models that take into account contextual and goal-relevant constraints.

4.1 Culture and science

Cognitive accounts of science usually do not consider the role of culture in science. Analyses of scientific thinking and cognitive models of science have been largely based upon the idea that cognition is not susceptible to cultural influences. However, Richard Nisbett demonstrates in his recent work on culture and cognition that culture can have significant effects on cognition. Nisbett and his colleagues have shown that Chinese participants in his experiments induce new concepts in a different way from Western participants (Nisbett *et al.*, 2001). This type of finding demonstrates that culture can influence cognition and leads to the prediction that we should see cultural differences in the ways that scientists conduct their research. If this is the case, this type of finding would have important implications for cognitive models of science as it would point to key places where culture influences science. A significant issue here is to take scientists working on similar problems, using similar techniques and having similar lab meetings to the North American labs. If we see differences in types of cognitive procedure, then this will allow us to understand some key aspects of the scientific mind.

We investigated the reasoning strategies and practices used in similar immunology labs in Canada, the USA and Italy. As in my research on the American labs, we taped the weekly lab meetings, supplemented by interviews with the director of the lab. The Italian lab was of similar size to the American labs, used the same types of materials, cells and equipment as the American labs, published in the same leading journals and the members of the lab attended the same conferences as the North American scientists. Thus, at the scientific level this lab was equivalent to the American labs. What we were interested in was whether the reasoning was equivalent, or whether the reasoning used in the science labs conformed to that of the wider culture.

Our initial analyses were of analogy. Over eight meetings we found few analogies made. In fact we found that induction and inductive reasoning strategies were rare. This was very different from what we observed in the North American Labs. In the North American labs we found anywhere from three to fifteen analogies in a meeting. We found that other forms of inductive reasoning such as generalizing over a set of findings were also infrequent at the Italian lab. Instead, we found that deductive reasoning was the norm, even in interpreting the data. In the North American labs, deduction was mainly used to make predictions for experiments or explain the rationale for experiments. In Italy, deduction was used for virtually all goals. Interestingly, this predominant use of deduction in an Italian lab is consistent with analyses of discourse in the wider Italian culture. For example, Triandis (personal communication) has found important cultural differences in reasoning and has suggested that cultures will have important effects on the reasoning. What is surprising here is that the scientists use the same materials and methods, speak at the same conferences and publish in the same journals as their North American counterparts, are often trained at American Universities, yet their styles of reasoning are different. While this conclusion fits with some historical analyses of differences between European and North American scientists (e.g. Crombie, 1994), it is nonetheless surprising to see such stark differences in the 'live' reasoning of scientists in a lab that uses the same theory, materials and methods as similar North American labs. What this points to is that one of the key roles of culture is to put together the same cognitive processes in different ways. Thus cognitive models of science can capture cultural differences in scientific thinking and point to the precise locus of the differences. This allows us to build a model of science showing how cognition, culture and science interact which I will articulate in the next section.

5 The nature(s) of the scientific mind: the *Science as Category* framework

We have investigated reasoning in scientific laboratories in Canada, the USA and Italy. We have also investigated the reasoning of politicians at political meetings (Blanchette and Dunbar, 2001) and the development of scientific reasoning skills in children (Dunbar and Klahr, 1989; Klahr, 2000). By investigating reasoning in these different contexts a picture of the scientific mind is beginning to emerge. Scientific thinking involves a whole gamut of cognitive processes such as *Induction*, *Deduction*, *Analogy*, *Causal Reasoning*, *Categorization*, *Problem-Solving* and *Socio-cognitive Interaction*. However, a brief look at babies, baseball players, politicians, artists, carpenters and playwrights reveals that the same mental processes are at work in these domains, too. Baseball players make predictions, categorize types of throws, have theories and test their hypotheses. How do we deal with these similarities? Many have noted that there are fundamental similarities between science and other human activities and have drawn three main types of conclusion:

1. One common conclusion has been to take science off its throne and make it just another error-prone and biased human activity. Much of the sociology and anthropology of science has taken this approach. Theorists using this approach stress the cultural embeddedness of science and see science as no different from any other cultural activity (e.g. Latour and Woolgar, 1986).
2. Another approach has been to note a similarity between the cognitive activity or mental representations of scientists and non-scientists, and to label the non-scientist a scientist. Thus, many developmental psychologists have noted that babies appear to be testing hypotheses and have proposed the 'child-as-scientist' view or the 'scientist-in-the crib' view (e.g. Gopnik, Meltzoff and Kuhl, 1999a). This view has resulted in an enriched view of the cognitive capacities of children and has been an important shift away from the blank-slate and developmental-stages views of children's thinking.
3. The third view has been to cling on to certain cognitive activities as being essential defining features of science and science only. Popper's falsificationist agenda is an example of this third way of characterizing science.

Much of the cognitive research on scientific thinking has focused on particular cognitive activities such as falsification of hypotheses and noted that even scientists often fail to reason in a normatively correct manner (that is assuming the norms are correct!). (See, for reviews of this literature, Klayman and Ha, 1987; Tweney and Chitwood, 1995.)

Each of these views of science captures important aspects of science, culture and the scientific mind. All of these theoretical stances attempt to capture certain important features of the scientific mind, but also ignore other features. Much of the debate among proponents of the different perspectives concerns the question of what the alternative accounts fail to capture. In this section, I will propose an account of science that attempts to encompass all three approaches to the scientific mind.

The underlying premise of the *Science as Category* framework is that science uses a number of basic thinking and reasoning skills that all human beings, and some non-humans, possess – such as *Categorization*, *Induction*, *Analogy*, *Causal Reasoning* and *Deduction*. Science puts these thinking strategies together in particular ways to subserve very particular goals. Other human activities such as games or sports can take the same mental operations and cognitive activities, put them together in different ways to subserve different goals, and result in different types of human activity. Thus, both science and non-scientific activities can use the same cognitive building blocks. The main difference between science and these other activities will be their goals and the way that the cognitive building blocks are put together.

A key insight is to regard science as a humanly constructed category like *vehicles*, *clothing*, *money*, or *food*. Over the past thirty years cognitive scientists have developed models and accounts of categories that can be applied to science and bring clarity to the diversity of views of science and the scientific mind (see Medin, Lynch and Solomon, 2000, for a fascinating review of this literature). The *Science as Category* view proposes that science is the combination of different cognitive components rather than being one type of cognitive activity such as having a theory or having a causal reasoning module.

By regarding science as a humanly constructed category of different cognitive processes, four important ideas emerge. First, like all categories, science can have internal structure. Second, goals will be an important cognitive glue that will aid the construction of the category. Third, it is possible to highlight a particular cognitive activity such as having a theory and draw analogies between theory-building in science and any other human activity that also involves theory-building. Fourth, like many human categories there can be multiple ways of considering the category *science*: there is more than one way of carving up the category *science*. Each of these different aspects of the *Science as Category* framework will now be considered.

5.1 *Internal structure of the category* science

Most categories possess graded structure. That is, there are typical and atypical members of categories. There are prototypes which are typical members of the category. For example, cars are typical vehicles, and in the year 2001, SUVs are also typical vehicles. However, the category of vehicles can have much less typical members such as skateboards and stealth bombers. Science, like all categories, has prototypes – such as chemistry, biology and physics – and less prototypical sciences, such as ceramics and economics. What makes something a prototype? The consensus among cognitive scientists has been that prototypical members of a category share many features with each other and that atypical category members share fewer features with either each other or the prototypes (Smith and Medin, 1981). There is no one feature that is essential or defining for something to be a member of a category. Instead, category members have constellations of features, none of which are defining. Having argued that categories consist of constellations of certain features, we can ask, what are the *cognitive features* that are important in science? Our *In Vivo* analyses indicate that *Induction*, *Deduction*, *Causal Reasoning*, *Analogy*, *Distributed Reasoning* and *Categorization* are key cognitive features. Furthermore, cognitive–historical analyses such as those of Nersessian and Thagard have also identified sequences of these types of cognitive activity as being crucial to many scientific discoveries (see chapters 7 and 12 in this volume).

Most importantly, these cognitive activities do not exist in science alone. What distinguishes different types of human activity are not the specific cognitive building blocks, but the sequences of cognitive building blocks. There are key constellations of reasoning activities that scientists engage in, such as *Analogy–Deduction–Induction–Causal Reasoning* chains. These sequences are frequently used in science and are composed of basic cognitive building blocks that all human beings possess. Just as the genes that code for the human eye are made up of one sequence of nucleotides and the genes that code for the human hand are made up of a different sequence of the same nucleotides, the scientific mind is made up of specific sequences of cognitive activities and a baseball player's mind is made up of a different sequence of these cognitive activities.

Different sciences will not necessarily use the same sequences of cognitive activities, and sub-disciplines of a field such as Theoretical and Experimental Physics may use different sequences of cognitive activity. What is important to note is that the cognitive features of science are organized in ways that are similar to those of any other humanly constructed category. This allows cultural and social factors to be a key element of the SAC account of science. In the next section we will consider the role of goals in the construction and maintenance of the category *science*.

5.2 *Goals and the category* science

A further aspect of the *Science as Category* framework is that many categories are goal-derived. Larry Barsalou (1983, 1991) has elegantly demonstrated that people will construct categories when given particular goals. For example, when asked to construct a list of ways to escape from the Mafia, participants have little difficulty generating categories that have prototypes and internal structure, with typical and atypical members. More recent research by Ross and Murphy (1999) also shows that categories such as food can be organized taxonomically (e.g. meats, vegetables, cereals, etc.), or around goals such as breakfast foods, movie foods or junk foods. Goals are an important way of organizing and categorizing the world in which we live, and help organize the long-term representation of our knowledge. Like all other humanly constructed categories, goals are a key feature of science and particular disciplines have overarching goals and vast networks of underlying sub-goals that help generate questions and establish particular methods as being appropriate for that field.

Goals are the glue that bind together various cognitive activities in scientific fields and because the overarching goals are frequently instantiated, particular combinations of cognitive activity are engaged in and become accepted ways of thinking and reasoning in that field. For example in molecular biology in the 1990s, scientists frequently used analogy to propose hypotheses, fix experimental problems and categorize their data and give causal explanations for particular results. These reasoning activities are focused around particular genes and particular gene products. Now that entire genomes have been mapped, there has been a shift to looking at vast sets of genes (20,000 genes at the same time). This has resulted in totally new ways of designing experiments and theories of gene function, and consequently the types of thinking and reasoning strategies that scientists use in their day-to-day research are changing. Rather than focusing on just one gene, scientists are now developing different ways of interpreting and understanding the genomes that they investigate. This necessitates new ways of putting together the cognitive building blocks used in molecular biological research. Given that goals can change in particular fields, there can be changes in the types of cognitive and social activities that members of the field use. Thus, by seeing sciences as 'goal constrained cognitive categories' we can see that the ways that scientists think and reason will not be constant but will vary both over time and across disciplines. The SAC framework makes it possible to propose cognitive models of science that are sensitive to the changing contexts of science and makes it possible to go beyond simplistic accounts of the role of particular cognitive processes in science.

5.3 Analogies to science

Beginning with the work of Inhelder and Piaget (1958), many researchers have noted that children bear certain striking similarities to scientists: they have theories, conduct experiments and engage in the whole gamut of thinking processes that are also involved in science, ranging from deduction, induction, to analogy. This is the 'child-as-scientist' analogy. Many researchers have argued that there is little difference between a three-year-old and a scientist in terms of underlying reasoning ability. That is, while scientists clearly have considerably more knowledge of domains than children, their underlying competencies are the same. Other researchers have expanded the 'child-as-scientist' analogy and argued that infants are basically scientists (Gopnik, Meltzoff and Kuhl, 1999a). Yet other researchers have argued that there are fundamental differences between children and scientists and that scientific thinking skills follow a developmental progression. (An overview of this debate can be found in Klahr, 2000.) This wide variety of viewpoints on the child-as-scientist hinges upon the nature of the analogy between a child and a scientist. At its most basic level, the controversy surrounds whether one focuses on the similarities or the differences between children and scientists. However, we also have to ask whether the analogy between babies and scientists captures the sequences of cognitive activities that scientists engage in.

While one could pluck any one of these cognitive activities out and say that this is *the* key feature of science, or these are the defining features of science, what we have seen is that it is the combination of the cognitive activities that allows us to see the category of science. Just as juiciness is an important feature of fruits, it is juiciness in conjunction with other features such as *soft*, *grows above ground*, etc. that make these juicy things fruits. Taking one feature of typical sciences such as making predictions, or having theories, does capture some important features; however, it fails to take note of how different cognitive activities are brought together in science. Without question, human beings, including children, possess each of these cognitive skills. The way that these skills are brought together is what makes the thinking of a baseball player different from a molecular biologist. We have seen this in our investigations of scientific reasoning in children where we have found that children have few problems in generating hypotheses, or in designing experiments, but do have difficulty in co-ordinating their hypotheses with their experiments. Only in situations where both hypotheses and experiments are highly constrained can children design appropriate experiments to test hypotheses.

Overall, while it is useful to draw an analogy between scientists and children, because it shows that children have complex knowledge, this does not make children scientists. When researchers such as Gopnik, Meltzoff and Kuhl (1999a) have claimed that infants are scientists they have merely plucked out one feature of the scientific mind and made that equivalent to the whole discipline. While there is nothing wrong with their claims *per se*, the claims are too general to help us understand how children's scientific minds change over time. To properly address the nature of the trajectory of the scientific mind we must address the development of the patterns of scientific reasoning that scientists use and the social context in which these activities are embedded. By merely focusing on one activity such as having a theory or using causal models it is possible to point to an important similarity to what scientists do. However to understand scientific thinking it is necessary to examine how the different cognitive processes are combined rather than focusing on one process.

5.4 Other approaches to science: cross-classifications of science

Cognitive views of science are not the only views. Many accounts exist which are apparently inconsistent with, and often hostile to, cognitive views of science, such as the social studies and anthropological studies of science (Latour and Woolgar, 1986; Pickering, 1992). These views have regarded cognitive approaches as bankrupt and misleading. The two main reasons for dismissing cognitive approaches are first, that cognitive researchers have often ignored the social and cultural practices in which science is embedded; and second, that cognitive researchers are seen as being reductionist crusaders, boiling down science to a set of basic cognitive processes and again have missed the role of social and cultural factors which, according to this viewpoint, are the driving force of science. There is an important element of truth in these two criticisms of cognitive views of science. Cognitive researchers have ignored social and cultural influences and we have been reductionist in our approach. However, our *In Vivo* work on science demonstrates that cognitive, cultural and social factors are all involved in science. Rather than attempting to reduce social and cultural components of science to cognition, these components must be combined with cognitive components to truly understand the nature of science.

However, there is a more interesting way in which cognitive and social studies of science have diverged, which is captured by the *Science as Category* framework. Most categories have multiple ways of dividing the world up. As discussed in sub-section 5.2, we can divide foods up taxonomically into fruits, meats, vegetables, etc. Alternatively, we can easily divide foods up into breakfast, dinner, lunch and snack foods. People can readily divide up foods in both ways. Thus one way that the human mind tends to divide the world up is *taxonomic* and another is *functional*. Both ways of dividing up the world have their virtues and human beings effortlessly switch from one view to the other. Cognitive approaches to science have tended to stress the cognitive attributes of science in a very taxonomic fashion. For example, computational models of scientific thinking have neat categories of cognitive activities that can be organized in taxonomically ordered ways. Social studies of science have tended to stress situations and functionally important components of science. Both the cognitive and social studies of science views have categorized the same domains in different, but important ways. Both ways of investigating science provide important insights into science. However, we are both engaged in building categories of science. By seeing these as two different ways of categorizing science, we can see that each has its own uses. A question then becomes whether, like a visual illusion, we can only see one interpretation at a time; or whether it is possible to build a unified view of science that can account for social and cognitive aspects of science in the same theory.

9 Theorizing is important, and collateral information constrains how well it is done

Barbara Koslowski and Stephanie Thompson

In the psychological literature, people are often deemed to be poor scientists. We argue that this is because many psychological measures are framed (at least tacitly) in a positivist tradition that emphasizes abstract formal rules rather than content. However, realist conceptions of science point out that formal or abstract rules can be used successfully only when content, or collateral information, is also taken into account. When psychological measures do take account of collateral information, then people perform appropriately. An especially important type of collateral information is information about mechanism, or theory. However, theories must also be evaluated to identify those that are plausible and to avoid being misled by those that are dubious. Realist philosophers note that theories or explanations can also be evaluated only by taking account of collateral information. We identify some of the types of collateral information that scientists rely on to evaluate explanations and examine the psychological function that such information plays in the reasoning of non-scientists. Furthermore, we note that the collateral information people search for is limited by the collateral information that is initially available to them and argue that it is limited as well by the social context of science.

1 Overview

If a large body of the psychological literature is to be believed, then people lack many of the dispositions that are prerequisites for sound thinking in general, and for scientific reasoning in particular. They have difficulty co-ordinating theory and evidence. They do not consider alternative accounts. They are reluctant to seek disconfirming evidence and, when confronted with it, do not take it into account. They fail to realize that when two factors are confounded and both co-vary with an effect, the actual cause of the effect is indeterminate. Of course, one possibility is that these characterizations are accurate, that there is something that scientists do (or at least that the scientific community does) that

Many thanks to commentators at the Cognitive Basis of Science Workshop at Sheffield in April 2000, and at the Cognitive Basis of Science Conference at Sheffield in June 2000 for their many helpful suggestions; and to Richard Boyd.

non-scientists fail to do. However, another possibility is that the measures that psychologists often rely on to assess scientific reasoning are seriously flawed. We argue that the latter is the case, because psychological measures are often based on positivist rather than realist conceptions of science.

Psychological measures that reflect a positivist view emphasize the importance of relying on various abstract principles. The principles include, for example, *Identify the cause of an effect by identifying events that co-vary with it*, and *Treat causation as indeterminate when two (or more) possibly causal events are confounded*. Other principles include, *When testing a theory, rule out or control for alternative hypotheses*, and *Prefer theories that yield good predictions*.

Psychological measures framed in a positivist tradition treat such principles (at least tacitly) as though they were formal in the sense of being content-free. Thus, in terms of how psychologists actually measure scientific reasoning, the principle of identifying causes by relying on co-variation is treated as good scientific reasoning irrespective of which variables are correlated. The result is that, when psychological measures are couched in a positivist tradition, the correlation of hair length and susceptibility to disease cannot be distinguished, in terms of how likely it is to be causal, from the correlation of nutrition and susceptibility to disease. This, despite the fact that our background information suggests strongly that the first correlation is simply an artefact or that hair length is merely an index of some underlying difference that is the actual cause.

By treating abstract principles as though they were content-free, psychological measures systematically ignore a central point made by realist philosophers of science – namely, that collateral information plays a crucial role in scientific inquiry (Putnam, 1962; Miller, 1987; Boyd, 1989). When psychological researchers rely on formal measures, they often (at least tacitly) stipulate that good scientific reasoning actually ignores or overrides collateral information in favour, typically, of co-variation or some other abstract principle. The result is that, when people in such tasks do take account of collateral information, they are deemed by psychological researchers to be poor scientists.

When psychological measures reflect a realist acknowledgement that relying on collateral information is scientifically legitimate, then one can see that people (including children) do have many of the dispositions that make for sound scientific reasoning. In section 3, we illustrate this argument with respect to some of the psychological research on causal reasoning. Much of this research examines causal reasoning by measuring whether people rely on co-variation to identify causal agents. However, as realists note, for co-variation to be used successfully, it must be used inter-dependently with collateral information. We summarize studies documenting that non-scientist adults and adolescents do rely on collateral information in deciding whether co-variation is causal or artefactual. An especially important type of collateral information is information about mechanism, or theory.

However, relying on mechanism or theory is a two-edged sword – useful when the theory is approximately accurate, but potentially misleading when it is either inaccurate or incomplete. Indeed, one of the (at least tacit) motivations that lead positivists and psychologists following in a positivist tradition to want to construct a set of principles that are content-free is that theories can so often be misleading. Therefore, an important aspect of good science also consists of evaluating explanations or theories, to distinguish those that are plausible from those that are dubious. As realist conceptions of science point out, this, too, requires acknowledging that theories are embedded in and evaluated with respect to collateral information. Sections 4, 5 and 6 identify some of the types of collateral information that people take into account when they evaluate explanations, and examine the psychological role that they play in how people reason about and evaluate explanations. These sections also present some data that explores how collateral information affects whether people decide to reject or to modify an explanation or theory.

Furthermore, when people evaluate explanations, they do not restrict themselves to information that they are given; often, they seek out information on their own. If theories are – and if non-scientists treat theories as being – embedded in a broad network of collateral information, then the collateral information that a person initially has access to might be expected to constrain the sorts of additional collateral information the person subsequently seeks. In section 7, we present some evidence that it does.

Finally, acknowledging that theories or explanations are embedded in and evaluated with respect to collateral information has several consequences that psychological measures of scientific reasoning would need to take into account. One is that formal principles from the positivist tradition do not constitute a magic formula or set of algorithms that guarantee good science. Because such principles can be applied successfully only when collateral information is taken into account, good scientific thinking is limited by the way in which age, culture and historical time period restrict the collateral information that is available and therefore the sorts of explanations that will be judged plausible. That is, collateral information is constrained by the social context of science. This has implications for education as well as for scientific practice. These are discussed in section 8.

2 Causal agents and causal mechanisms

Psychological research on causal and scientific reasoning has focused on two questions: how people identify causal agents and how people reason about theories or mechanisms. The distinction between causal agents and causal mechanisms is not a sharp one (Koslowski and Masnick, 2001). Often, identifying a causal agent is the first step in explaining a phenomenon. For example, the first step in explaining recent deaths might be to identify treatment with penicillin as a potential causal agent. Subsequent steps in explaining the death might include an allergic reaction to moulds and the histamines produced by the allergy, the resulting constriction of the lungs and so on, as mechanisms by which the penicillin operated. As this example illustrates, the distinction between agents and mechanisms is also slippery. The histamines that are the mechanism at one level of explanation can be construed as the agents causing lung constriction at the next level.

In addition, in the above example, treatment with penicillin was identified as a potential causal agent because of our background knowledge, or collateral information, that many people are allergic to moulds. However, formal rules can also help discover additional information that is not currently known. For example, although there was no known mechanism to explain the correlation, among pre-school children, of playing on recently cleaned carpets and death, the correlation helped point the way to the discovery that cleaning fluid can activate the viruses that cause Kawasaki's syndrome.

Section 3 summarizes literature demonstrating that, when reasoning about causal agents, people treat formal rules and collateral information as interdependent, relying on collateral information to decide which correlations are likely to be causal and relying on correlations to suggest information that has not yet been discovered. The studies demonstrate that an important type of collateral information consists of information about mechanism, or explanation, or theory. Subsequent sections will examine how mechanisms themselves are evaluated.

3 Psychological studies of reasoning about causal agents

Psychological research in a positivist tradition emphasizes the role of formal or abstract rules in identifying causal agents, the typical formal rules being the Humean indices of which co-variation or correlation is the most prominent. Thus, X causes Y if X is present when Y is present and X is absent when Y is absent. However, as realist philosophers have noted, because the world is rife with correlations, we must rely on collateral information to decide which correlations are likely to be causal and which are likely to be artefacts. For example, given currently available collateral information, we treat the correlation of level of ice cream consumption and incidence of violent crime as artefactual. Were we to learn that consumption of fats increases testosterone production, we might give the correlation a second look as potentially causal. As this example illustrates, a type of collateral information that is often relied on in distinguishing causal from artefactual correlations is information about mechanism. In both the philosophical and the psychological literature, 'mechanism', 'theory' and 'explanation' are often used interchangeably.

First, then, consider evidence that when non-scientists, including adolescents, employ formal rules to identify causal agents, their decisions treat formal rules and collateral information as inter-dependent. The formal rules in question have all been operationalized by psychological researchers as versions of the co-variation rule.

3.1 Co-variation and collateral information are treated as inter-dependent

In the research on causal reasoning, one of the most basic formal rules in the positivist tradition argues that, to identify the causes of a phenomenon, one needs to identify the variables that co-vary with it. However, non-scientist adults and adolescents take a realist tack, treating the co-variation rule and collateral information as inter-dependent.

Sixth-graders (roughly eleven-year-olds) as well as college students are more likely to treat a co-varying factor as causal rather than coincidental when there is a possible mechanism that can explain the process by which the factor might have brought about the effect. This is especially salient when the factor is an implausible cause to begin with. For example, an implausible cause of better gas mileage (such as colour of car) is more likely to be treated as causal when there is a possible mechanism (such as the boost that red gives to attentiveness) that could explain how it might operate. Conversely, sixth- and ninth-graders, as well as college students, are less likely to treat the co-variation as indicating causation when given collateral information that plausible explanatory mechanisms have been checked and ruled out (Koslowski, 1996). That is, the strategy of relying on co-variation to identify causal factors is not applied mechanically; it is applied judiciously, in a way that takes account of collateral information.

Furthermore, co-variation information is often relied on as a way of discovering new information. For example, if a factor and an effect co-vary a sufficiently large number of times, then even if no mechanism is currently known by which the factor might cause the effect, people as young as eleven years of age treat the factor as warranting a second look (Koslowski, 1996). That is, like the Center for Disease Control, they consider that the co-variation might indicate a causal relation that has not yet been discovered. Note that the belief in the possibility of undiscovered mechanisms is continuous with younger children's ability to recognize the existence of invisible causal processes (Siegal, chapter 16 in this volume).

In addition, when evaluating possible causal agents, not only do people take into account alternatives that they are presented with, but they also spontaneously generate possible alternatives that need to be considered. For example, when asked how they would find out if having parents stay overnight with their hospitalized children would increase recovery rates, virtually all of a sample of college students and college-bound sixth- and ninth-graders noted that one would need to consider the general alternative hypothesis that maybe the recovery rate would have been the same if the parents had not stayed. All but one of 48 participants noted that, for example, 'You'd need to have some kids stay with their parents and some kids not.' In addition, a third of the sixth-graders and almost all of the college students also spontaneously proposed a specific alternative to be considered, for example, 'You'd need to make sure the kids with their parents weren't just less sick, or didn't have better doctors' (Koslowski, 1996).

(Sometimes, it appears that one would *not* need to take either mechanisms or alternatives into account to identify causal agents. For example, one might be concerned with whether vitamin C prevents colds without being concerned with the process by which it does so. To find out, one could simply randomly assign half of the participants to the vitamin C group and half to a placebo group and measure the resulting incidence of colds in the two groups. However, the reason random assignment is so useful is precisely because of the assumption that, with a large enough sample, random assignment will control for alternative hypotheses and enable them to be ruled out. However, even random assignment does not deal with the thorny problem of which population is the representative one from which the sample should be drawn. Whether a sample is representative is a judgement that depends on collateral information about the phenomenon under study. And, as already noted, the procedure described above would tell us little about the mechanism by which the vitamin C works.)

3.2 Confounding yields indeterminacy depending on whether the confounded factors are both plausible causes

When it is operationalized in terms of co-variation, the indeterminacy principle is that, when two possibly-causal factors are confounded, and both co-vary with an effect, it is not possible to determine which, if either, is the actual cause. This principle, too, is not applied mechanically. Both college students and adolescents take content or collateral information into account in deciding whether to conclude indeterminacy.

Specifically, when adolescents and college students are presented with two confounded variables (such as colour and texture of sports balls) and one of the variables is a very implausible cause of the effect (bounciness of the balls), then the confounding is ignored, and causation is ascribed to the plausible variable (Kuhn, Amsel and O'Loughlin, 1988). In contrast, when the confounded variables are *both* plausible causes (such as size and model of car as possible causes of reduced gas mileage), then both adolescents and college students *do* note that causation is indeterminate (Koslowski, 1996).

A devil's advocate might argue that people are reasoning incorrectly when they take account of confounding (or alternative causes) in some situations but ignore it in others, because doing so runs the risk of overlooking a variable that is, in fact, causal. However, recall that, as mentioned earlier, people do not *always* behave as though implausible co-variations are unlikely to be causal. When plausible causes have been ruled out, both adolescents and college students are increasingly likely to treat the implausible covariate as causal if a possible mechanism exists that could explain the process by which the factor might have produced the effect. More striking, and as already noted, when alternative causes have been ruled out, adolescents and college students are increasingly likely to treat an implausible covariate as causal if the co-variation occurs a sufficiently large number of times (Koslowski, 1996). These tendencies provide a safeguard against overlooking a possibly causal factor that appears, at first glance, to be non-causal.

3.3 Information about causal mechanism affects whether a hypothesis will be modified or rejected in light of anomalous information

Another principle of scientific inquiry is that anomalies to an explanation reduce its credibility by calling it into question. When researchers have operationalized anomalies within a co-variation framework, the participant is given some initial data that factor X co-varies with and thus causes effect Y. After this hypothesis has been established, the participant is then presented with some observations, in which X and Y do *not* co-vary, that constitute anomalies. For some psychologists, tasks have been defined so that hypotheses ought always to be rejected in the face of anomalous data (Mynatt, Doherty and Tweney, 1977). However, in scientific inquiry, not all anomalous data is treated as requiring hypothesis rejection; sometimes it is seen as warranting a modification to take account of anomalies to what is essentially a working hypothesis. For example, in light of the information that penicillin had no effect on viral infections, it was appropriate to modify rather than reject the initial hypothesis that penicillin kills germs. The resulting modification (that penicillin kills only bacterial germs) was an appropriate way of taking account of the anomalous information. The problem is that, in principle, a hypothesis can always be modified to take account of anomalies. However, in practice, only some modifications are seen as scientifically legitimate, because they are theoretically motivated, while others are not. This point is marked by the distinction between theory modification, which is seen as appropriate, and *ad hoc* modifications, which are seen as inappropriate patch-work.

For college students as well as college-bound ninth-graders, the decision about whether to reject rather than modify an explanation in light of anomalous data is also made judiciously rather than mechanically, because it depends on content. Specifically, modification rather than rejection is more likely to occur if there is a mechanism that might account for why the causal factor did not operate in the anomalous situation (Koslowski, 1996).

For example, imagine that, in several samples, treatment with drug X co-varies with a cure of illness Y, leading to the working hypothesis that X cures Y. Anomalies to the hypothesis would consist of learning that drug X failed to cure Y in two populations. Now imagine learning that one population was anaemic (and that too little iron prevents absorption of the drug) and that the other population was suffering from high blood pressure (and that the negative effects of high blood pressure can outweigh the beneficial effects of the drug). The presence of possible mechanisms that could account for why the drug failed to operate in the two populations would make us more likely to modify rather than reject the working hypothesis. By adding to our understanding of when the drug fails to operate, the mechanism information would also add to our understanding of when and why the drug does operate.

Analogously, imagine that, after generating the working hypothesis that X cures Y, we subsequently learn that X was not effective in two populations: a population of the elderly living alone, and a population of people living in poverty. Imagine further that we learn that the populations had something in common, namely, poor nutrition. Even if we had no understanding of the mechanism by which poor nutrition might compromise the effectiveness of the drug, we might nevertheless be inclined to modify rather than reject the initial working hypothesis. The pattern in the anomalies would suggest a possible mechanism, as yet undiscovered, by which poor nutrition compromises the drug's effectiveness.

College students and college-bound ninth-graders make comparable judgements when asked whether to reject or modify working hypotheses in the face of anomalous evidence. They do not follow a blanket, abstract rule according to which anomalies call for rejection of a hypothesis. Instead, they decide whether to reject or to modify by considering collateral information.

3.4 Section summary

When causal reasoning has been couched in a positivist framework, researchers have typically examined how people identify causal agents by relying on principles that are fairly straightforward variations on the basic co-variation principle. Although such principles seem to be content-free, they are actually more or less likely to be applied depending on collateral information, the most salient type of which is information about mechanism or explanation. People rely on collateral information to decide which co-variations to treat as causal, whether confounding yields causal indeterminacy, and whether explanations should be revised or rejected in light of anomalous information. Thus, if one of the cognitive bases of science includes relying on collateral information, as realists would argue, then people are able to do this by at least the sixth grade.

4 Theories and collateral information

4.1 Theories are a two-edged sword

As the research summarized above demonstrates, a type of collateral information that plays a prominent role in identifying causal agents is information about theory or mechanism. In the past few years, the importance of theory (or mechanism, or explanation) has received much attention in the psychological literature (for example, Gopnik and Glymour, chapter 6 in this volume). Researchers have argued that theories provide the underpinning for categorization (Murphy and Medin, 1985), affect how we understand words (Carey, 1985), appropriately influence our causal reasoning (Koslowski, 1996) and affect the way in which we make causal attributions (Ahn *et al.*, 1995). Theories also drive the sorts of information that we search for as we seek to evaluate hypotheses (Kuhn, Amsel and O'Loughlin, 1988; Schauble, 1990). A recent textbook (Goswami, 1998) is structured around the general role that children's theories play in cognitive development. Clearly, theories are crucial in reasoning in general and in scientific reasoning in particular. As Evans (chapter 10 in this volume) notes, prior beliefs (including prior theories) influence our reasoning, and it is not clear that this is always a bad thing. The importance of considerations of theory, explanation, or mechanism is not at issue.

However, the emphasis on theories has a potential downside, because relying on theories can be a two-edged sword. Useful though many theories are, some theories are simply wrong and ought to be rejected. Theories that are wrong can, at best, lead science down fruitless paths and, at worst, can have quite pernicious consequences (as theories about the genetic inferiority of various ethnic groups have demonstrated). Other theories are only approximately correct; they are incomplete and need to be refined or augmented. Therefore, for theories to be maximally useful, they need not only to be evaluated but also to be, in many cases, refined and elaborated with additional information.

4.2 Formal rules must be used in conjunction with collateral information to distinguish plausible from dubious theories

A predisposition that permits good scientific reasoning and indeed good thinking in general is the tendency to embed theories in, use them with and evaluate them with respect to networks of collateral information. Thus, just as collateral or background information is relied on to identify causal agents, it is also relied on to evaluate explanations or mechanisms.

In both the psychological (Murphy and Medin, 1985) and the philosophical literature, the point has been made that we can evaluate theories by assessing the extent to which they are congruent with collateral information, or what else we know about the world. For example, in trying to explain speciation, we find it more plausible to invoke evolution than to take seriously the possibility that visiting Martians created different species to perplex us. We make this judgement because an explanation based on evolution is congruent with collateral information about population genetics, plate tectonics, etc.

One way to see the realist point about the importance of collateral information in evaluating theories is to consider whether theories could be evaluated by making do with only formal, positivist strategies such as relying on Humean co-variation. For example, consider a possible theory or mechanism by which drug X cures illness Y, namely that it dissolves the cell walls of the bacteria causing the illness. A positivist Humean might argue that one could evaluate this mechanism by simply reducing it to a series of co-variations. Thus, one might 're-write' the mechanism by noting that drug X co-varies with the production of a certain set of toxins, which in turn co-varies with the dissolution of the cell walls, which in turn co-varies with the demise of the bacteria, etc.

However, one would still need to rely on collateral information to decide which of the many possible mechanisms to reduce to co-variations in the first place. For example, it is *possible* that drug X cures illness Y, not because it dissolves bacterial cell walls, but because it increases production of white blood cells to such an extent that they crush the harmful bacteria. The principle of co-variation could not, by itself, account for why we do not treat this alternative as worth pursuing. After the fact, one might be able to reduce virtually any mechanism to a series of co-variations. The question is, why is it that, *before* the fact, one possible mechanism rather than another would loom as worthy of reduction? We argue it is because some possible mechanisms accord with collateral information about the phenomenon being explained.

Another formal strategy would be to evaluate an explanation by generating predictions that follow from the mechanism and seeing whether the predictions obtain. However, collateral information would nevertheless be required to distinguish informative predictions from those that are uninformative. For example, if the possible mechanism is the dissolution of cell walls, then one prediction is that the drug will not be effective against germs that lack cell walls. However, another prediction is that, for germs that *do* have cell walls, the drug should be effective at 2.00 p.m. as well as at 2.15 p.m. Although both predictions are compatible with the mechanism in question, we rely on collateral information to treat the first as more informative than the second.

In the same vein, it would also not be possible to evaluate an explanation simply by attending to the amount of evidence for it. One would also need to assess whether the evidence was more or less informative. This, too, would require collateral information.

Finally, to evaluate a possible causal mechanism one would also need to rule out or control for competing, alternative mechanisms. Again, however, one would need collateral information to decide which particular alternative mechanisms would be worth taking seriously. In terms of the above example, it would not be reasonable to give patients a tranquillizer to rule out the possibility that drug X works, not by dissolving the cell walls, but by scaring the germs to death. Such an alternative would be a straw person, because of our collateral information about the emotional life (or lack thereof) of germs.

This is not to argue that formal rules are useless. Relying on co-variation information is a fine strategy to use when theoretically driven explanations fail. For example, when faced with a set of symptoms that cannot be explained by relying on known causes and known mechanisms, the Center for Disease Control looks for variables that co-vary with the symptoms, even if mechanisms are not now known that could explain how the variables function. In the case of Kawasaki's syndrome, after ruling out known possible causes, the only variable that emerged as distinguishing victims from non-victims was that the victims' families had had their carpets cleaned in the previous three months. Despite the counter-intuitive nature of this variable as a possible cause (counter-intuitive, because cleanliness is next to godliness), it was pursued, because other more standard possible causes had been ruled out, and because the correlation was noted in several samples. Once pursued, the correlation led scientists to discover the mechanism (a virus activated by cleaning fluid) by which the co-variation obtained.

As this example illustrates, pursuing unlikely co-variations can lead to discovery of new content. But, again, pursuing every unlikely co-variation would be a pragmatic disaster. Unlikely co-variations tend to be most worth pursuing when theory-driven approaches have failed, and when the co-variations are reliable occurrences.

4.3 Section summary

Good thinking in general and good scientific thinking in particular requires that we distinguish dubious from plausible mechanisms. To do this, we cannot rely exclusively on formal strategies framed within a positivist tradition. Rather, as realist philosophers argue, we need to treat collateral information and formal rules as inter-dependent.

5 Psychological studies of reasoning about mechanisms, explanations, or theories

The research summarized below suggests that non-scientists also rely on collateral information when reasoning about mechanisms or explanations. As already noted, the notion of a mechanism is ill defined. The literature covered in the present section deals with mechanism in the sense that the tasks presented to participants require going beyond merely identifying a causal agent. In general, the tasks in these studies were not necessarily aimed at demonstrating people's reliance on collateral information, but the results can be re-cast to do so.

5.1 Children rely on collateral information to decide which prediction is useful

Sodian, Zaitchik and Carey (1991) asked children to determine whether an unseen mouse was large or small. To do this, the children had to decide whether to leave mouse-food in a box with a large opening (which a mouse of either size could pass through) or a box with a small opening (which could accommodate only a small mouse). They correctly chose the latter.

Note that, to make the correct choice, the children needed to rely on the background assumptions that the mice would not change their girth to become narrower and that the small opening would not stretch to become larger. One way of making the role of collateral information salient is to consider whether the same predictions would have been informative with different collateral information. For example, if the children had been told that the walls of the box were made of rubber sheeting or that mice can reduce their girth (as some snakes can change their girth by disconnecting their joints), then the same predictions would have been quite uninformative.

5.2 Collateral information helps children recognize anomalies

Studies of the spontaneous questions of children suggest that, when they identify anomalous information, they do so by relying on collateral information. For example, we asked the parents of 30 preschool children to keep a two-week diary of the sorts of explanations and requests for explanations that the children provided (Koslowski and Winsor, 1981). Approximately 40% of the children identified anomalous data and did so by relying on collateral information. They provided statements such as the following: (while watching a movie set during the Second World War) 'If an army is running to battle and someone gets sick, what happens? Do they leave him behind or do they carry him or what?' This child was noting the tension between not being able to run when sick, and needing to run when in battle. Consider the statement provided by another child, 'Johnny says when his cat died, his cat went *up* to heaven, but it couldn't go *up* because cats have bones and bones are heavy and clouds couldn't hold up heavy bones, so they'd fall right through the clouds.'

Some of the questions compiled by Harris (2000a) demonstrate the same point. For example, consider the six-year-old reported by Isaacs (cited in Harris, 2000a) who asked, 'Why do angels never fall down to earth when there is no floor to heaven?'

5.3 People rely on collateral information when responding to anomalies

As noted in section 3, when people aim to identify causal agents and rely on co-variation to do so, an anomaly consists of an expected co-variation that does not obtain. As already noted, anomalies are treated as more or less problematic for an explanation depending on collateral information. Similarly, when people reason about mechanisms rather than co-variation, collateral information also affects how anomalies are judged.

The aim of Chinn and Brewer's (1993, 1998) work has been to document the strategies that people use when responding to anomalies. However, although not the focus of their research, their studies also illustrate that theories, anomalies to the theories and the strategies used for dealing with anomalies, all involve collateral information.

Chinn and Brewer provided their participants with evidence for various theories about dinosaurs (for example, that the dinosaurs were cold-blooded). In this example, the evidence included such things as the anatomical similarity between dinosaurs and extant, cold-blooded reptiles and the small brain size, also typical of extant, cold-blooded animals.

Anomalous data were aspects of the world that would not have been expected if the theory were true. In this example, the anomaly was that, for the museum specimens studied, the dinosaurs had the pattern of bone density produced by the fast growth that characterizes extant animals that are warm-blooded. For this example, one way of resolving the anomaly was to note that the process of fossilization might have affected patterns of bone density in the specimens, or that bone density while the animals were alive might also have been affected by basic nutrients, such as calcium, instead of or in addition to warm-bloodedness.

In terms of the present chapter, we note that in Chinn and Brewer's work, the evidence for the theories, the anomalies themselves and possible ways of resolving the anomalies are all embedded in collateral information about, for example, extant animals, the effects of fossilization and the role of nutrients in bone development. That nutrition, for example, rather than warm-bloodedness might have affected bone density is plausible, because we know that nutrition affects bone growth in other ways as well. (To be sure, one could, after the fact, re-write this information as a series of formal correlations, for example, between calcium consumption and various features of bone development. However, as already noted, if one were to do that, it would be collateral information that would suggest which correlations to look for and to take seriously in the first place, as realist philosophers note.)

5.4 *To call a belief into question, one must disconfirm the collateral information in which it is embedded as well as the belief itself*

In the cognitive as well as the social psychological literature on disconfirmation, the modal conclusion is that people neither seek nor take account of disconfirming evidence (for example, Mynatt, Doherty and Tweney, 1977; Lord, Ross and Lepper, 1979). Some researchers (Koslowski, 1996; Kuhn and Lao, 1996) have suggested several reasons why the modal conclusion might not be warranted. The present argument about the importance of collateral information suggests an additional reason.

If explanations are evaluated with respect to collateral information, then explanations are not circumscribed. Rather, they are embedded in broader contexts of information. Furthermore, if this is true, then attempts to disconfirm an explanation should not be expected to be successful unless the broader context of information, in which the explanation is embedded, is also called into question.

In many cases, the belief in question includes two components. One is the belief that a factor causes an effect because the factor co-varies with the effect. The network of collateral information in which the belief is embedded often includes some notion (however non-specific) of the mechanism by which the causal factor operates. This is the second component. Therefore, to disconfirm a belief that is based on co-variation, it is not sufficient simply to note that the co-variation was a fluke and, in fact, does *not* obtain; one ought also to call into question the mechanism component of the belief.

A very preliminary study (Koslowski, 1996) examined the causal belief that type of candy (high-sugar vs. low-sugar) makes it difficult for children to sleep and the non-causal belief that type of milk (high-fat vs. low-fat) has no effect. When participants held the belief that type of candy did co-vary with sleeplessness, their co-variation belief was integrated with a belief about the mechanism that rendered the co-variation causal, namely, that eating sugar 'revs up' children. In one condition, the researchers disconfirmed only the co-variation component of the belief (by noting that type of candy did *not* co-vary with sleeplessness, and that type of milk did). In another condition, in addition to disconfirming the co-variation component, the researchers also disconfirmed the mechanism component by offering an alternative mechanism to explain the co-variation or non-co-variation that actually obtains. Participants who believed that sugar causes sleeplessness were told that most sugar is ingested as chocolate candy and that it is caffeine in chocolate, not the sugar, that 'revs up' children. Participants who believed that fat content was causally irrelevant to sleepiness were told that high-fat milk causes sleepiness by increasing 'serotonin, a chemical in your brain'.

The results were suggestive. For adolescents as well as college students, causal (as opposed to non-causal) beliefs were *more* likely to change when the disconfirming evidence undermined the mechanism as well as the co-variation components of the belief (instead of undermining the co-variation component alone). The results suggest that one reason it has been difficult to disconfirm (causal) beliefs may be that the disconfirming evidence has called into question only the co-variation component of the belief and has left unchallenged the collateral information with which the belief was integrated – in this case, information about mechanism.

In a much more sophisticated study, Swiderek (1999) examined beliefs about capital punishment – beliefs embedded in and integrated with a broad context of collateral information. Swiderek identified six major components of participants' belief networks, such as whether capital punishment was or was not a deterrent, applied in a non-biased way, cost-effective, etc. She asked participants to rate not only their general agreement or disagreement with capital punishment in general, but also their agreement or disagreement with each of the six major belief components. Irrespective of whether the participants agreed or disagreed with a particular belief (such as whether capital punishment is a deterrent), Swiderek then challenged each of the six component beliefs. Afterwards, she asked participants to rate again not only their beliefs about the six components but also their general belief about capital punishment.

For none of the six component beliefs was it possible to use change on that particular belief to predict change on the general belief. However, more interestingly in terms of the present argument, it *was* possible to predict change on the general belief by relying on the *sum* of the changes on the individual components. That is, it was the changes on the individual component beliefs *taken together* that predicted change on the general belief. This is exactly what one would expect if the model of causal reasoning suggested by the present studies is approximately correct. If a particular belief is integrated with what else is known (or at least believed to be true) about the world, then disconfirming the belief can only be done if the supporting context of beliefs is also called into question.

5.5 Section summary

Just as collateral information plays a role in identifying causal factors, it also plays a role in how people reason about explanations or mechanisms. Specifically, collateral information plays a role in how people detect and respond to data that are anomalous to an explanation. Furthermore, since an explanation is often embedded in and evaluated with respect to collateral information, disconfirming an explanation often requires also undermining the collateral information that is evidentially relevant to it and in which it is embedded.

6 The network of evidentially relevant collateral information includes several types

The research summarized above demonstrates the general point that collateral information is important to thinking in general, and to scientific explanation in particular. The point would be more useful if one could specify some of the types of collateral information that are treated as evidentially relevant. One way of identifying these types of information is to use scientific inquiry as a model. However, the fact that information functions a certain way in scientific inquiry does not guarantee that it will serve the same function in the thinking of individuals, especially individuals who are non-scientists. Thus, identifying possible types is only a first step; assessing their role in individual psychology is also necessary. We draw attention to five types of collateral information that we have identified and tested with college students (Masnick *et al.*, 1998).

One could view these types as reflecting different aspects of information about the mechanism or explanation. We have found that, at least for college students, the various types are treated as evidentially relevant, though not necessarily across the board. That is, in at least some situations, they affect whether people judge an explanatory mechanism to be plausible. The types include the following:

- *Details of the mechanism.* A possible explanatory mechanism becomes increasingly plausible when information becomes available that can flesh out the details of how the mechanism operates. For example, evolutionary theory became increasingly plausible as information from genetics spelled out the details of how the mechanism of natural selection might operate on genetic variability to produce change in a population. Genetics provided a deeper level of description of the process by which the mechanism of natural selection works.
- *Information that yields anomalous predictions.* Anomalies are events that we infer should *not* obtain if the explanatory mechanism is true, and given what else we know about the world. For example, if the mechanism of evolution is right, then animals with common evolutionary histories ought to share certain traits. Thus, an anomaly to evolutionary theory is that some Australian mammals lay eggs. By itself, this anomaly creates a problem for evolutionary theory.
- *Information that resolves anomalies.* However, the anomaly is resolved in a theoretically motivated way with the collateral information that Australia was

the earliest land mass to separate from Pangaea – suggesting a fairly isolated niche in which the evolution of some Australian mammals diverged from the path followed by other mammals. This resolution is theoretically motivated in that it fine-tunes our understanding of how evolution operates. Thus, the resolution makes the anomaly less problematic for the explanation.

- *Information that yields congruent predictions.* Such predictions might be considered to be the flip side of anomalies. They are events that ought to obtain if the explanatory mechanism is true, and given what else we know about the world. Consider another example from genetics. Mountain dwellers are often shorter, on average, than people who live on the plains. One possible mechanism is that the two populations come from different gene pools. Another is that reduced oxygen supply at high elevations stunts growth. If the oxygen explanation is true, then a corollary effect would be that, when people leave the mountains and move to sea level, their children are taller than would otherwise be expected.
- *Alternative accounts.* In our study, the presence of a competing, alternative mechanism did not directly affect the credibility of the target mechanism. However, the presence of an alternative mechanism did make it more likely that other sorts of collateral information (such as anomalies) would be treated as affecting the credibility of the target.

That competing alternative accounts did not directly affect the credibility of the target mechanism was surprising, especially in light of previous research (Koslowski, 1996) in which competing alternatives did reduce the credibility of a target. We note that the previous research asked subjects to reason about situations that were less esoteric than were the situations just described. Given the participants' justifications, we suggest that, when reasoning about situations that are commonplace, people are able to assess the likelihood of competing accounts, so that having a viable alternative present reduces the credibility of the target. In contrast, when reasoning about situations that are less familiar, people are unsure of how to evaluate an alternative and are thus unsure of whether it in fact renders the target less questionable. However, even when reasoning about unfamiliar situations, learning that there is some alternative does make people think more critically about the evidence for the target.

One way of thinking about the above types of information is to note that they permit various empirical inferences to be drawn, inferences that can then be tested as one way of assessing the explanation being evaluated. For example, if one is assessing oxygen deprivation as the explanation for shorter stature in the mountains, relevant collateral information is that the effects of limited oxygen are not passed on from parent to child. This, in turn, yields the inference or prediction that, when mountain dwellers move to sea level, their subsequently born children ought to be taller than would otherwise be expected.

The importance of evaluating a theory by examining inferences that follow from it has long been recognized as important by philosophers working in a formal, positivist tradition. What the above examples illustrate is the more recent, realist, philosophical point that inferences that are *informative* require considering a theory *in conjunction with* collateral information. The inferences are empirical, rather than formal, precisely because they are not content-free; they depend on taking account of collateral information.

7 Differential access to the network of collateral information structures the search for other facets of the network

The research summarized in the preceding sections examined the ways in which people take account of collateral information that they are either given or already in possession of. However, when people reason about explanations, they often seek out information on their own. Furthermore, different people often have initial access to different facets of the network. Therefore, if explanations are embedded in a network of collateral information, one might expect that different people would search for different information depending on the sorts of information to which they initially have access.

We have been finding exactly this with college students. We presented college students with various events to be explained, such as why it is that people who live in the mountains of Mexico have a shorter stature than do people who live at sea level. Depending on the condition, participants were told only about a possible Target explanation (the two populations come from different gene pools); or about a possible Target and a possible Alternative (lower levels of oxygen lead to diminished growth); or about a possible Target along with information anomalous to the Target (the two populations frequently inter-marry, which is problematic for the different gene pools hypothesis); or about a possible Target, information anomalous to the Target and a possible Alternative. About each story problem, we asked participants: 'If you were trying to explain this phenomenon, what additional information would you like to have in order to help you explain it?'

The kind of information people were initially presented with certainly did have an effect on the sorts of information they subsequently asked to search for. Worth noting at the outset is that the different conditions did not affect the total number of pieces of information that participants requested. Rather, what differed across conditions was the distribution of types of information that were requested.

Consider three types of information that participants could request. One type asked about possible alternative accounts. (These alternatives were in addition to the Alternative provided by the experimenter in some of the conditions.) Another type asked about mechanisms that were either provided by the experimenter or proposed by the participant. A third type were requests for general information, often about the parameters of the event (for example, when the event was noticed). Thus, the first two types of information were fairly specific and focused, while the third was quite general.

The information requested by the participants depended on the information the experimenter had provided them with. A striking finding involved the effect of Alternative explanations. The distribution of requests when an Alternative was present, and the distribution when an Alternative was absent were rough mirror images of each other, with the presence of an Alternative making participants' requests more general and less focused. Specifically, when told about both a Target and an Alternative account, roughly a third of participants' requests asked for general information, while roughly only a fifth were split evenly between asking about other possible alternative accounts or about possible mechanisms. In contrast, when information about a possible Alternative was absent, and participants were told only about a possible Target explanation, then participants requested quite specific information. Over two-thirds of their requests were split evenly between asking about possible alternative explanations and possible mechanisms. Only about one-seventh of the requests were for general information.

One possibility is that presenting a Target explanation along with an Alternative made participants' requests less focused than presenting them with a Target alone, simply because the Alternative resulted in information overload that interfered with a focused approach. This is unlikely because, in a third condition, we gave participants information about the Target and also about an Anomaly to the Target. Thus, this condition also provided participants with an additional piece of information. However, the distribution of requests in this condition was roughly comparable to the distribution when the Target explanation was presented alone. That is, when the additional information was an anomaly, rather than an alternative, participants' requests remained fairly focused.

Recall also that, when the participants' requests had to do with alternative possible causes, the alternatives were different from the Alternative provided by the experimenter. Thus, it is not likely that, for each situation, there is a particular alternative which, if the experimenter does not mention it, is hypothesized by the participant.

What does this tell us about how people reason about explanations? As noted in section 3, at least by early adolescence people spontaneously generate alternative causes and possible mechanisms when trying to evaluate a possible causal factor. We suggest that the combination of target explanation, alternative account and possible mechanisms constitutes a kind of explanatory unit or capsule. When presented with only a Target explanation, people try to complete the capsule by searching for possible alternatives and for mechanism information. In contrast, when the experimenter provides a possible Alternative in addition to the Target, the search for other alternatives and for mechanisms to flesh them out is short-circuited. Similarly, when the Target explanation is accompanied by an Anomaly to the Target, participants are once again left with an incomplete explanatory capsule and thus search for other possible explanations and possible mechanisms to flesh them out.

In earlier parts of this chapter, we presented data suggesting that people rely on collateral information in deciding whether to take possible alternative hypotheses seriously. The results just described suggest that collateral information also affects the likelihood that people will search for alternative hypotheses on their own. That is, the scientific principle, *Consider alternative hypotheses*, is applied in a way that is content-bound rather than formal.

These results complement Hilton's point (chapter 11 in this volume) about the search for evidence. Hilton notes that sometimes group membership truncates the search for evidence. The results just described suggest the evidence that is initially available can also truncate the search for additional evidence. Furthermore, only a small leap is required to note that group membership can also constrain the evidence that is initially available. That is, there is a social dimension to science.

7.1 Section summary

Collateral information is not always simply presented; sometimes, people search for it on their own. However, the collateral information that is initially available to someone structures or constrains the additional collateral information that is sought. We suggest that a fairly strong tendency to consider alternative accounts is interrupted when people are presented with both a Target and an Alternative possible explanation. The result is a search for general information. The search becomes increasingly specific when only a Target is presented or when the Target is accompanied by an Anomaly to it. In these conditions, participants try to complete the explanatory capsule by searching for other explanations and for mechanisms to flesh them out.

8 General summary and conclusions

On a formal, positivist view of science, which places emphasis on the importance of content-free rules such as relying on co-variation, people in general and children in particular are not very adept at scientific reasoning. Instead of treating various rules as content-free, people rely on collateral information both when identifying causal agents and when reasoning about causal mechanisms. However, as realist philosophers have argued, relying on collateral information is not only scientifically legitimate, it is also crucial; it is not possible to do science (or causal reasoning) without taking collateral information into account.

Nevertheless, as realist philosophers also acknowledge, not all mechanisms have already been discovered, and some possible mechanisms are dubious. Therefore, on a realist view of science, formal rules and collateral information ought to be treated as inter-dependent, with collateral information constraining when formal rules ought to be taken seriously, and formal rules pointing the way to mechanisms that are as yet undiscovered or that are dubious. People, including children, do treat formal rules and collateral information as inter-dependent. That is, on a realist view of science, people are not as ill equipped to do science as a positivist approach would suggest.

The importance of collateral information to both identifying causal agents and to reasoning about causal explanations or mechanisms has several general consequences. One is that there is no magic formula that guarantees good science. As we have tried to argue, strict application of formal rules does not guarantee good science. At best, formal rules are heuristics rather than algorithms. Formal rules can be successfully deployed only if collateral information is also taken into account.

The problem is that the entire network of information that is actually relevant to a question is not always available. That is why people continue to do research. Thus, it is one thing to note that successful implementation of formal principles requires relying on collateral information. It is quite another to have available all the collateral information that is relevant. And this point speaks to the social dimension of science.

The collateral information that is actually available can be limited by individual differences (in age and knowledge). Not every child could be expected to know that Johnny's dead cat could not have gone to heaven because its bones would have fallen through the clouds. As Harris (chapter 17 in this volume) notes, children often learn from the testimony of adults. Not every child has access to adults who would (or could) provide her with information about the relative densities of bones and clouds. This is one aspect of the social dimension of science.

Another aspect is that the available collateral information can also be limited by culture, whether it is the culture of a particular time or a particular group (Hilton, chapter 11 in this volume). For example, during one of the plague periods of the Middle Ages, some scientists considered and ruled out the alternative hypothesis that plague was caused by poverty by noting that the wealthy and the poor died in comparable numbers. Lacking the relevant collateral information about germs and how germs are transmitted, they did not consider the alternative causal factor of proximity to rats and fleas. More recently, it is only in the last few years that the Tuskegee Airmen came to be known in the USA as the only group of fighter pilots who never lost a bomber during the Second World War – and that all of the Tuskegee Airmen were Black. Restricted access to this sort of information makes it less likely that institutionalized racism will be offered as an alternative to the hypothesis of genetic inferiority when trying to explain, for example, racial differences in IQ scores. That is, the social dimension of science includes culturally restricted access to evidence.

In addition, the collateral information to which one initially has access can constrain the subsequent information that is sought. That is, the initial effects of having restricted access to information can be magnified by influencing the sorts of information that are subsequently sought.

One of the reasons it is useful that science is a social enterprise is that it increases the amount of collateral information that is available. Even when the motivation for mustering additional information is to shoot down a rival's theory, the effect can nevertheless be salutary (Evans, chapter 10, Hilton, chapter 11 and Kitcher, chapter 14 in this volume).

Finally, acknowledging the importance of collateral information also has consequences for future research. It shifts the emphasis from studying how people acquire formal rules to a focus on the way a network of collateral information is acquired and structured and the way it affects the deployment of the formal rules.

10 The influence of prior belief on scientific thinking

Jonathan St B. T. Evans

I should start by defining the kind of scientific thinking that I will be talking about in this chapter. It is not, for example, the thinking that scientists mostly do. Most scientific activity is mundane and repetitious, so most thinking is probably the day-dreaming that goes on while waiting for instruments to produce data or computer programs to run their analyses. It is not, in general, the creative process which leads to sudden breakthroughs, although I shall speculate a little on this matter. What I will really focus on is the processes of hypothetical and deductive reasoning that underlie the testing of hypotheses and the interpretation of evidence. This kind of reasoning has been extensively studied by psychologists, particularly over the past thirty years or so. (For recent reviews and discussions see Evans, Newstead and Byrne, 1993; Evans and Over, 1996; Manktelow, 1999; Stanovich, 1999; Schaeken *et al.*, 2000.)

Experimental research on human reasoning generally consists of giving people tasks in the laboratory for which a correct answer can be computed according to a standard normative system such as propositional logic. Typically such experiments may manipulate both the logical structure of problems and a variety of non-logical factors such as problem content and context, instructional set, and so on. This kind of research has led broadly to three kinds of finding. First, intelligent adults (university students) untrained in logic make many errors on these tasks. Second, the errors people make are not random, but reflect a whole host of systematic biases that have been identified in the literature. Third, people's reasoning is strongly affected by the content and context used to convey logically equivalent problems.

The discovery of systematic errors and biases has led to a major debate about the implications of this research for human rationality, in which philosophers and psychologists have both participated enthusiastically. In recent years, a distinction has been drawn between what we might term *normative* and *personal* rationality (see Evans and Over, 1996; Stanovich, 1999). Experiments certainly show a high degree of violation of normative rationality, which is failure to comply with the normative system applied to the structure of the problems set. What is much more contentious is whether they also provide evidence for violation of personal rationality, which is the ability of individuals to achieve their ordinary goals. For example, the experimenter may have applied the wrong normative system, or the norms might not apply well to the real world in which the individual normally operates. Biases might reflect the operation of heuristics carried over from everyday reasoning, where they are effective, to a contrived laboratory task, where they fail – and so on. The scope for interpretation and debate is immense. With regard to expert thinking, including scientific thinking, we can add the extra qualification that where formal thinking is required such as in the control of confounding variables in experimental design or the interpretation of statistical evidence, then extensive training will have been given to the individuals concerned.

Of more direct concern to the present chapter is the finding that content and context profoundly affects reasoning even when logical structure is held constant. So pervasive are the effects of prior knowledge and belief on reasoning, that Stanovich (1999) has recently described the tendency to contextualize all information given as the *fundamental computational bias* in human cognition. While I agree with Stanovich about the importance and pervasiveness of the tendency, I am a little uncomfortable with terming this a *bias* as this presupposes the viewpoint of normative rather than personal rationality. It looks like a bias in psychological experiments where the logical problems are framed in contexts of no formal relevance to the logical task in hand. In real life, however, it is not only rational to introduce all relevant belief into our reasoning about a given situation, but it is a marvel of the human brain that we are able to do so. Indeed, we could turn the phrase on its head when we observe the repeated failures of artificial intelligence programs to deal with context and meaning in their attempts to process natural language or perceive the everyday world. The fundamental computational bias in machine cognition is the *inability* to contextualize information.

1 Content dependent reasoning: an example

Lest this discussion appear too abstract, I will introduce an early example of the kind of phenomenon that appears in the psychological laboratory. The notorious four card, or 'selection task' problem devised by Peter Wason (1966) can be presented in a form which is easily solved. For example in the 'drinking age problem' of Griggs and Cox (1982), participants are told to imagine that they are police officers observing people drinking in a bar. They have to ensure that people are conforming to a rule such as:

> IF A PERSON IS DRINKING BEER THEN
> THAT PERSON MUST BE OVER 18 YEARS OF AGE

We can denote the logical form of this rule as the conditional: *If p then q*. Participants are shown four cards and told that these represent a drinker, with their age on one side and the beverage consumed on the other. The exposed values are Beer (*p*), Coke (*not-p*), 20 years of age (*q*) and 16 years of age (*not-q*). Most people (around 75% on the original study) given this task turn over the card marked Beer and the card marked 16 years of age (*p and not-q*). These cards are the ones which could detect an underage drinker violating the rule, and so are generally regarded as the correct choices.

In contrast, the apparently similar abstract form of the selection task is very difficult to solve, with typically only about 10% of student participants providing the correct *p and not-q* choice. In the Letter–Number version, participants are told that the four cards have a letter on one side and a number on the other and that they have to choose cards to turn over in order to see if the rule is true or false. A typical rule would be:

> IF THE CARD HAS AN 'A' ON ONE SIDE THEN
> IT HAS A '3' ON THE OTHER SIDE

The cards presented might be A (*p*), D (*not-p*), 3 (*q*) and 7 (*not-q*). A and 7 (*p and not-q*) is normally regarded as the correct choice because only these cards could reveal the falsifying combination of A-7. Most participants, however, choose A and 3, or just A. Why is this so difficult while the drinking age problem is so easy?

The sceptical reader, unfamiliar with the vast psychological literature on this task, might raise a host of objections on the basis of these two examples such as:

- The tasks are logically different because the Drinking Age rule concerns deontic logic – discovering whether or not a rule is obeyed, whereas the Letter–Number problem is an exercise in indicative logic – deciding whether a claim was true or false. Curiously, this went unnoticed in the literature for several years before being pointed out by Manktelow and Over (1991).
- The Letter–Number problem refers directly to cards, whereas the Drinking Age problem has cards that represent something else referred to in the rule.
- People might not be reasoning on the Drinking Age rule at all, but introducing their direct real world experience of drinking laws.

And so on. As an alternative to a lengthy and tedious review of the literature, I ask the reader to trust my assurance that all these issues and many more have been studied exhaustively by experimental psychologists. For example, thematically rich but indicative versions of the task have been shown to facilitate performance; deontic tasks involving permission rules of which people could have no previous experience have been demonstrated to facilitate; abstract but deontic versions of the task do not facilitate, and so on.

What research on this task does clearly show is that both the content and the context of the problem are critical. For example, in the Drinking Age rule, if the short introductory paragraph is removed, so that people are not asked to imagine that they are a police officer checking the law, then the Drinking Age rule becomes almost as hard as the Letter–Number rule (Pollard and Evans, 1987). Without this pragmatic cue, the relevant contextualization (which *assists* performance on this task) fails to occur. It has also been shown that subtle changes in context can reverse card selections and produce a pattern almost never observed on an abstract version of the task: *not-p and q*. For example, Manktelow and Over (1991) investigated a Shop task with the following rule:

> IF YOU SPEND MORE THAN £100 THEN
> YOU MAY TAKE A FREE GIFT

The cards for this task showed on one side whether or not a customer had spent £100 and on the other whether or not they had taken a free gift. For all participants there was a context indicating that this was an offer being run in a major store. They were, however, given different perspectives with regard to the motives of the individual checking the rule. One perspective is that of a sceptical customer who wants to make sure the store is keeping its promise. With this perspective, people turn over the cards for customers spending the £100 (p) and those not receiving the gift (*not-q*). This looks similar to the facilitation on the Drinking Age rule. However, other participants are given a different perspective, with an otherwise identical set of instructions. They adopt the view of a store detective who is on the lookout for cheating customers. With this perspective people turn the cards showing customers who did *not* spend £100 (*not-p*) and those who took the gift (q). This makes perfect sense from the viewpoint of personal rationality. People make the choices which could lead to detection of a cheating store or a cheating customer, depending upon the goals they assume.

Let us now think a bit more about the 75% of people who solve the Drinking Age problem and the 10% who solve the Letter–Number problem. What exactly is being facilitated in the former case? Surely not a process of abstract logical reasoning. We know this is not the case, because a number of studies have shown that when an abstract problem is presented immediately after solving such a facilitatory version, it remains as difficult as ever. There is no transfer of performance whatsoever. This is striking evidence in itself of the context dependent nature of human reasoning. Even when two structurally isomorphic problems are presented in succession, people fail to map these structures. What of the 10% who solve the Letter–Number problem? Are *they* engaging in abstract logical reasoning? The answers to these questions have emerged quite recently and relate to what are known as 'dual process theories' of reasoning. A brief diversion to explain the basics of dual process theory is needed before we examine the influence of belief on scientific thinking.

2 Evidence for dual thought processes

A simple form of dual process theory was proposed originally by Wason and Evans (1975) to account for findings on the selection task. In this experiment, Wason and Evans led participants to give correct or incorrect choices on the task by manipulating a factor known as 'matching bias' (see Evans, 1998). Essentially, people tend to choose the cards which match the lexical content of the rules, even when negatives are introduced which alter the logical significance of the cards. For example, if the rule reads, 'If there is an A on one side of the card then there is not a 3 on the other side' most people still choose A and 3 but now they are logically correct. Wason and Evans asked people to write justifications for their choices and discovered something very interesting. On the negative rule, people would give logical sounding arguments: for example: 'I am turning over the A because a 3 on the back would prove the rule false.' Despite this apparent insight, however, the same individual subsequently given an affirmative rule (with different lexical content) would typically match again, giving the usual wrong answer. The justifications now seemed to lack insight: 'I am turning over the A because a 3 on the back would prove the rule true.'

Wason and Evans argued that the matching bias was an unconscious choice and that the verbal justifications offered were rationalizations produced by a different kind of thinking. They called these type 1 and type 2 thought processes, respectively. After various intermediate developments, the dual process theory was recently reformulated by Evans and Over (1996) to refer to two distinct cognitive systems: implicit processes which are computationally powerful, context-dependent and not limited by working memory capacity; and explicit processes which permit general purpose reasoning but are slow, sequential and constrained by working memory capacity. In review of a wide range of psychological research on reasoning and decision making, the explicit system was seen in much more positive light than the rationalizations of Wason and Evans had suggested, although the account of that particular experiment is essentially unaltered. Other authors have proposed dual process theories also, most recently Stanovich (1999) who describes them as System 1 (implicit) and System 2 (explicit). We agree with Stanovich that System 2 thinking is needed to achieve normative rationality while System 1 is often sufficient for personal rationality. For example, many of our everyday goals can be achieved by applying routine learned solutions that require little if any conscious thinking and reasoning. However, Stanovich has added important new insights and data by linking the dual process theory with his work on individual differences in reasoning and decision making tasks.

Stanovich adds to our list of characteristics the idea that System 2 thinking is correlated with measures of general intelligence while System 1 thinking is not. (This idea was also proposed earlier in an analogous dual process theory in the implicit learning field by Reber, e.g. 1993.) To illustrate this, we can now return to the issue of how people solve the Drinking Age and Letter–Number versions of the selection task. Stanovich proposes that if, as argued by Evans and Over, solving the Drinking Age problem requires only pragmatic processes of the System 1 kind, then performance should be unrelated to measures of general intelligence. On the other hand, the 10% of people who solve the abstract task should be high in general intelligence if System 2 resources are required for abstract logical reasoning. In a large-scale empirical study, Stanovich and West (1998) provided evidence that strongly supported this analysis. Thus it seems that there exists a minority of individuals who can decontexualize (in this case resisting matching bias) and solve problems involving abstract logical reasoning. Stanovich (1999) reviews evidence that he has accumulated for this theory across a wide range of cognitive tasks.

At first sight this work might seem to suggest that scientific thinking will not be contextualized, constrained and biased by prior knowledge and expectations. After all, scientists are pretty intelligent people, are they not? To draw such a conclusion would be to profoundly misunderstand the psychological literature. The pragmatic influences of System 1 thinking are universal and pervasive. Some individuals succeed in suppressing these some of the time. Because they *can* do so, it does not mean that they generally do. We know that the instructions given on reasoning tasks can strongly influence the extent to which people base their reasoning on prior belief as opposed to restricting their attention to premises given (see, for example, Evans *et al.*, 1994; George, 1995; Stevenson and Over, 1995). We also know from such studies that in the absence of clear instructions to reason logically most people will reason pragmatically. We also know that even in the presence of such instructions a lot of people will continue reasoning pragmatically.

3 Belief and rationality in science

Let us turn now to the issue of whether it is rational or desirable for our thinking to be so heavily contextualized by prior belief and knowledge. Conflicting views seem to abound on this. In the Bayesian approach, it is seen as rational and appropriate to consider all new evidence in the light of prior belief. Bayesian conditionalization is far more than a statistical technique for computing posterior probabilities in the light of prior probabilities and likelihoods. It provides the foundation for a whole philosophy of science and one that stands in stark contrast with Popperian and classicist traditions (see Howson and Urbach, 1993). In the Bayesian tradition a formally identical probability calculus is given an entirely different interpretation than in the classical tradition. Probabilities are subjective: they represent no more and no less than a degree of belief that something will happen. To a classicist, probabilities are relative frequencies. In the Bayesian tradition it is meaningless to consider a single hypothesis in isolation or without regard to its prior likelihood of being correct. In the classical tradition, it is meaningless to assign a probability to a hypothesis, because it belongs to no frequency distribution.

British legal procedure – and other systems derived from it – contains a curious anti-Bayesian bias. In a criminal case juries are instructed to require the prosecution to prove their case beyond 'reasonable doubt', since the error of convicting the innocent is considered more serious that that of acquitting the guilty. So far, so decision-theoretic. But the rules of evidence specifically prevent the prosecution from introducing evidence of the prisoner's prior criminal record which may establish as strong prior probability of guilt and may have often have been the main cause of detection and arrest. Jurors are frequently dismayed to learn of the record of a defendant whom they have just acquitted. The origin of this and other anti-Bayesian arguments appears to be that prior belief is a source of bias and prejudice. Rational dispassionate observers (like, presumably, scientists) should go only on the evidence before them. Of course, the psychological evidence, as indicated above is that people find it almost impossible to disregard prior knowledge when instructed to do so. Hence, the rules of evidence which prevent the knowledge being acquired in the first place.

The psychological literatures on reasoning and judgement include a similar ambivalence on the rationality of prior belief (see Evans and Over, 1996, chapter 5). Participants are berated on occasions by psychologists for being influenced by their prior beliefs – the literatures on belief bias and confirmation bias (see below) typify this approach. On the other hand, there are literatures where people are dubbed bad Bayesians for ignoring prior probabilities, such as in the base rate neglect literature. As Evans and Over point out, however, base rate neglect has almost invariably been demonstrated in experiments where base rates are presented as arbitrary statistics and do not correspond to actual prior beliefs of the participants. In our own recent research, we have shown that people are much more strongly influenced by base rates which accord with, or even are supplied implicitly by their actual beliefs (Evans *et al.*, 1999). Hence, we agree with Stanovich's view that the pervasive tendency is for prior beliefs to influence reasoning. That still leaves the issue of whether it is rational or not that they should do so.

Stanovich (1999) addresses this issue in his chapter on the fundamental computational bias, a term which – as noted earlier – seems to beg the question. Stanovich argues that the demands of a modern technological society and the explicit aims of many of our educational practices require decontextualized thinking skills. He comments with admirable rhetoric that:

> Tests of abstract thought and of the ability to deal with complexity . . . will increase in number as more niches in post-industrial society requires these intellectual styles and skills . . . For intellectuals to argue that the 'person in the street' has no need of such skills of abstraction is like a rich person telling someone in poverty that money is really not that important. (1999, p. 205)

There is certainly an important need for abstract and decontextualized thought in some areas of science some of the time. However, we must keep this notion in proportion. Scientific training consists largely in imparting as much knowledge as possible to the students about prior research on the topics that they plan to investigate. Journal editors are swift to reject manuscripts which show ignorance or misunderstanding of the relevant literatures. In this respect, our presumption seems to be that expert research requires the researcher to be strongly equipped with relevant prior knowledge and belief. Against this, however, is the suggestion that creativity in science may be inhibited by too much prior knowledge leading to greater discovery of theoretical breakthroughs in early career scientists.[1]

At this point, some readers may wish to dwell upon the distinction between knowledge and belief, regarding the former as beneficial and the latter as often suspect and prejudicial. Unfortunately, a philosophical distinction – say, defining knowledge as justified true belief – is of little psychological value. We have in general no way of knowing whether our beliefs are true or justified and so we can't distinguish subjectively what we know from what we believe. For example, we may feel that we are knowledgeable enough of the prior literature on some topic to cite it confidently in writing our research papers. However, there is strong evidence that researchers frequently mis-cite findings in the literature, with famous findings being idealized and schematized in reconstructive recall (Vicente and Brewer, 1993). This example shows two things of importance. First, the cognitive processes of experts are no different from anyone else's. Experts have specialized knowledge but they do not in general have better memories and superior processes of reasoning and judgement to anyone else, and they are not immune to cognitive biases. Second, the guarantee of knowledgeability in our sciences that we rely upon – the peer review and editorial processes – is far from fail-safe. The numerous citation errors described by Vicente and Brewer were mostly published in journals which operated critical peer review policies.

Another distinction of interest – particularly for those who would required objective knowledge in science – is that between knowledge and information.

[1] My own PhD supervisor was the highly creative Peter Wason, who founded the modern study of the psychology of reasoning. Wason notoriously advised his PhD students to 'conduct the experiments first and to read the literature afterwards' in order to avoid having their creativity constrained by too much prior knowledge. As a long-term strategy, of course, this won't work as one is bound to acquire knowledge of the literature in order to publish papers.

Consider the question: Where is the knowledge in a scientific discipline to be found? My answer is – in the heads of the scientists, and *nowhere else at all.* The journals do not contain knowledge – they provide information from which a human being with a considerable amount of training and effort may construct knowledge. If a freak accident wiped out all the members of a particular research field leaving the journal publication intact, it would probably take many years to train a replacement set of researchers and recommence sensible investigation on the topic. From this standpoint, the knowledge in a given field is the collection of understandings that its participants roughly share, but with no unique or definitive versions to be found. There is also – in my experience of scientists – no such animal as the disinterested observer.

Of course, some sciences are more exact than others, although I believe my comments would apply to any science at the cutting edge of new research. In my own discipline – experimental cognitive psychology – it would be pretty safe to assert that there is rather little that counts as hard objective fact and that the current state of knowledge consists mostly of a set of partially justified and often conflicting beliefs on the part of its various proponents. In such a state of affairs, the *influence* which scientists exert on the thinking of others rests upon many factors other than hard evidence. For example, the quality and quantity of the written output which proponents of particular positions can produce may have a large effect. I became aware of this when reviewing the highly contentious debate between the mental logic and mental model theories of human reasoning (see Evans and Over, 1996, 1997). Proponents of the mental models account have been much more prolific in their conduct of experimentation and scientific papers than proponents of mental logic, but the truth of a scientific proposition should hardly be decided by the energy level of its advocates. The status of the institutions for which people work and the journals in which they published are clearly another factor. The editors of *Psychological Review*, for example, bear an awesome responsibility given the influence of papers published in that journal.

In summary, what I am saying is this. Even in so-called ‘exact sciences’ it may take many years of retrospection to agree the objective factual knowledge of the discipline. In currently active research fields we have only partially justified and conflicting beliefs whose success depends upon the argumentation skills, status and influence of their proponents and not simply upon ‘objective’ research methods viewed by ‘dispassionate’ observers. This process has profound cognitive and social psychological aspects to it. The social psychology of science is beyond the scope of this chapter. What I will focus for the rest of this chapter is the way in which prior beliefs influence our reasoning in the testing of hypotheses and the evaluation of evidence. For reasons explained above, I make no attempt to distinguish belief from knowledge.

4 Confirmation bias in the testing of hypotheses

There are two major cognitive biases studied in the psychological literature which are of relevance to the current chapter: *confirmation bias* and *belief bias*. Although some definitions of confirmation bias are broad enough to include belief bias (see Klayman, 1995), I prefer the following non-overlapping specifications:

- *Confirmation bias* – the tendency to seek evidence which will confirm rather than falsify prior beliefs and hypotheses.
- *Belief bias* – the biased evaluation of evidence in order to favour prior beliefs and hypotheses.

As we shall see, the evidence produced by psychologists of these biases often falls short of establishing the above definitions when carefully examined.

The term 'confirmation bias' suggests the negative interpretation of prior belief discussed above, and readers will not be surprised to know that such research is largely driven by a Popperian definition of normative rationality in science. This was certainly and explicitly the case in Peter Wason's pioneering studies of what he called 'verification bias'. Wason asserted that people were in general bad Popperians who were motivated to confirm rather than refute their hypotheses. One of the experiments on which he based this claim was the famous '2 4 6' task (Wason, 1960). Participants are told that the experimenter has a rule in mind which classifies triples – groups of three integers. An example of a triple which conforms to the rule is '2 4 6'. The task is to discover the rule by generating triples of one's own. The experimenter will in each case indicate whether the triple generated conforms to the rule or not. The rule is actually a very general one: any increasing sequence of integers.

What happens on this task is that people usually adopt a more specific hypothesis, as suggested by the 2 4 6 example, and then test it only with positive instances. This leads to repeated confirmation of the hypothesis and participants become convinced that it is correct. Wason was fascinated by the perseverance shown and the tendency to reformulate the same hypothesis in different words when told that it was wrong. For example, in one protocol quoted by Wason (1960) the participant persisted in testing triples such as

8 10 12
1 50 99
2 6 10

all of which were positive examples, but announced successively the following rules:

- The rule is that the middle number is the arithmetic mean of the outer two.

- The rule is that the difference between numbers next to each other is the same.
- The rule is adding a number, always the same, to form the next number.

Perseverance with what Wason considered a false hypothesis was so marked among his undergraduate student participants that he was moved to consider the task as providing strong evidence for a general confirmation bias in human thinking. This view persisted for many years in the literature, despite an early criticism (Wetherick, 1962) pointing out a basic flaw in Wason's interpretation. This flaw was not widely recognized until much later (see Klayman and Ha, 1987; Evans, 1989). The problem is that the phenomenon – powerful and interesting though it is – demonstrates only a positive testing bias and not a confirmation bias. That is to say, it appears that people rarely think of testing negative examples of their hypothesis. However, it is a peculiar feature of Wason's task that only negative tests can lead to falsification. Hence, we cannot infer that people are motivated to verify their hypothesis, as Wason claimed.

Positive testing is the preferred strategy in science, and one which will normally work well (see Klayman and Ha, 1987). That is to say, we normally construct an hypothesis H and deduce a prediction P from it. We then conduct an experiment designed to test whether P occurs or not. If this positive prediction fails, then falsification of the hypothesis will occur. According to Popper (1959) the best predictions to test are *risky* ones, i.e. predictions that could not easily be derived if the hypothesis were not held. Some spectacular examples of risky predictions being confirmed (or in Popper's terminology, corroborated) were provided by Einstein's theory of relativity. For example, why else would one predict that clocks would run slower at the equator than the north pole, or that light would curve when passing the sun during an eclipse? Negative predictions are generally not good from a Popperian point of view, because they are too broad. Most negative claims have high *a priori* probability.

What the Wason 2 4 6 task actually models is an insufficiently generalized hypothesis. Suppose, for example, one presents the claim:

Iron expands when heated

One could run many experiments involving the heating of iron objects, all of which would confirm this claim. The hypothesis is not wrong, but it is insufficiently generalized. The real rule, as Wason would put it, is:

Metals expand when heated

This could be discovered from the first hypothesis only by checking a negative test – whether things other than iron also expand when heated. However, it is general practice in science to make one's theories as general and as powerful as possible. I rather doubt that the Wason error will occur very much in real science.

In spite of these criticisms of the Wason task, I am sure that people, including scientists, do engage in a lot of confirmation-seeking in their hypothesis testing – a conclusion reached by Klayman (1995) in a survey of a broad range of psychological tasks that have addressed this issue. Like the 'fundamental computational bias' discussed earlier, however, there is a question as to whether the pejorative term 'bias' should really be applied here. One can argue easily that Popper's proposals have little psychological reality. David Over and I have suggested that hypothetical thinking has three typical properties (Evans and Over, 1996; Evans, 1999). First, we tend to consider only one hypothesis or mental model at a time. Second, we by default consider the most probable or plausible model, given our background beliefs. Finally, we use a satisficing strategy. That is we accept the current hypothesis as long as it meets broad criteria of satisfaction. We abandon or revise it only when required to do so by evidence. This is our general view of hypothetical thinking that people engage in in ordinary life. It is also a good model for actual scientific thinking.

In an uncertain world, it is not a practical psychological possibility to believe only things for which there is incontestable, or even very strong, evidence. Our natural mode of thinking is to accepting working hypotheses and to reason in a defeasible manner (see Oaksford and Chater, 1998). In order to build such a hypothesis, we do of course seek to find confirming evidence so that we can build up some practical system of belief. However, it is also intrinsic to our natural way of thinking to revise beliefs – or even abandon them altogether – in the light of the evidence encountered. In general, people – scientists included – behave more like Bayesians than Popperians. That is, we tend to increase and decrease our degree of belief in a gradualist manner as evidence is encountered.

Studies of expert thinking are clearly relevant here. We carried out an extensive study of engineering design using both diary methods for project conducted over a lengthy period and verbal protocol analysis of the solution of short (two-hour) design problems (Ball, Evans and Dennis, 1994; Ball *et al.*, 1997). Engineering design takes place in very large and complex problem spaces where exhaustive search and optimization are not practical possibilities. We found that satisficing strategies were dominant. That is, designers find components of the design which are good enough to meet their criteria and leave them in place until or unless they are invalidated by some later aspect of the design.

A fascinating case study of the role of confirmation and disconfirmation in scientific thinking has been presented by Gorman (1995). His research was based on qualitative analyses of the notebooks and correspondence of Alexander Graham Bell in the period of research leading to the invention of the telephone. Bell approached the problem by taking an analogy with nature and studying the mechanics of the human ear. His basic idea was to use a diaphragm, vibrating reed and electromagnet to mimic the mechanism of the human ear and hence to translate sound waves into electric current. A great deal of experimentation with the component materials of this set-up was required before his telephone would work, including a large set of studies of liquid transmission which was eventually abandoned. Gorman finds evidence of confirmation bias in this phase of the work, in that Bell was very alert to any confirming findings.

Gorman's discussion mirrors the debate concerning the Wason 2 4 6 problem. That is, while it is easy to distinguish positive from negative testing, it is much harder to say what constitutes confirmatory and disconfirmatory behaviour, because this depends upon the motivations and expectations of the researcher. As he points out, particular experiments could count as confirmation for one scientist and disconfirmation for another. It is important also to remember that science is a collaborative enterprise, engaged in by many individuals with different perspectives and motivations. It is clear that individual scientists are often motivated by a desire to build belief in their particular theoretical system. Such motivation can lead to considerable productivity. For example, the range of experimentation offered in support of their theory of mental models in deductive reasoning by Johnson-Laird and Byrne (1991) is quite staggering. Whether the theory is right or not, such ambition can lead to much new knowledge. If the theory has weaknesses, then it is not essential that these are discovered by the theories' proponents. Other scientists, with their own differing perspectives and ambitions, will be only too pleased to provide the falsifying evidence.

In conclusion to this section, it is clear that we tend to test hypothesis in a positive manner, but this is typically appropriate and not evidence of confirmation bias in itself. However, we do tend to focus on particular hypotheses or mental models and stick with them until evidence causes us to abandon them. Strong belief in their own theories or career motivation might lead individual scientists also to adhere to theoretical positions and pursue them beyond justification by the evidence. However, research so motivated will still lead to the discovery of new knowledge and the soundness of the theoretical claims will be subject to constant examination by other scientists in the same field. Popperian prescriptions do not map comfortably on to our natural ways of thinking and on the whole scientists behave more like Bayesians, whatever their explicit philosophy of science may be. For all of these reasons, I do not regard confirmation bias as a particular problem.

5 Pseudo-diagnostic reasoning

A task which has been claimed to provide evidence of confirmation bias is the 'pseudo-diagnosticity' (PD) problem (Doherty *et al.*, 1979; Mynatt, Doherty and Dragan, 1993). It is interesting to consider this task briefly as it concerns probabilistic reasoning and statistical evidence to which the deductive

Popperian framework is less obviously applicable. An example of the standard PD paradigm is taken from Mynatt *et al.* (1993, pp. 757–68).

> Your sister has a car she bought a couple of years ago. It's either a car X or a car Y but you can't remember which. You do remember that her car does over 25 miles per gallon and has not had any major mechanical problems in the two years she's owned it.
>
> You have the following information:
>
> A 65% of car Xs do over 25 miles per gallon.
>
> Three additional pieces of information are also available:
>
> B The percentage of car Ys that do over 25 miles per gallon.
>
> C The percentage of car Xs that have had no major mechanical problems for the first two years of ownership.
>
> D The percentage of car Ys that have had no major mechanical problems for the first two years of ownership.

Assuming you could find out only one of these three pieces of information (B, C or D) which would you want in order to help you to decide which car your sister owns? Please circle your answer.

The conventional normative analysis for this task states that B is the correct answer, because it gives a diagnostic information – a complete likelihood ratio allowing comparisons of the two hypotheses with regard to one piece of data. Most people, however, choose option C – which gives more information about the *focal hypothesis* X (focal because it is mentioned in the initial information given, also known as the anchor). Doherty and colleagues regard this as a form of confirmation bias. People have focused on X as the likely hypothesis and are trying to confirm it.

Girotto, Evans and Legrenzi (2000) have shown that this phenomenon can be *de-biased* (accepting the normative analysis for the moment) in various ways. For example, if no anchor is provided and people are asked simply to choose two pieces of information from A, B, C and D then they strongly prefer to choose one of the two diagnostic pairings that compare the two hypotheses (this was replicated in Feeney, 1996). More relevant to the present chapter is the influence of the rarity of the information given. Suppose you are told that your sister's car has four doors (a common feature) and does over 160 miles per hour (a rare feature). If participants are now told that car X does over 160 mph they switch choices from C to B, seeking more evidence about this rare feature and apparently choosing diagnostically (Feeney, 1996). We also investigated a variation on the task where people are not asked to choose information, but simply to rate their belief in the hypothesis after seeing first one and then a second piece of evidence (Feeney *et al.*, in press). If the two pieces of information are pseudo-diagnostic (that is, both concern the same hypothesis), people end up with higher confidence in the hypothesis if the supporting information given is rare rather than common.

My interpretation of the PD task is as follows. The use of an anchor generally focuses people on a single plausible hypothesis, in accordance with our general principles of hypothetical thinking. This leads to a confirmatory choice in the second piece of data. However, if the focus is removed, as with the unanchored version, people do appreciate the value of a diagnostic pairing. The effect of feature rarity is of particular interest with regard to the current discussion as it shows that people spontaneously introduce background belief into these experimental tasks (compare Koslowski and Thompson, chapter 9 in this volume). The preference given to rare over common information can be seen as appropriate because it reflects the *implicit diagnosticity* of rare evidence, relative to background belief. If you know that car X has four doors, your expectations is that Y will also have four doors – since most cars do – and hence you will expect to gain little from this 'diagnostic' choice. On the other hand, since very few cars do over 160 mph, learning that one does provides strong evidence for your hypothesis in itself. The diagnosticity of rare evidence is judged relative to the low prior probability provided by background belief. Even Popper should approve here, since evidence based on rare features is akin to that derived from risky predictions.

I think that pseudo-diagnostic reasoning is a genuine phenomenon that occurs under some circumstances and to which scientists may be prone. Essentially, it is not enough to show that the evidence is probable given one's favoured theory. One has to show that it is more probable than would be predicted by the rival theories available in the literature. Based on my personal experience of reviewing hundreds of experimental papers over the years, I would say that experimental psychologists quite frequently neglect to demonstrate that their data are better accounted for by their own theories than by other people's. Until, that is, referees like me get to them. This is where the same form of protection comes in that applies to other forms of confirmation bias discussed above. Science fortunately is a social activity, where knowledge advances via the processes of peer review and debate and ultimately consensus.

6 Bias in the evaluation of evidence

A final issue to consider is whether our prior belief in theories and hypotheses biases our evaluation of data. Do we see what we hope and expect to see in our findings and those of others reported in the literature? Are we much more critical of the design of studies whose findings conflict with our theories? In general, psychological research as well as everyday observation provides an affirmative answer to these questions. For example, Lord, Ross and Lepper (1979) presented participants with reports of research studies purporting to find evidence for or against capital punishment – a topic on which the the participants' attitudes were polarized in advance. Each participant was shown a study supporting capital punishment using methodology A and one opposing capital punishment using methodology B. Of course, this was counter-balanced across the participants so that A and B were reversed for half of them. Nevertheless, participants were much more critical of the methodology of the study whose conclusions conflicted with their prior belief – whichever method that happened to be. There are numerous similar findings in the social cognition literature showing that people's evaluation of evidence is biased by their prior beliefs.

In the cognitive literature, the analogous finding is that of 'belief bias' in deductive reasoning. It has been known for many years that when people are asked to evaluate the validity of logical arguments, they tend to endorse those whose conclusions accord with prior belief far more often than those which conflict with prior belief (see Evans, Newstead and Byrne, 1993, chapter 8). As I have argued more recently, however (e.g. Evans, in press) the effect of belief on these tasks is generally misunderstood, because most studies have omitted to include arguments with belief-neutral conclusions. In general, people have a strong tendency to endorse fallacious conclusions with neutral materials, and when this control is introduced belief bias is seen to be mostly a negative effect. That is to say, people do not increase their endorsement of fallacies when conclusions are believable; rather, fallacies are suppressed when conclusions are *un*believable. Hence, the effect of belief on these tasks could be seen as a de-biasing effect.

This finding can be explained with regard to the principles of hypothetical thinking discussed earlier. The default reasoning mechanism focuses our attention on a single most plausible model of the premises (strong evidence for this tendency in syllogistic reasoning has recently been provided by Evans *et al.*, 1999). This model is maintained provided that it satisfies. An unbelievable conclusion does not satisfy, so here a search for an alternative model of the premises takes place, leading to refutation of the fallacy. The evidence for biased evaluation of scientific studies – as in the study of Lord, Ross and Lepper, discussed above – can be explained in a similar way. When the findings of a study are congenial with our beliefs, we accept it uncritically. However, when we don't believe the conclusion, we look very critically at the methodology.

Recent evidence suggests that people are not only influenced by the believability of the conclusions but by the believability of the premises of deductive arguments. They are reluctant to endorse arguments with false premises, even when the logic is simple and transparent, and appear to confuse validity with soundness (see George, 1995; Stevenson and Over, 1995; Thompson, 1996). This is a good example of a case where normative and personal rationality diverge. Relative to the context of the experimental instructions it is normatively incorrect to be influenced by the believability of the premises and the conclusion. In real life, of course, it is rational to take account of all relevant beliefs in our reasoning, and a waste of time and effort to draw conclusions from premises we believe to be false. In fact, relative to real-world reasoning it is arguably adaptive to be influenced by beliefs generally in the way that experiments on belief bias reveal.

It is clearly necessary for us to have an organized and broadly consistent belief system in any area of our lives, which may include science if we happen to be scientists. Of course, we need to revise and update our belief system in the light of evidence, but we should not lightly discard a core belief upon which foundation much else may rest. Hence, if a scientist has built up a theoretical understanding of a research area and then encounters a falsifying result, how should she or he react? Is it rational to reject one's hard earned belief-system in one fell Popperian swoop? Surely, the first step is to examine the methodology of the study in great detail in order to check that the result is actually sound. The next step would be to replicate the finding and ensure its reliability. If the finding is unshakeable, then one would expect the scientist to make the smallest and least costly revision to the belief system required to accommodate the finding.

7 Conclusions

There is a highly pervasive tendency for people to contextualize all problems and to be influenced by prior belief in their hypothesis testing, reasoning and evaluation of evidence. In general, expert thinking, including that of scientists, is prone to the same tendencies and biases as can generally be demonstrated in the psychological laboratory. Whether the influence of prior belief is likely to be the cause of error and bias in scientific thinking, however, is highly debatable. I admit to a preference for the Bayesian over the Popperian perspective as a normative framework. I have no doubt at all that the Bayesian process corresponds much more closely to what scientists actually do.

To summarize some of the phenomena reviewed in this chapter:

- People often reason more effectively when problems are viewed in the context of prior knowledge and experience. In such cases, pragmatic processes can supply solutions without much conscious reasoning effort being required.
- Hypotheses tend to be tested in a positive manner, a strategy which is usually appropriate but has been confused with confirmation bias in the psychological literature.

- Genuine confirmation bias – the tendency to seek information to support a favoured theory – probably also exists, but it can be argued that confirmatory strategies can be helpful in the development of theories.
- Disconfirmation is not really a problem in practice, because science is a social and collective activity. If it suits the ego or career interests of one scientist to confirm a point of view, then it will suit similar interests of another to attempt to refute the same hypothesis.
- The interpretation of psychological experiments on hypothesis testing and reasoning is complicated by people's habitual introduction of prior belief. This may lead to error relative to the experimental instructions, but a similar process would often be adaptive in the real world.
- The way in which prior beliefs influence thinking in these experiments often makes sense from the perspective of personal rationality. For example, people prefer rare evidence – which is implicitly diagnostic – in hypothesis testing, and resist reasoning from false premises or toward false conclusions.

Perhaps I should emphasize in conclusion that we have little choice in this matter of how we think. The pervasive effects of belief – the fundamental computational bias of Stanovich (1999) – reflect the operation of a very powerful pragmatic system, operating at an implicit level, and providing the main foundation of our cognitive processes. These processes cannot be switched off by experimental instructions (psychologists have tried and failed to do this). People's critical thinking and reasoning can be improved by training, but the domination of implicit over explicit cognition remains. Exhorting people to be disinterested observers or Popperian falsificationists is unlikely to have much effect.

11 Thinking about causality: pragmatic, social and scientific rationality

Denis Hilton

While one can argue that full rationality is the adequate normative framework to evaluate the reasoning of *homo scientificus*, as science should spare no effort to validate causal hypotheses and to identify correct explanations, I shall argue that ordinary causal reasoning does in fact have other goals – particularly those of cognitive economy and social co-ordination. This translates into a concern with bounded rationality – using heuristics that are simple but smart; and social rationality – using assumptions about conversation which enable speakers and listeners to co-ordinate their efforts around the most relevant explanation. Understanding such concerns can help explain why *homo pragmaticus* deviates from scientific criteria for causal reasoning while being rational in his or her own terms. It can also help us understand what we need to do to help people reason 'scientifically' about causes, where we consider that to be important. Indeed, understanding the pragmatic rationality of human inference processes should help us better understand why they are 'sticky', and sometimes resist reformation into 'fully rational' modes of thinking which are principally adapted to doing science. I apply this pragmatic analysis to three domains of causal judgement: learning of causal relations; causal hypothesis testing; and discounting vs. backgrounding of causes in explanations.

1 Introduction

In this chapter I will be primarily interested in how scientific rationality may differ from pragmatic and social rationality with respect to causal thinking. Since adaptations of thought which favour pragmatic and social rationality are likely to be evolutionarily ingrained in human beings, they may sometimes prevent human beings from attaining full scientific rationality in thinking about causes. Conversely, examples of causal thinking that may appear irrational from a 'scientific' point of view may of course reflect very rational adaptations – say, to information-processing limitations or to conversational conventions. I

I would like to thank Jean-François Bonnefon, Peter Carruthers and David Mandel for helpful comments on earlier versions of this chapter.

will address these questions through examining examples of scientific (and unscientific) *causal thinking* in domains of professional concern such as finance, law, management and medicine. I will also review empirical research in cognitive and social psychology that can give us a clue as to why people have the biases in causal reasoning that they do, and in the final section I will review social factors that can either amplify or attenuate biases in individual reasoning. Knowing these social factors should help institute some remedies for errors (when measured against some normative criterion) induced by cognitive biases.

In attempting to better understand people's natural capacities to reason in order to devise ways of improving that reasoning, I will be continuing a project begun by Bentham and Mill. These philosophers were concerned to understand the sources of logical fallacies which – in their view – bedevilled rational debate about social policy. In their view, poor causal reasoning tends to lead to the formulation of inadequate social policy recommendations. For example, if in the mid-nineteenth century one induces from the fact that most Irish people are out of work that they are lazy (rather than inferring that this is due to the lack of jobs), then a British politician could rationally justify a policy of inaction towards the Irish famine. These problems are still with us, as evidenced by the French government's decision to withdraw support for almost a quarter of the drugs hitherto funded by the state health service on the grounds that there was no scientific support for their causal efficacy (see *Le Monde*, 18 September 1999, *La liste des 286 médicaments inutiles*). For such reasons, Bentham published a *Book of Fallacies,* and Mill his *System of Logic* in which he examined the thought processes of established science in order to make recommendations to the nascent social sciences. Like Popper later, Mill saw the experimental method as the royal road to refuting dogma.

However, it is crucial to determine what the goals of a causal inquiry are, as we need a clear idea of what normative properties ideal scientific or common-sense reasoning *should* have before we can characterize a particular piece of reasoning as being erroneous. In particular, I will argue that we need to have a clear idea of what kinds of question are being addressed in scientific and common-sense causal inquiries before we can decide whether a given response (or method of finding a response) is correct or not. Otherwise we risk mistaking causal reasoning that is in fact well adapted to its goals as being irrational because we are applying an inappropriate normative model.

In particular, it seems to me that there is a danger of supposing that a *full rationality* model, where agents have unlimited processing capacities, is always the appropriate normative standard for human beings' causal inferences. This risk is serious as many researchers on causal reasoning have compared the layperson to the scientist (e.g. Heider, 1958; Kelley, 1967), and sometimes unfavourably so (Ross, 1977; Nisbett and Ross, 1980). While one can argue that full rationality is the adequate normative framework to evaluate the reasoning of *homo scientificus*, as science should spare no effort to validate causal hypotheses, I shall argue that ordinary causal reasoning does in fact have other goals – particularly that of economy. This translates into a concern with bounded rationality – using heuristics that are simple but smart (cf. Gigerenzer, Todd and the ABC Research Group, 1999); and social rationality – using assumptions about conversation which enable speakers and listeners to co-ordinate their efforts around the most relevant information (Hilton, 1990, 1991, 1995b). Understanding these concerns can help explain why *homo pragmaticus* deviates from scientific criteria for causal reasoning while being rational in his or her own terms. It can also help us understand what we need to do to help people reason 'scientifically' about causal relations, where we consider that to be important. Indeed, understanding the pragmatic rationality of human inference processes should help us better understand why they are 'sticky', and sometimes resist reformation into 'fully rational' modes of thinking which are principally adapted to doing science.

2 The rationality of explanations: science, common-sense and pragmatics

I begin by noting two ways in which the questions addressed by scientific and common-sense explanation seem to me to differ, and thus how answers to these questions will therefore obey different criteria of rationality.

2.1 Causal induction vs. explanation: inferring relations between types of event vs. identifying causes of particular events

The pragmatic perspective suggests that causal inference may serve different purposes. Of particular importance is the recognition that whereas science is generally concerned with the establishment of universal causal *generalizations, such as oxygen causes fires, lack of oxygen causes death, light causes plants to grow, smoking causes cancer*, practical and common-sense causal inquiries are often concerned with why a *particular* event occurred when and how it did. This is well illustrated by Hart and Honoré (1985), who wrote that:

> The lawyer and the historian are both primarily concerned to make causal statements about *particulars*, to establish on some particular occasion some particular occurrence was the effect or consequence of some other particular occurrence. The causal statements characteristic of these disciplines are of the form, 'This man's death was caused by this blow'. Their characteristic concern with causation is not to discover connexions between types of events, and so not to *formulate* laws or generalizations, but is often to *apply*

generalizations, which are already known or accepted as true and even platitudinous, to particular concrete cases.

As an example of the uselessness of many causal generalizations for common-sense explanation, Hart and Honoré (1985) consider the case of a man who has died as a result of being shot. Here, the causal generalization that *lack of oxygen in the brain causes death* would not explain why this man died when and how he did, being true of all deaths. Rather, the pragmatic interests of the lawyer, doctor and layman would be to identify the *abnormal condition* that distinguishes this man, who is dead, from the normal state (being alive). The explanation that he died because he was shot gives new information that is relevant to the question about the cause of this particular death, which the answer that he died because of lack of oxygen in his brain would not, even though it is more likely to be true, and to have a higher co-variation with death than being shot.

Causal generalizations can of course figure in common-sense explanations. This can happen in at least two ways. The first is that they may be used as a rule for justifying categorization decisions if the causal generalization identifies a condition that is both necessary and sufficient for the effect to occur, as when a doctor certifies that *I declare this patient dead because of brain death consequent on lack of oxygen* (cf. Hilton's 1990, 1995a discussion of epistemic justifications). Causal generalizations which identify conditions which are necessary but not sufficient for the effect to occur (e.g. *oxygen causes fires*) do not seem able to support such categorization decisions.

A second case is when giving answers to causal questions about universal generalizations that are not situated in specific times and places, such as *What causes fires?* For example, as Cheng (1993) persuasively argues, using Salmon's (1984) notion of 'statistical relevance', the causal generalization *High heat causes fires* may seem a better explanation than *High energy causes fires* (because in many cases fire does not occur even in the presence of high energy) or *Lightning causes fires* (because many fires occur in the absence of lightning). Cheng (1993) correctly points out that this may be because high heat may, in general, have a higher co-variation with fire than either high energy (owing to its greater sufficiency) or lightning (owing to its greater necessity). However, note that her reasoning breaks down when applied to the explanation of a *particular* event that happened to a named individual situated in space and time. For example, it hardly seems natural to explain the explosion of the space shuttle Challenger in January 1986 by saying that, 'The Challenger caught fire because of high heat'. This would seem to be because the presence of high heat would, being a necessary condition for fires, be presupposed anyway. Although it would distinguish the Challenger (which caught fire) and the normal space shuttle (which should not, and until the launch of Challenger, did not catch fire), it would not add any new information which could not be inferred from the description of the event (the Challenger caught fire).

2.2 *Science vs. common-sense: the problems of causal connection and causal selection*

As Hesslow (1988) argues, the central task for causal explanation is to select one (or two) of a set of necessary conditions as *the* cause(s) of a particular event, whereas the aim of scientific causal induction is to establish causal *connections* between two types of event. The structure of such tasks follow the classic pattern of induction: from a series of particular observations, a generalization about a causal relation may be made such as *AZT cocktails retard the development of AIDS.* This inductive–statistical generalization (Hempel, 1965) may, of course, be used to support the explanation about a particular individual that, 'Tom did not develop AIDS because he took the AZT cocktail'.

I consider the causal generalizations that are *outputs* from inductive processes to be the part of the *inputs* into causal explanation processes. Typically, a common-sense causal explanation process is triggered by surprise or disappointment at an unexpected, undesired or incongruous outcome (Hastie, 1984; Weiner, 1985; Bohner *et al.*, 1988; Brown and van Kleeck, 1989). In such cases, we consider that the surprising event is then compared to suppositions about what may normally have been expected to happen in this context, which are retrieved from world knowledge (cf. Kahneman and Miller, 1986; Schank and Abelson, 1977). Consequently, the input into the explanation process is an *abnormal target event* and a set of relevant *causal generalizations* which describe the laws (or tendencies) of nature, and the output is some *abnormal condition* which shows why the target event had to occur, given these causal generalizations.

These considerations illustrate why I consider *normality* to be a pivotal feature of causal inference. First of all, the outputs from the causal induction process are causal generalizations which express our understanding of what should normally happen in the world. If an event (e.g. a space shuttle launch) goes off normally, this may simply strengthen an established causal generalization along the lines of *NASA organizes successful space shuttle launches.* Secondly, it is precisely violation of these expectations about what should happen in particular cases that triggers spontaneous search for causal explanations of surprising events, e.g. *Why did the Challenger explode?* Consequently, causal explanation typically involves a comparison of the particular abnormal event with the normal case, where the space shuttle does not explode during lift-off. In the case of the Challenger disaster, the result of the explanation process is to identify factors that made the difference between the Challenger launch and the 24 previous successful lift-offs (a statistical norm), such as cold weather, and that made the difference between the actual space shuttle design and its ideal design (an ideal norm), such as faulty rocket seals.

Although investigations into the causes of the Challenger disaster generated an enormous amount of causally relevant knowledge about the nature of the space shuttle, such as the functioning of its propulsion and control systems, which helped understand how the disaster happened, these were ruled out as *the* causes of the accident because they explained facts about the shuttle's normal functioning. Rather they seem to help explain *how* an abnormality is connected to the final disaster. Thus the unusually cold weather was thought to be a cause of the disaster because it caused the rubber O-ring seals to lose their flexibility. The disaster then became inevitable, even 'normal' in these circumstances, as the dysfunction in the seals allowed the hydrogen in the fuel tanks to come into contact with oxygen, thus setting off the explosion (Hilton, Mathes and Trabasso, 1992).

3 Biases in causal learning: rational adjustment or error?

Having established *what* the goals of causal inquiry are (general vs. particular, connection vs. selection) we can ask whether people do them well or not, or at least 'scientifically'. We therefore need some kind of 'gold standard' to evaluate causal claims by, and for this we turn to Mill's (1872/1973) 'System of Logic', where he sought to provide methods of *proof* that would enable the validity of causal claims to be evaluated. In particular, he proposed that causal hypotheses be verified using specific methods of induction, namely the methods of (1) difference, (2) agreement, (3) concomitant variation, (4) residues and (5) the joint method of difference and agreement. Of these, the methods of difference and agreement are the most important, the others essentially being derivative of the first two. Mill considered the method of agreement to be 'less sure' than the method of difference, and recommended its application only when there was not sufficient (contrastive) data about what happens in the absence of the putative cause to apply the method of difference. Mill's methods have become part of the philosophical canon (Mackie, 1980) and have inspired many psychological models of causal induction (e.g. Heider, 1958; Kelley, 1967).

As an example of how prone people are to errors in causal inference (even those who should know better), I take the example of illusory correlations in human resource management. Although one might like to think that human resource management in modern Western countries would be 'scientific', a recent survey shows that over 90% of French recruitment agencies use graphology whereas only some 60% of them use tests of intelligence, aptitude and personality (Amado and Deumie, 1991). However, a recent review of empirical research on graphology has concluded that the validity of graphologists' assessments of personality is 'virtually zero' (Neter and Ben-Shakhar, 1989). Another study showed that an 'expert' graphologist was unable to give the same analysis to specimens of hand-writing presented twice to her (Ben-Shakhar *et al.*, 1989). None of this research seemed to justify the faith placed in graphology in France and abroad.

The Amado and Deumie survey also shows that the most common selection method in France is the interview. Here again, it seems that the value of these selection methods may be overestimated. For example, one prominent French business school uses interviews as a means of selecting candidates. However, a recent study (conducted by Alain Bernard at ESSEC) which asked the same candidates to be interviewed twice by different juries, found no significant correlation between the evaluations given by the two juries for the same candidate. However, when asked, most people at the school affirm that these selection interviews do have some validity. Their reasoning illustrates how such illusory correlations might come into existence. The people we ask often justify the effectiveness of the selection procedure by arguing that most of the students selected are successful – for example, they have no problems finding good jobs once they leave the school.

As Mill would have pointed out, this involves the fallacy of supposing the method of agreement to be valid. That is, our interlocutors think about what the successful students have in common, and one thing that they all have in common is that they were interviewed at admission. They therefore conclude that the interview is at least a contributory cause to selecting good students. However, they do not have the full range of information necessary to make a proper causal induction using the method of difference. This would be provided by knowing how well students who failed the interview would do if the school admitted them anyway, to see how well they did when educated the same way (Einhorn and Hogarth, 1978). Only if there was a difference in ultimate performance between those who 'succeeded' or 'failed' the interview, could one conclude that interviews help select better students. Significantly, when studies using this design have been carried out in the USA, no difference in ultimate performance was observed between students who 'succeeded' or 'failed' the initial admissions interview (Dawes, 1994).

How then might intelligent, motivated decision makers with time enough to reflect carefully about their judgements be so prone to make decisions which lack scientific validity? To provide a partial answer to this question below, I consider two traditions of research on ordinary causal induction: the first, on associative learning, where people (and animals) are exposed on-line to pairings of events from which they form associations; and the second on contingency judgement where people are required to use summary information presented in the form of contingency tables to assess causal relations between two predefined variables. Research in both these traditions has identified characteristic biases and errors, which give us insight into the kinds of processes people actually use to make these inductions.

3.1 Associative learning processes and causality judgement

Associative learning processes have a distinguished philosophical pedigree, going back to the philosopher David Hume's account of how people perceive causation. More recently, models of associative learning in animal learning have been extended to describe how human beings learn about co-variations (Dickinson, Shanks and Evenden, 1984). There thus seems to be considerable evidence that people learn contingencies using simple associative processes when making causal judgements on-line (Lober and Shanks, 1999). It therefore seems that Hume's *psychological* supposition that we form beliefs about causal relations through associative processes such as experience of co-variation and spatial–temporal contiguity between putative causes and effects is largely correct, and supported by recent research on human causal learning.

In this contemporary view of associative learning, processes of causal judgement may be automatic and outside conscious control. For example, a student who wakes up in the morning with an unexpected headache may automatically search for the cause of his headache in the following heuristic way. He may, for example, think back to what he did the night before (he went to a bar, he drank a lot, he ate a lot of mandarins, he talked a lot) and consider these as somehow associated with his headache. A week later, he wakes up with another headache in the morning, and again thinks back (painfully) to what he did the night before. This time he remembers that he went to a bar and drank a lot, but did not eat any mandarins and talked little. As a result, going to a party and drinking a lot become more strongly associated with getting a headache, and eating mandarins and talking a lot become less strongly associated with his hangover. If he does this enough times, going to a bar and drinking too much will perhaps become so strongly associated with getting a headache in the morning, that he will not even be surprised when he gets a hangover after drinking in a bar, and will not even think about it. He will have learnt the rule *drinking too much in a bar leads to a hangover*, a kind of causal generalization.

The associative account predicts that when an unpredicted event occurs, the organism scans the immediate context for factors that might be associated with it (Rescorla and Wagner, 1972; Pearce and Hall, 1980). No such processing is assumed to take place when the event is predicted by its context. If the event is already predicted by the context (e.g. the rat has already learnt that pressing a food lever will produce food), then it will not find the arrival of food surprising and will not attend to it, just as the seriously party-going student will be unsurprised if he wakes up after a drunken party with a hangover (Shanks and Dickinson, 1988).

The process whereby we scan the environment for precursors of an unpredicted event reflects Hume's principle that we tend to associate events which are close together in space and time. Consistent with this principle, Shanks and Dickinson (1988) show that temporal contiguity between a shell hitting a tank and the tank's explosion leads to stronger ratings of causal strength for the shell. In scanning the environment for precursors of an unpredicted event, people are implicitly using a naïve version of Mill's method of agreement, where factors which are always present when the effect is present are judged to be the cause.

While these processes can lead us into error (when compared to a normative system for proof of causal relations such as that advanced by Mill in his 'System of Logic'), these simple associative processes can be considered as 'simple but smart' heuristics for learning about causal relations (cf. Payne, Bettman and Johnson, 1993; Gigerenzer, Todd and the ABC Research Group 1999) in that that they can often attain considerable accuracy at minimal cognitive cost. It should therefore be no surprise that we find such adaptive heuristics in humans and other animal species. Below we evaluate the bounded (pragmatic) rationality of associative learning processes by comparing them with a co-variational model of causal induction.

3.2 *The bounded rationality of associative learning processes*

One way to evaluate the computational advantages of these simple associative learning rules is to consider the cost of implementing a co-variational model of causal induction as a process model. One such model has been proposed by Cheng and Novick (1992; see also Cheng, 1997), which implements a probabilistic form of Mill's method of difference and is thus an *a priori* candidate to be a model of 'scientific' causal induction. As well as claiming the *Delta-P* rule to be a normative model of induction, they also propose it as a model of 'natural' causal induction, arguing that people use its rules even in basic causal judgement tasks. If this model were implemented as an algorithmic (i.e. process) model, it would require the following things to be done: every observed pairing of events must be compiled as exemplars into memory; this memory must comprise all four cells obtained by pairing the (non-)occurrence of p with the (non-)occurrence of q; this memory must be updated every time a new pairing is observed; the whole matrix will then be used to calculate a *Delta-P* coefficient. (This is obtained by subtracting the ratio of occurrences to non-occurrences of q when p is present from the ratio when p is absent. In terms of the contingency table presented in figure 1, *Delta-P*

	Effect present	**Effect absent**
Cause present	(Cell A)	(Cell B)
Cause absent	(Cell C)	(Cell D)

Fig. 1 Contingency table of observations of a putative cause and effect

would correspond to $a/(a+b) - c/(c+d)$.) Whereas associative learning processes use Mill's method of agreement at each trial, co-variational rules such as Cheng's use a probabilistic form of Mill's method of difference on the complete, stored and retrieved data-set.

Clearly, the associative learning models would seem to have advantages of computational simplicity over co-variational models such as that of Cheng and Novick (1992) and Cheng (1997). First, associative learning models fire only when surprised (i.e. when confronted with an unpredicted event); otherwise they do not activate in the presence of non-events (i.e. *not* q) or predicted events (q when predicted by its context). Second, associative learning models only sample what is present when the surprising 'event' is present (i.e. for the presence of p when q is surprising, or for the presence of p when *not* q is surprising). Third, associative learning models store the index of the relationship between p and q as a simple association, rather than storing and updating exemplars in four different cells. In Millian terms, associative learning models iteratively apply Mill's method of agreement (sampling the a and b cells) whereas co-variational models apply Mill's method of difference (sampling all four cells). For these reasons, it is easy to imagine that associative learning processes have evolved in a wide range of species owing to their comparative simplicity, as well as effectiveness at detecting co-variations where the cause and effect are spatially and temporally contiguous.

However, associative learning processes risk certain kinds of errors characteristic of Mill's method of agreement, such as the failure to detect third variables. For example, the student who attributes his morning headaches to drinking heavily in bars may not have noticed that if he drinks as much in a dinner party he does not wake up with a headache the following morning. The real cause of his headache may be the smoke in the bars, which is absent in the dinner parties he attends. He may not notice this because he never launches a causal search after waking up without a headache. In addition, when he does wake up with a hangover, he fails to learn the co-variation between the morning headache and the previous night's smoke because this is 'blocked' by his prior association of the headache to drinking. To properly test his hypothesis that drinking too much causes hangovers, he would have to do an experiment where he spends some evenings in bars without drinking alcohol, which of course he never does!

The operation of associative learning processes may thus help explain why people are very good at learning spatially and temporally contiguous co-variations (Shanks and Dickinson, 1988), but are very poor at learning lagged co-variations (Sterman, 1994). In addition, they help explain why people form many illusory correlations. Thus the fact that these heuristics sample factors that are present only when the effect-to-be-explained is present may help explain why people fail to spontaneously generate relevant counter-factuals when evaluating causal claims. For example, a doctor may spontaneously associate a patient's recovery to the preceding psychotherapy without asking the counter-factual question of what would have happened if the patient had had no psychotherapy. As Dawes (1994) suggests, in many cases the outcome (recovery) would have happened anyway. Unfortunately, people often omit even to generate this counter-factual question, let alone collect the relevant data.

4 Causal induction from information in contingency tables

Unlike rats, humans use higher symbolic processes such as language, diagrams and mathematics to communicate information. Whereas the associative learning processes discussed above operate on sequentially presented events, inferences about whether causal relations exist can also be made from summary information that is *communicated* (e.g. in the form of a contingency table) rather than from on-line event-by-event learning. For example, we might want to know whether a particular type of person tends to watch early evening TV programmes. To do this, we might divide people into relevant categories (e.g. male vs. female, young vs. old) and then assess their viewing habits to see whether there was any causal relationship between, say, age and watching early evening TV programmes. The result of such an inference process is an induction about a *class* of events, such as *Young people tend to watch early evening TV* or *Protease inhibitors prevent the development of AIDS*. In this section, I review psychological research on this problem and show how biases in contingency judgement can be understood as being not only boundedly rational, but conversationally rational.

Let us begin with the example of a medical researcher who might want to know whether a particular symptom was associated with a disease or not by taking a sample of patients who had the symptom and a sample that did not, to see whether the disease was more present in the first case than the second.

	Disease present	**Disease absent**
Symptom present	37 (Cell A)	33 (Cell B)
Symptom absent	17 (Cell C)	13 (Cell D)

Fig. 2 Contingency table used in Smedslund's (1963) experiment

This information might be compiled into a 2×2 contingency table such as in figure 2.

Smedslund (1963) presented this contingency table to Swedish nurses, 85 percent of the nurses judged that there was indeed a relationship between the symptom and disease. However, a statistical test such as a Chi-square would reveal that there is in fact no statistical relationship between the symptom and the disease. This can be shown by reflecting that the proportion of sick and healthy people is approximately the same (54%) both when the symptom is present and when it is absent. In other words, $a/(a + b) = c/(c + d)$, hence yielding a *Delta-P* of zero.

Another way of arriving at the same conclusion is by counting observations in Cells A and D as evidence for a causal relationship, and observations in Cells B and C as evidence against. This would be consistent with Kelley's (1967, p. 154) co-variational definition of a cause as 'that condition which is present when the effect is present and which is absent when the effect is absent'. If we do this, then we end up with exactly 50 votes 'for' (37 + 13) and 50 'votes' against (33 + 17) the existence of a causal relation. The 'judgement rule' that we are implicitly using in this case is that a cause exists when the sum of Cells A and D $(a + d)$ is significantly greater than the sum of Cells B and C $(b + c)$. We can express this formally as:

A cause is present when $(a + d) > (b + c)$

Why, then, do people make errors on contingency judgement tasks? One possibility is that people use a co-variational rule, but find information in some cells more salient than others, or at least attribute more importance to them. Thus research has suggested that people weight information in Cell A the most and information in Cell D the least when making judgements about causal relationships from contingency tables (Schustack and Sternberg, 1981). We can apply this weighting system informally to Smedslund's example in the following way, by giving Cell A a weight of 3, Cells B and C each a weight of 2 (representing their intermediate importance) and Cell D a weight of 1. Implicitly, this amounts to using the following weighted co-variational judgement rule:

A cause is present when $(3a + d) > (2b + 2c)$

When applied to Smedslund's contingency table, this weighted co-variational rule does in fact yield 124 'votes' in favour of a causal relationship because it yields 124 'votes' for a causal relationship and only 100 'votes' against:

A cause is present because $(3 \times 37 + 13) > (2 \times 17 + 2 \times 33)$,
i.e. $124 > 100$

This judgement process would result in the erroneous judgement that there is indeed a causal relation between the symptom and the disease in the above example, resulting from a tendency to overweight information in Cell A and to underweight information in Cell D.

Many researchers have sought to identify the rules that people actually use when making causal judgements from contingency information. In general, it is considered that people often tend to pay more attention to the occurrence than the non-occurrence of events. This may explain the tendency to weight Cell A more than Cells B and C, which in turn are more weighted than Cell D, when making such contingency judgements. Thus Mandel and Lehman (1998) suggest that the tendency to focus on positive cases (where either the putative cause or effect is present) may lead people to make comparisons between Cells A and B (considering cases where the putative cause is always present to see if the effect always occurs) and Cell A vs. C comparisons (considering cases where the effect is always present to see whether the putative cause is always present, too). Mandel and Lehman term these, respectively, *sufficiency test* and *positivity test* (Klayman and Ha, 1987) strategies. Favouring these strategies would naturally lead people to lend most weight to Cell A information (involved in both favoured comparisons) and to ignore information from Cell D, because they were less likely to make Cell B vs. D or Cell C vs. D comparisons.

Mandel and Lehman (1998) obtained data supporting their analysis that people showed a bias towards using the sufficiency (A vs. B) and positivity (A vs. C) test strategies in causal inference from contingency tables. Interestingly, Mandel and Lehman observed this pattern even in subjects who *said* that Cell D information was important in making causal judgements. This suggests that the processes that people *actually* use when making causal judgements often escape conscious awareness and that their beliefs about how they make causal inferences may not be accurate.

4.1 Conversational and bounded rationality in contingency judgement

The A > B or C > D tendency is often taken as evidence of confirmation bias, as events which occur are more heavily sampled than events which do not occur. However, it is only rational to sample all cells if the tested relation is considered to be *necessary and sufficient*; but Mandel and Lehman (1998) obtained evidence that their subjects tended to consider the causal relation they were asked to test as one of *sufficiency* alone. If subjects made this interpretation, the C and D cells are no longer relevant tests, and only sufficiency tests contrasting cell A to cell B are relevant. A sufficiency interpretation of the conditional will therefore explain the bias towards 'confirmatory' hypothesis testing.

We can also envisage an explanation for why people should shy away from sampling cell D even when the conditional relation to be tested is interpreted as *necessary and sufficient*. McKenzie *et al.* (2000) have shown that when given the option, people have a distinct preference to phrase hypotheses in terms of relations between rare events than between common ones. For example, when presented with a table showing relations between grades and admissions to a university, they are more likely to formulate the hypothesis-to-be-tested as *If the student gets high grades then he will be accepted* if high grades and acceptance are rare. Thus if we treat getting high grades as p and being accepted as q, a tendency to weight *p and q* combinations more than *p and not q* and *not p and q*, which in turn are sampled more than *not p and not q* will necessarily involve preferential sampling of smaller cells. Even when the causal relation is interpreted as *necessary and sufficient* (as seems likely in this example), and thus all cells should be explored, a preference for sampling explicitly mentioned elements will therefore lead the judge to explore cells with fewer members, which is less time-consuming.

McKenzie and Mikkelsen (in press) showed that when asked to test an abstract conditional, people tend to select observations mentioned in the hypothesis. However, when they had either implicit or explicit knowledge suggesting that observations unmentioned in the hypothesis were rare, the tendency to prefer observations mentioned in the hypothesis disappeared. The tendency to focus on mentioned elements when a hypothesis is stated in an abstract but not a 'concrete' form has also been found in studies of 'matching bias' in studies of conditional reasoning (Evans, 1998).

Overall, people do seem to be sensitive to the informational value of rare observations in hypothesis testing (cf. Oaksford and Chater, 1994). Thus if rare events tend to be the focus of conditional assertions, hypothesis testing strategies which focus on explicitly mentioned elements are Bayesian as they will naturally result in sampling of the most informative cells. In this perspective, it is more efficient to focus on the probability of the effect given that the cause is present (given by $a/(a+b)$ ratio), since a high value of a will make a big difference to this probability, as b will generally be small. On the other hand, since most observations will be in the d cell (cause absent and effect absent) the $c/(c+d)$ ratio will be slight whether the causal relation exists or not (cf. Green and Over, 2000).

In the absence of other information about the frequency or rarity of observations, people's default strategy of attending to items mentioned in the hypothesis thus not only follows rules of conversational relevance (cf. Sperber, Cara and Girotto, 1995), but is also efficient from the perspective of optimal data search (Oaksford and Chater, 1995). Such a strategy will thus be both conversationally and boundedly rational, even if it does not meet the requirements of 'full rationality' which would require analysis of information from all four cells (cf. Cheng and Novick, 1992) to evaluate a causal claim of this type. In addition, people may not routinely search all four cells of a contingency table, as other research suggests that cells are differentially searched as a function of the way a causal question is asked. For example, McGill and Klein (1995) report that questions that specify the putative cause of an event (e.g. '*Do you think that something about being female caused Mary to fail at this task?*') tends to trigger evaluations of the necessity of the candidate cause for the effect (i.e. a vs. c comparisons), whereas causal questions that did not specify a particular putative cause for the effect (e.g. '*What do you think caused Mary to fail?*') tended to trigger evaluations of the sufficiency of the causal relation (i.e. a vs. b comparisons). We therefore need to take the idea seriously that conversational relevance assumptions govern information search from contingency tables.

Nevertheless, it is important to note that the tendency to focus only on mentioned aspects of a stated hypothesis can lead to error. A case in point is the discussion of the hypothesis that launching the Challenger space-shuttle in cold weather caused the booster rocket seals to burn through. In this discussion, which took place on the evening before the fatal launch in January 1986, NASA's engineers considered only the eight previous launches of the space-shuttle in cold weather. When they observed that the seals burnt through badly in four cases, but not so in four other cases, they concluded that there was no causal relation ('There's only 50–50 chance of burn-through after a cold weather launch, so what can you conclude from that?'). Had they considered the other sixteen launches in warm weather, and observed that there was burn-through in only one of these cases and no burn-through in the other fifteen cases, then they would surely have accepted the hypothesis that *launching in cold weather causes burn-through*. However, it seems that they truncated their search to test the hypothesis when the sufficiency test appeared inconclusive.

5 Conversational models of causal explanation: when should alternative explanations hurt a focal explanation?

Causal induction of the kind that I have discussed above (a *whether*-question, as to whether a causal relationship exists between two types of event) differs from causal explanation which seeks to explain *why* a particular event happened, or why a general pattern of events exists. Careful attention to the kind of causal question being asked can help us avoid mis-classifying some explanations as irrational. Below I illustrate how a 'conversational' approach to explanation (cf. Hilton, 1990, 1991, 1995a) can help elucidate issues of particular interest to scientific explanation. Paying careful attention to the kind of causal question being asked can help us see why and how a particular explanation should be changed when another candidate is present, rather then seeing the (non-) effect of alternatives on target explanations as being 'capricious' or erratic. Recognizing the conversational point ('aboutness') of explanations thus helps us see orderliness and rationality in what might otherwise seem to be chaos and inconsistency.

5.1 Alternative explanations when explaining particular events or general patterns

Although the creation of universal generalizations is a distinguishing feature of science, science also seeks to explain historically situated events. A good example is the theory of evolution, which is in part composed of generalizations such as *the fittest adaptations are most likely to survive* which help explain in a general way why some species survive and others become extinct, but also, as Koslowski and Thompson (chapter 9 in this volume) show, seeks to explain specific historical events such as the extinction of saurischian dinosaurs. Here, insofar as an event can only have one cause, rival historical hypotheses are exclusive; thus evidence which supports one hypothesis such as *saurischian dinosaurs became extinct because the earth cooled* necessarily should damage a rival hypothesis such as *saurischian dinosaurs became extinct because they evolved into birds*.

We may note that the same is not true of hypothesized universal generalizations. Suppose I am testing two hypotheses about the causes of cancer: *smoking causes cancer* and *working in asbestos factories causes cancer*. Intuitively, collecting support for the one does not seem to damage my beliefs in the other. Here we simply collect evidence for and against hypotheses, and Bayesian updating of beliefs of one hypothesis should not affect our beliefs in a rival hypothesis; in the end, both hypotheses may be considered to be true.

However, suppose that we are trying to explain why a particular individual has contracted cancer. His lawyer may argue that working in an asbestos factory is the cause, thus trying to get his employers to pay compensation. In order to falsify this argument, his employers' lawyers may counter that this worker has smoked for many years, and that this is likely to be the cause of his cancer (Einhorn and Hogarth, 1986). Experimental research does in fact confirm that the activation of a rival hypothesis does indeed lead people to consider the initial hypothesis to be less probably true (Hilton and Erb, 1996). This kind of causal discounting for particular events has often (but not invariably) been observed (Morris and Larrick, 1995; McClure, 1998).

5.2 *Explanatory relevance: discounting vs. backgrounding of explanations*

A final distinction that is of use is that between causal discounting of the kind described above and *causal backgrounding*, where an initial explanation is dropped in favour of an alternative explanation, not because it is considered to be less likely, but because it is considered to be less relevant. For example, suppose that you learn that a hammer strikes a watch, and the watch-face breaks. Most people here assume that the watch broke because the hammer hit it. However, suppose that you learn that this happens as part of a routine testing procedure in a watch factory. In this case, people tend to prefer the explanation that the watch broke because of a fault in the glass (Einhorn and Hogarth, 1986). However, in this case, people still believe that the hammer hit the watch and that this was necessary for the watch to break; however, they now consider the first explanation to be less informative and relevant than the second explanation (Hilton and Erb, 1996). The fault in the glass is an abnormal condition that makes the difference between this watch that breaks and other watches that do not break, and is therefore dignified as the cause.

However, in the case of particular events situated in space and time, causal discounting should occur as alternative explanations are genuine rivals for focal explanations since each event normally has only one causal history. However, *within* a given causal history, certain factors may be backgrounded not because they are disbelieved but because they are instead *presupposed*, becoming part of the causal field (Mackie, 1980) of backgrounded necessary conditions. If anything, even though backgrounded, these factors may be believed *more* strongly as they now form part of a coherent causal story (Hilton and Erb, 1996).

We may therefore conclude that the capacity of people to change their explanation when background information changes does not reflect changes in underlying beliefs about which causal mechanisms are responsible for the production of an event, so much as changes about which aspect of the mechanism people consider to be informative and relevant to mention in an explanation.

Rather than being 'capricious' changes in which causes are selected to mention in an explanation, these changes seem to reflect a pragmatic rationality – some conditions are more interesting than others, and are thus elevated to the rank of 'cause' (McGill, 1989; Hilton, 1990, 1991; Hilton, Mathes and Trabasso, 1992; Slugoski *et al.*, 1993).

6 Cognitive and social sources of bias, and remedies

We are now in a position to diagnose some sources of bias and error in causal decision making by managers and to suggest remedies. Below I consider the cases of associative learning of causal relations, testing of verbally stated hypotheses concerning causal relations and alternative explanations for particular events.

6.1 Associative learning

On-line causal learning is likely to be accurate when the causes and effects are spatially contiguous; otherwise when there are lagged relationships people are very unlikely to be able to detect them (Sterman, 1994). People are also likely to have cognitive load problems when causal induction problems are complex and there are multiple candidate hypotheses. Hilton and Neveu (1996) show that similar populations of management students are better able to solve a multiple-hypothesis causal induction problem if it is presented in verbal summary form than when it is presented in on-line form with lag between causes and effects. This suggests that strategies for improving decision making will be essentially cognitive in nature, having to do with aids for overcoming cognitive limitations.

Biases due to 'hard-wired' automatic processes will be difficult to correct. However, it is worth noting even here that social pressure may lead people to work harder. Slugoski, Sarson and Krank (1991) showed that requiring people to justify their decision led to reduced formation of illusory correlations on a learning task. Another strategy for improving decision making is to present co-variation information in pre-packaged summary form. However, as we saw above, testing verbally stated hypotheses on co-variation or contingency information presented in pre-packaged form runs its own risks.

6.2 The social context of hypothesis testing and contingency judgement

I have argued that confirmation bias in response to verbally stated hypotheses may not be exclusively cognitive in origin. It can be explained in terms of assumptions about conversational relevance, where one assumes that a co-operative and competent interlocutor explicitly mentions what is relevant in the hypothesis. In line with this reasoning, Slugoski and Wilson (1998) found that conversationally skilled people were *more* likely to produce matching bias on Wason's (1966) four-card selection task.

Another consequence of the conversational position is that if one undermines assumptions about the interlocutor's co-operativeness then confirmation bias may be attenuated or disappear. A striking demonstration of such an effect has been found using Wason's (1960) 2-4-6 task by Legrenzi *et al.* (1991), who show that people become much more disconfirmatory when told that the rule they have to test has been generated by a minority rather than a majority source. This result suggests that confirmation bias is in large part due to conformity pressure rather than any intrinsic cognitive bias.

Another line of research indicating the close relationship between social factors and cognitive bias comes from research on overconfidence. Some people are characteristically more confident than others on various trivia tasks (Klayman *et al.*, 1999). While this trait does not correlate with a cognitive variable such as intelligence (Stanovich and West, in press), it does correlate with membership of highly self-contained social networks (Klayman and Burt, 1999). These are just the sorts of social networks which one may expect to be most prone to group-think (Janis, 1971), where members of a decision making group collectively repress disagreeable ideas that threaten the consensus of the group.

Consistent with this analysis, Leyens *et al.* (1999) show that people who ask confirming questions in social interaction are generally more liked than those who ask disconfirming questions. It would therefore be unsurprising that where there is evidence of high group-think as in the Challenger decision process (Marx *et al.*, 1987), there would be a truncated information search in testing a disagreeable hypothesis such as *launching in cold weather causes seal burn-through.* Finally, collectivist societies who value social consensus have been shown to be more conformist on tasks such as the Asch line-judgement task (Bond and Smith, 1996). These societies thus may be less likely to foster the independent critical spirit that many consider a key characteristic of the scientific approach (Faucher *et al.*, chapter 18 in this volume).

At another level, science already seems to be organized in terms of competition between individuals or groups who wish to prove that their hypothesis is the best through falsifying the hypotheses of rival groups; hence individual biases to confirmation may be counteracted through social organization (cf. Evans, chapter 10 in this volume). These seem to capitalize on a basic tendency to valorize one's own group. Further support for this general position has been found by Scaillet and Leyens (2000) who show that people are more likely to seek to falsify hypotheses that have unfavourable implications for their in-group. This line of reasoning suggests that one way of eliminating confirmation bias is through social organization.

6.3 *Causal explanation of particulars: discounting and backgrounding*

I have argued that explanation of particular events has some important differences to causal induction from multiple observations, and has different criteria of adequacy. Thus evidence for an alternative hypothesis in causal induction should not hurt a focal hypothesis where the hypotheses are not exclusive, but should lead to discounting a focal hypothesis where explanation of a particular event is concerned. On the other hand, backgrounding of a cause when the context makes another cause more relevant to mention does not reflect capriciousness but rather the rational operation of conversational relevance.

7 Conclusions

In this chapter, I have reviewed some ways in which scientific and common-sense explanation differ. Scientific explanations typically aim at the induction of causal generalizations connecting types of event, whereas common-sense explanation typically aims at elucidating why an unexpected particular event happened when and how it did. Consequently, common-sense explanation typically has different goals to scientific explanation, and it can be a mistake to apply the same normative criteria to common-sense explanation as to scientific explanation. In particular, common-sense explanation may follow criteria of bounded rationality that do not apply to scientific explanation, where an idealized model of full rationality may be applied.

Recognition of the distinctive rationality of common-sense explanation enables us to see that some criticisms of common-sense explanation follow from the inappropriate application of a normative model, as when Mill criticized the context-sensitive way in which common-sense selects causes from conditions in ordinary conversations as 'capricious'. However, information-processing strategies that might be quite adaptive in terms of bounded or conversational rationality may become maladaptive when on-line causal reasoning problems become complex, or because assumptions about the conversational relevance of communicated information cannot be relied upon. Confirmation bias may in part have social roots in in-group favouritism, and may be eliminated through specifying the source of the hypothesis as being a member of the out-group.

Consequently, some 'cognitive' biases may have 'social' remedies. Already, work on socio-technical systems suggests that firms that use management structures which are adapted to the kinds of environment they operate in are more likely to prosper and survive. For example, Woodward (1965) showed that 'organic' organizations in which jobs are very general, there are few rules and decisions can be made by lower-level employees perform better on small-batch production (satellites, customized dresses) and continuous process production (oil refining and chemical companies), whereas 'mechanistic' organizations in which people perform specialized jobs, many rigid rules are imposed and authority vested in a few top-ranking officials are better for large-batch production (e.g. cars, spare parts). It seems plausible, for example, that organic organizations are more likely to foster creative, critical thinking. An interesting question for science management could be to see whether certain kinds of organization are better for all kinds of science, or whether the success of a kind of organization was contingent on the kind (e.g. physics or biology) or phase (normal vs. extraordinary) of science involved.

Cognitive and social psychologists have begun to analyse the ways in which organizational structures may both attenuate and amplify biases in individual decision making (Tetlock, 1992; Payne, 1997). Given the difficulty of de-biasing using 'cognitive' strategies (Fischhoff, 1982), those who wish to induce 'scientific' thinking may likewise wish to also look more closely at the influence of socio-structural design factors on thinking. Management texts (e.g. Greenberg and Baron, 1997) already recommend heterarchical rather than hierarchical organizations as ways of fostering creativity and innovation in research and development units. Science managers (e.g. research group leaders) may therefore find that social organization is an important tool for facilitating the kind of work they would like to see done.

The interchange between the cognitive and social levels of analysis of scientific thinking is likely to result in a two-way exchange. One the one hand, through emphasizing the social adaptiveness of certain cognitive biases, we can avoid undue condemnation of these biases as 'irrational'. On the other hand, where after due consideration we consider these biases likely to lead to undesirable and indeed dangerous errors, science managers may profit from knowledge of the cognitive and social sources of these errors to design organizational structures that can counteract them.

Part three

Science and motivation

12 The passionate scientist: emotion in scientific cognition

Paul Thagard

This chapter discusses the cognitive contributions that emotions make to scientific inquiry, including the justification as well as the discovery of hypotheses. James Watson's description of how he and Francis Crick discovered the structure of DNA illustrates how positive and negative emotions contribute to scientific thinking. I conclude that emotions are an essential part of scientific cognition.

1 Introduction

Since Plato, most philosophers have drawn a sharp line between reason and emotion, assuming that emotions interfere with rationality and have nothing to contribute to good reasoning. In his dialogue the *Phaedrus*, Plato compared the rational part of the soul to a charioteer who must control his steeds, which correspond to the emotional parts of the soul (Plato, 1961, p. 499). Today, scientists are often taken as the paragons of rationality, and scientific thought is generally assumed to be independent of emotional thinking.

Current research in cognitive science is increasingly challenging the view that emotions and reason are antagonistic to each other, however. Evidence is accumulating in cognitive psychology and neuroscience that emotions and rational thinking are closely intertwined (see, for example: Damasio, 1994; Kahneman, 1999; Panksepp, 1999). My aim in this chapter is to extend that work and describe the role of the emotions in scientific thinking. If even scientific thinking is legitimately emotional, then the traditional division between reason and emotion becomes totally unsupportable.

My chapter begins with a historical case study. In his famous book, *The Double Helix*, James Watson (1969) presented a review of the discovery of the structure of DNA. Unlike the typical dry and scientific biography or autobiography, Watson provided a rich view of the personalities involved in one of the most important discoveries of the twentieth century. I will present a survey and analysis of the emotions mentioned by Watson and use it and quotations from other scientists to explain the role of the emotions in scientific cognition.

I am grateful to Peter Carruthers and Cameron Shelley for comments, and to the Natural Sciences and Engineering Research Council of Canada for financial support.

My account will describe the essential contributions of emotions in all three of the contexts in which scientific work is done: investigation, discovery and justification. Initially, emotions such as curiosity, interest and wonder play a crucial role in the pursuit of scientific ideas. Moreover, when investigation is successful and leads to discoveries, emotions such as excitement and pleasure arise. Even in the third context, justification, emotions are a crucial part of the process of recognizing a theory as one that deserves to be accepted. Good theories are acknowledged for their beauty and elegance, which are aesthetic values that are accompanied by emotional reactions.

2 The discovery of the structure of DNA

In 1951, James Watson was a young American post-doctoral fellow at Cambridge University. He met Francis Crick, a British graduate student who was already in his mid-30s. Both had a strong interest in genetics, and they began to work together to identify the structure of the DNA molecule and to determine its role in the operations of genes. The intellectual history of their work has been thoroughly presented by Olby (1974) and Judson (1979). My concern is with emotional aspects of the thinking of Watson and Crick and of scientific thinking in general; these aspects have been largely ignored by historians, philosophers and even psychologists. The primary question is: What role did emotions play in the thinking of Watson and Crick that led to the discovery of the structure of DNA and the acceptance of their model?

In order to answer this question, I read carefully through *The Double Helix*, looking for words that refer to emotions. Watson described not only the ideas and the hypotheses that were involved in the discovery of the structure of DNA, but also the emotions that accompanied the development of the new ideas. In Watson's short book, whose paperback edition has only 143 pages, I counted a total of 235 'emotion words'. Of the 235 emotion episodes referred to, more than half (125) were attributed by Watson solely to himself. Another 35 were attributed to his collaborator Francis Crick, and 13 emotions were attributed by Watson to both of them. There were also 60 attributions of emotions to other researchers, including various scientists in Cambridge and London. I coded the emotions as having either positive valence (e.g. happiness) or negative valence (e.g. sadness), and found that more than half of the emotions (135) had positive valence. Of course, there is no guarantee that Watson's reports of the emotions of himself and others are historically and psychologically accurate, but they provide a rich set of examples of possible emotional concomitants of scientific thinking.

In order to identify the kinds of emotions mentioned by Watson, I coded the emotion words in terms of what psychologists call *basic* emotions, which are ones taken to be culturally universal among human beings (Ekman, 1992).

The usual list of basic emotions includes happiness, sadness, anger, fear, disgust and sometimes surprise. To cover classes of emotion words that occurred frequently in the book but did not fall under the six basic emotions, I added three additional categories: interest, hope and beauty. Figure 1 displays the frequency with which these categories of emotions appear in Watson's narrative. Happiness was the most frequently mentioned emotion, occurring 65 times with many different words referring to such positive emotional states as excitement, pleasure, joy, fun, delight and relief. The next most frequently mentioned classes of emotions were: interest, with 43 emotion words referring to states such as wonder and enthusiasm; and fear, with 38 emotion words referring to states such as worry and anxiety.

The point of this study, however, was not simply to enumerate the emotion words used by Watson in his story. To identify the role that emotions played in the thinking of Watson and Crick, I coded the emotion words as occurring in three different contexts of the process of inquiry: investigation, discovery and justification. Most scientific work occurs in the context of investigation, when scientists are engaged in the long and often difficult attempt to establish empirical facts and to develop theories that explain them. Much preparatory experimental and theoretical work is usually required before scientists are able to move into the context of discovery, in which new theoretical ideas and important new empirical results are produced. Finally, scientists enter the context of justification, when new hypotheses and empirical results are evaluated with respect to alternative explanations and the entire body of scientific ideas.

The distinction between the contexts of discovery and justification is due to Reichenbach (1938). I use it here not to demarcate the psychological and subjective from the philosophical and rational, as Reichenbach intended, but merely to mark different stages in the process of inquiry, all of which I think are of both psychological and philosophical interest. I further divide Reichenbach's context of discovery into contexts of investigation and discovery, in order to indicate that much work is often required before actual discoveries are made. In scientific practice, the contexts of investigation, discovery and justification often blend into each other, so they are best viewed as rough stages of scientific inquiry rather than as absolutely distinct.

Most of Watson's emotion words (163) occurred in the context of investigation. 15 words occurred in the context of discovery, 29 in the context of justification and 28 words occurred in other more personal contexts that had nothing to do with the development of scientific ideas. In the rest of this chapter, I will provide a much more detailed account of how the many different kinds of emotion contribute to scientific thinking in the three contexts. I will offer not only a correlational account of what emotions tend to occur in what contexts, but also a causal account of how emotions produce and are produced by the cognitive operations that occur in the different stages of scientific inquiry.

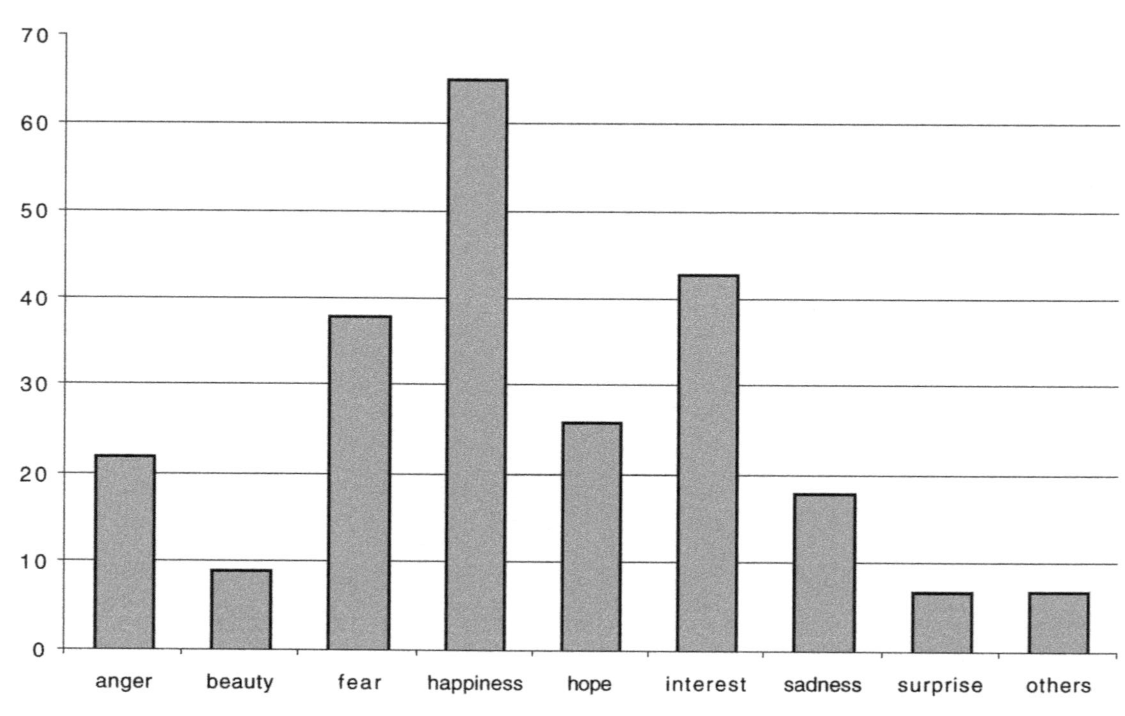

Fig. 1 Frequency of classes of emotion words in Watson (1969)

3 Emotions in the context of investigation

Discussions of scientific rationality usually address questions in the context of justification, such as when is it rational to replace a theory by a competing one. But scientists engage in a great deal of decision making that precedes questions of justification. Students who are interested in pursuing a scientific career must make decisions that answer questions such as the following: What science (e.g. physics or biology) should I concentrate on? What area of the science (e.g. high-energy physics or molecular biology) should I focus on? What particular research topics should I pursue? Which questions should I attempt to answer? What methods and tools should I use in attempting to answer those questions?

On the traditional view of rational decision making familiar from economics, scientists should attempt to answer these questions by calculating how to maximize their expected utility. This involves taking into account scientific goals such as truth and understanding, and possibly also taking into account personal goals such as fame and financial gain, as well as considering the probabilities that particular courses of research will lead to the satisfaction of these goals. Some philosophers (e.g. Kitcher, 1993; Goldman, 1999) describe scientists as choosing projects based on maximizing epistemic goals such as truth.

I would like to propose an alternative view of scientific decision making in the context of investigation. Decisions concerning what topics to research are much more frequently based on emotions than on rational calculations. It is rarely possible for scientists to predict with any accuracy what choice of research areas, topics and questions will pay off with respect to understanding, truth, or personal success. The range of possible actions is usually ill specified, and the probabilities of success of different strategies are rarely known. Because rational calculation of maximal utility is effectively impossible, it is appropriate that scientists rely on cognitive emotions such as interest and curiosity to shape the direction of their inquiries. See Thagard (2001) for a model of decision making as informed intuition based on emotional coherence.

Watson's narrative makes it clear that he and Crick were heavily motivated by interest. Watson left Copenhagen where his post-doctoral fellowship was supposed to be held because he found the research being done there boring: the biochemist he was working with there 'did not stimulate me in the slightest' (Watson, 1969, p. 23). In contrast, he reacted to questions concerning the physical structure of biologically important molecules such as DNA with emotions such as excitement: 'It was Wilkins who first excited me about X-ray work on DNA' (p. 22); 'Suddenly I was excited about chemistry' (p. 28). Crick similarly remarks how he and Watson 'passionately wanted to know the details of the structure' (Crick, 1988, p. 70). Much later, Watson (2000, p. 125) stated as one of his rules for succeeding in science: 'Never do anything that bores you.'

Once interest and curiosity direct scientists to pursue answers to particular questions, other emotions such as happiness and hope can help motivate them to perform the often laborious investigations that are required to produce results. It was clearly important to Watson and Crick that they often became excited that they were on the right track. Watson (1969, p. 99) wrote that, 'On a few walks our enthusiasm would build up to the point that we fiddled with the models when we got back to our office.' Both took delight in getting glimpses of what the structure of DNA might be. They were strongly motivated by the hope that they might make a major discovery. Hope is more than a belief that an event might happen – it is also the emotional desire and joyful anticipation that it will happen. I counted 26 occurrences where Watson mentioned the hopes of himself or others for scientific advance.

In addition to positive emotions such as interest and happiness, scientists are also influenced by negative emotions such as sadness, fear and anger. Sadness enters the context of investigation when research projects do not work out as expected. Watson and Crick experienced emotions such as dismay when their work faltered. Such emotions need not be entirely negative in their effects, however, because sadness about the failure of one course of action can motivate a scientist to pursue an alternative and ultimately more successful line of research.

Fear can also be a motivating emotion. Watson and Crick were very worried that the eminent chemist Linus Pauling would discover the structure of DNA before they did, and they also feared that the London researchers, Rosalind Franklin and Maurice Wilkins, would beat them. Watson wrote that when he heard that Pauling had proposed a structure, 'my stomach sank in apprehension at learning that all was lost' (p. 102). Other worries and anxieties arose from setbacks experienced by Watson and Crick themselves. Watson was initially very excited by a proposal that was shown by a crystallographer to be unworkable, but tried to salvage his hypothesis: 'Thoroughly worried, I went back to my desk hoping to salvage the like-with-like idea' (p. 122).

The other negative emotion that Watson mentions frequently in the context of investigation is anger. To himself, Watson ascribes only weak forms of anger, such as annoyance and frustration, but he describes Crick as experiencing fury and outrage when his senior professors wrote a paper that failed to acknowledge his contributions. According to Oatley (1992), people experience anger when accomplishment of their goals is blocked by people or events. Most of the anger-related episodes mentioned by Watson are directed at people, but some concern facts, as when Watson and Crick both become annoyed at the complexity of DNA bonds. I do not get the impression from Watson's chronicle that anger was ever a motivating force in the scientific work he described, but we can see it as an effect of undesirable interactions with other people and the world.

A general model of the role of emotions in the context of investigation of scientific investigations is shown in figure 2. Interest, wonder, curiosity and

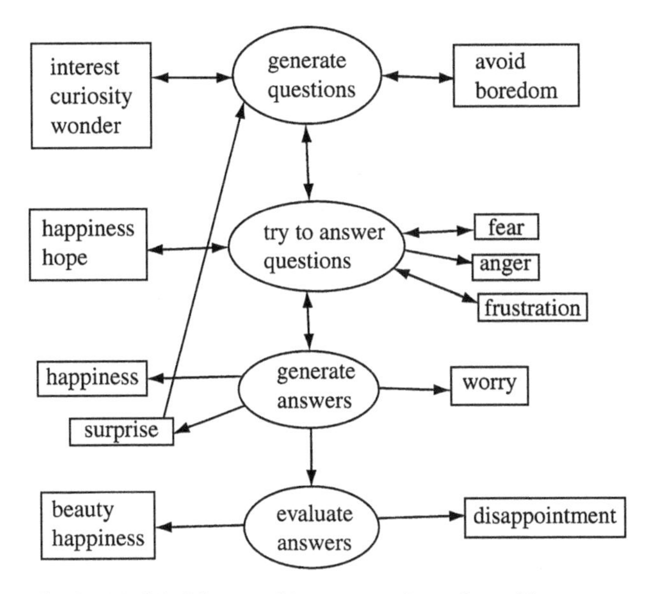

Fig. 2 Model of the causal interconnections of cognitive processes of inquiry and emotional states
Note: Arrows indicate casual influences.

the avoidance of boredom are key inputs to the process of selecting scientific questions to investigate. I have made the causal arrows between emotions and the process of question generation bi-directional in order to indicate that emotions are outputs as well as inputs to the generation of questions. Coming up with a good question can increase curiosity and interest, and produce happiness as well. Once questions have been generated, the cognitive processes involved in trying to generate answers to them can also interact with emotions such as interest and happiness, as well as with negative emotions such as fear. For convenience, I have organized positive emotions on the left side of figure 2, and negative emotions on the right side. The causal arrows connecting processes of question generation, question answering and answer evaluation are also bi-directional, indicating inter-connections rather than a linear operation. For example, the attempt to answer a question can generate subordinate questions whose answers are relevant to the original question. For a discussion of the process of question generation in science, see Thagard (1999).

The emotions that Watson attributed to himself and his colleagues are commonly found in other scientists, as evident in a series of interviews conducted by the biologist Lewis Wolpert for BBC Radio (Wolpert and Richards, 1997). Wolpert elicited frank descriptions of leading researchers' intense emotional involvement in their work. For example, the distinguished biologist Gerald

Edelman stated: 'Curiosity drives me. I believe that there is a group of scientists who are kind of voyeurs, they have the splendid feeling, almost a lustful feeling, of excitement when a secret of nature is revealed . . . and I would certainly place myself in that group' (Wolpert and Richards, 1997, p. 137). Similarly, the eminent physicist Carlo Rubbia said of scientists: 'We're driven by an impulse which is one of curiosity, which is one of the basic instincts that a man has. So we are essentially driven not by . . . how can I say . . . not by the success, but by a sort of passion, namely the desire of understanding better, to possess, if you like, a bigger part of the truth' (Wolpert and Richards, 1997, p. 197).

According to Kubovy (1999, p. 147), to be curious is to get pleasure from learning something that you did not previously know. He contends that curiosity has its roots in animal behaviour, having evolved from the need to search for food. Many mammals prefer richer environments over less complex ones. Humans can be curious about a very wide range of subjects, from the trivial to the sublime, and scientists direct their intense drive to learn things that are unknown not just to them, but to people in general. Loewenstein (1994) defends a more cognitive account of curiosity as the noticing of knowledge gaps, but underplays the emotional side of curiosity. There are many knowledge gaps that scientists may notice, but only some of them arouse an emotional interest that spurs them to the efforts required to generate answers.

4 Emotions in the context of discovery

The discovery of the structure of DNA occurred after Maurice Wilkins showed Watson new X-ray pictures that Rosalind Franklin had taken of a three-dimensional form of DNA. Watson reports: 'The instant I saw the picture my mouth fell open and my pulse began to race' (p. 107). This picture moved Watson from the context of investigation into the context of discovery, in which a plausible hypothesis concerning the structure of DNA could be generated. While drawing, Watson got the idea that each DNA molecule might consist of two chains, and he became very excited about the possibility and its biological implications. Here is a passage that describes his resulting mental state; I have highlighted in boldface the positive emotion words and in italics the negative emotion words.

> As the clock went past midnight I was becoming more and more **pleased**. There had been far too many days when Francis and I *worried* that DNA structure might turn out to be superficially very *dull*, suggesting nothing about either its replication or its function in controlling cell biochemistry. But now, to my **delight** and **amazement**, the answer was turning out to be profoundly **interesting**. For over two hours I **happily** lay awake with pairs of adenine residues whirling in front of my closed eyes. Only for brief moments did the *fear* shoot through me that an idea this good could be wrong. (Watson, 1969, p. 118)

Watson's initial idea about the chemical structure of DNA turned out to be wrong, but it put him on the track that quickly led to the final model that Watson and Crick published.

Most of the emotions mentioned by Watson in his discussion of discoveries fall under the basic emotion of happiness; these include excitement, pleasure and delight. The chemist who helped to invent the birth control pill, Carl Djerassi, compares making discoveries to sexual pleasure. 'I'm absolutely convinced that the pleasure of a real scientific insight – it doesn't have to be a great discovery – is like an orgasm' (Wolpert and Richards, 1997, p. 12). Gopnik (1998) also compares explanations to orgasms, hypothesizing that explanation is to cognition as orgasm is to reproduction.

The pleasure of making discoveries is effusively described by Gerard Edelman:

> After all, if you been filling in the tedium of everyday existence by blundering around a lab, for a long time, and wondering how you're going to get the answer, and then something really glorious happens that you couldn't possibly have thought of, that has to be some kind of remarkable pleasure. In the sense that it's a surprise, but it's not too threatening, it is a pleasure in the same sense that you can make a baby laugh when you bring an object out of nowhere . . . Breaking through, getting various insights is certainly one of the most beautiful aspects of scientific life. (Wolpert and Richards, 1997, p. 137)

Carlo Rubbia reports that 'The act of discovery, the act of being confronted with a new phenomenon, is a very passionate and very exciting moment in everyone's life. It's the reward for many, many years of effort and, also, of failures' (Wolpert and Richards, 1997, p. 197).

François Jacob (1988, pp. 196–7) describes his first taste of the joy of discovery as follows: 'I had seen myself launched into research. Into discovery. And, above all, I had grasped the process. I had tasted the pleasure.' Later, when Jacob was developing the ideas about the genetic regulatory mechanisms in the synthesis of proteins that later won him a Nobel Prize, he had an even more intense emotional reaction: 'These hypotheses, still rough, still vaguely outlined, poorly formulated, stir within me. Barely have I emerged than I feel invaded by an intense joy, a savage pleasure. A sense of strength, as well, of power' (Jacob, 1988, p. 298). Still later, Jacob describes the great joy that came with experimental confirmation of his hypotheses. Scheffler (1991, p. 10) discusses the joy of verification, when scientists take pleasure when their predictions are found to be true. Of course, predictions sometimes turn out to be false, producing disappointment and even gloom.

It is evident from Watson and the other scientists that I have quoted that discovery can be an intensely pleasurable experience. In figure 2, I have shown surprise and happiness as both arising from the successful generation of answers to pursued questions. The prospect of being able to experience such emotions is one of the prime motivating factors behind scientific efforts. Richard Feynman proclaimed that his work was not motivated by a desire for fame or prizes such as the Nobel that he eventually received, but by the joy of discovery: 'The prize is the pleasure of finding a thing out, the kick of the discovery, the observation that other people use it [my work] – those are the real things, the others are unreal to me' (Feynman, 1999, p. 12). Discoveries are usually pleasant surprises, but unpleasant surprises also occur, for example when experiments yield data that are contrary to expectations. Disappointment and sadness can also arise when the evaluation stage leads to the conclusion that one's preferred answers are inferior or inadequate, as shown in figure 2.

Kubovy (1999) discusses virtuosity, the pleasure we have when we are doing something well. He thinks that humans and also animals such as monkeys and dolphins enjoy working and take pleasure in mastering new skills. Scientists can achieve virtuosity in many diverse tasks such as designing experiments, interpreting their results, and developing plausible theories to explain the experimental results. According to physicist Murray Gell-mann, 'Understanding things, seeing connections, finding explanations, discovering beautiful, simple principles that work is very, very satisfying' (Wolpert and Richards, 1997, p. 165).

5 Emotions in the context of justification

Although many would concede that the processes of investigation and discovery are substantially emotional, it is much more radical to suggest that even the context of justification has a strong emotional component. In Watson's writings, the main sign of an emotional component to justification is found in his frequent description of the elegance and beauty of the model of DNA that he and Crick produced. Watson wrote:

> We only wished to establish that at least one specific two-chain complementary helix was stereochemically possible. Until this was clear, the objection could be raised that, although our idea was aesthetically elegant, the shape of the sugar-phosphate backbone might not permit its existence. Happily, now we knew that this was not true, and so we had lunch, telling each other that a structure this pretty just had to exist. (1969, p. 131)

Thus one of the causes of the conviction that they had the right structure was its aesthetic, emotional appeal. Other scientists also had a strong emotional reaction to the new model of DNA. Jacob (1988, p. 271) wrote of the Watson and Crick model: 'This structure was of such simplicity, such perfection, such harmony, such beauty even, and biological advances flowed from it with such rigor and clarity, that one could not believe it untrue.'

In a similar vein, the distinguished microbiologist Leroy Hood described how he takes pleasure in coming up with elegant theories:

> Well, I think it's a part of my natural enthusiasm for everything, but what I've been impressed with in science over my twenty-one years is the absolute conflict between, on the one hand, as we come to learn more and more about particular biological systems there is a simple elegant beauty to the underlying principles, yet when you look into the details it's complex, it's bewildering, it's kind of overwhelming, and I think of beauty in the sense of being able to extract the fundamental elegant principles from the bewildering array of confusing details, and I've felt I was good at doing that, I enjoy doing that. (Wolpert and Richards, 1997, p. 44)

Many other scientists have identified beauty and elegance as distinguishing marks of theories that should be accepted (McAllister, 1996).

From the perspective of traditional philosophy of science, or even the perspective of traditional cognitive psychology that separates the cognitive and the emotional, the fact that scientists find some theories emotionally appealing is irrelevant to their justification. My own previous work is no exception: I have defended the view that scientific theories are accepted or rejected on the basis of their explanatory coherence with empirical data and other theories (Thagard, 1992, 1999). But my new theory of emotional coherence shows how cognitive coherence judgements can generate emotional judgements (Thagard, 2000, chapter 6). I will now briefly review the theory of emotional coherence and indicate how it can be applied to explain emotional scientific judgements concerning the acceptance of theories.

My coherence theory of inference can be summarized in the following theses (Thagard and Shelley, 2001; see also Thagard, 2000):

1. All inference is coherence-based. So-called rules of inference such as *modus ponens* do not by themselves license inferences, because their conclusions may contradict other accepted information. The only rule of inference is: Accept a conclusion if its acceptance maximizes coherence.
2. Coherence is a matter of constraint satisfaction, and can be computed by connectionist and other algorithms.
3. There are six kinds of coherence: analogical, conceptual, explanatory, deductive, perceptual and deliberative.
4. Coherence is not just a matter of accepting or rejecting a conclusion, but can also involve attaching a positive or negative emotional assessment to a proposition, object, concept, or other representation.

On this account, a theory is justified if inferring it maximizes coherence, but assessment can also involve an emotional judgement (see also Hookway, chapter 13 in this volume). Theories consists of hypotheses which are comprised of concepts. According to the theory of emotional coherence, these representations not only have a cognitive status of being accepted or rejected, they also have an emotional status of being liked or disliked. In keeping with Bower's (1981) account of the psychology of emotions, I use the term *valence* for the emotional status of a representation. A representation receives its valence as the result of its connections with other representations. The valence of a theory will flow from the valences of the hypotheses that constitute it, as well as from the overall coherence that it generates.

Overall coherence requires a judgement about how everything fits together. Such judgements can be made by the computational model HOTCO ('hot coherence'), which not only simulates the spread of valences among representations, but also simulates how the coherence of the representations with each other can generate a 'meta-coherence' inference which is associated with happiness. I will not repeat the computational details (Thagard, 2000), but merely want here to give the general flavour of how the model works. A proposition is highly coherent with other propositions if its acceptance helps to maximize the satisfaction of constraints, such as the constraint that if a hypothesis explains a piece of evidence, then the hypothesis and the evidence should be either accepted together or rejected together. If a group of propositions can be accepted together in such a way that the individual propositions each tend to satisfy a high proportion of the constraints on them, the overall system of proposition gets a high meta-coherence rating.

Figure 3 shows how judgements of coherence and other emotions can arise. Units are artificial neurons that represent propositions, and their assessment by neural network algorithms for maximizing constraint satisfaction leads to them being either accepted or rejected. In addition, the extent to which the constraints on a unit are satisfied affects how much it activates a unit representing

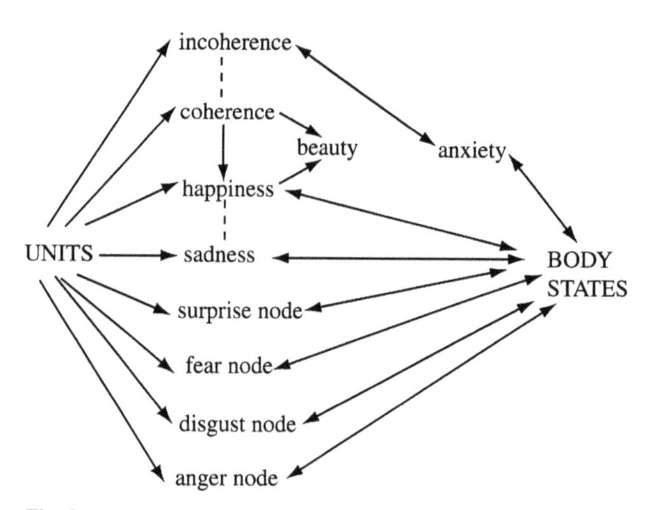

Fig. 3 Interconnections of cognitive units (neurons or neuronal groups) with emotions and bodily states
Source: Reprinted from Thagard (2000), by permission of MIT Press.

a judgement of coherence, which produces happiness. For example, if the acceptance of a hypothesis makes possible satisfaction of numerous constraints that tie the hypotheses with the evidence it explains, then a unit representing the hypothesis will strongly activate the coherence node and thence the happiness node. At the other end of the emotional distribution, incoherence tends to produce unhappiness or even fear-related emotions such as anxiety.

This sketch of the theory and computational model of emotional coherence shows how it is possible for a mind simultaneously to make cognitive judgements and emotional reactions. Watson and Crick's hypothesis concerning the structure of DNA was highly coherent with the available evidence and with the goals of biologists to understand the genetic basis of life. This high coherence generated not only a judgement that the hypothesis should be accepted, but also an aesthetic, emotional attitude that the hypothesis was beautiful. Coherence made Watson, Crick, Jacob and others very happy.

6 Conceptual change as emotional change

Discovering the structure of DNA was a major scientific breakthrough, but it was not a major case of conceptual change. The discovery brought about additions to the concepts of a gene and DNA, but did not require revisions of previous components of the concepts. New hypotheses were adopted, but there was no rejection of previously believed hypotheses. In contrast, scientific revolutions involve both major alteration of concepts and revision of previously held hypotheses (Thagard, 1992). I will now argue that conceptual changes also often involve emotional change.

Consider, for example, Darwin's theory of evolution by natural selection. When he proposed it in the *Origin of Species* in 1859, some scientists such as Huxley quickly realized the strength of the theory and adopted it. But others continued to maintain the traditional view that species arose by Divine creation, and viewed Darwin's theory as not only false but repugnant. Even today, there are creationists who view the theory of evolution as not only false but evil, since they see it as undermining religious beliefs and values. Most biologists, in contrast, feel that Darwin's theory is profound and elegant. Thus for a creationist to become a Darwinian, or for a Darwinian to become a creationist, thinkers must undergo emotional as well as cognitive changes. In addition to rejecting some hypotheses already held and accepting new ones, thinkers who are changing theories are also changing their emotional attitudes towards propositions, concepts and even people: there are different emotions associated with Darwin-as-scientific-hero and Darwin-as-heretic.

Consider also the more recent historical case of the initial rejection and eventual acceptance of the bacterial theory of ulcers (Thagard, 1999). When Marshall and Warren (1984) proposed that most peptic ulcers are caused by the newly discovered bacterium now known as *Helicobacter pylori*, their hypothesis was initially viewed as not only implausible, but even crazy! Marshall's persistent advocacy of the unpopular hypothesis led people to view him as brash and irresponsible. But annoyance turned to respect as evidence mounted that eradicating the bacteria brought many people a permanent cure of their ulcers. Not only did attitudes change toward Marshall but also towards the bacterial theory of ulcer and towards the concept of ulcer-causing bacteria. What was ridiculed became appreciated as an important part of medical theory and practice.

I propose that these changes come about through the process of achieving emotional coherence. At the same time that decisions concerning what to believe are made by a process that maximizes explanatory coherence (fitting hypotheses with data and each other), attitudes are formed by a process that spreads valences through the representational system in accord with existing valences and emotional constraints. For example, scientists who come to appreciate the explanatory power of Darwin's evolutionary hypotheses will come to believe them and to attach a positive emotional valence to them, to the concept of evolution, and to Darwin. On the other hand, scientists (if there are any) who adopt religious fundamentalism and decide that the Bible is to be revered will reject the hypothesis and concept of evolution and view it as odious. Our beliefs are revised by a holistic process of deciding what fits with what, and our emotions are similarly adjusted.

If conceptual change in science is often emotional change, what about science education? Often, science education will not require emotional change because students do not have any emotional attachments to pre-scientific ideas. For example, learning to use Newton's laws of motion to understand the behaviour of projectiles does not seem to require abandoning any emotional attachment on the part of the students to Aristotelian ideas. But learning in biology and the social sciences may involve much more emotional adjustment. Biology students learning about genetics and evolution may have to abandon beliefs that they not only hold but value about the nature of the universe. Similarly, social and political beliefs may be factual, but they are closely intertwined with strongly emotional values. A micro-economist who is convinced that *laissez-faire* capitalism is the best economic system differs in beliefs and also emphatically in values from a macro-economist who favours socialism. Hence educators should be alert for cases where resistance to conceptual change derives from emotional as well as cognitive factors. Normally, emotional change is more the province of psychotherapists than science teachers, but teaching emotionally laden scientific issues such as the theory of evolution may require therapeutic techniques to identify valences as well as beliefs. Appreciation of radically different conceptual frameworks, for example traditional Chinese medicine from the perspective of western medicine, may require emotional as well as conceptual flexibility (Thagard and Zhu, forthcoming).

7 Should scientists be emotional?

But should it not be the job of science educators to help students aspire to what scientists are supposed to be: objective and non-emotional reasoners? Even if emotions are associated with scientific thinking, perhaps the aim of scientific education, from grade school through the PhD, should be to inculcate in students the necessity of eradicating subjective emotions in order to improve their scientific thinking by insulating it from bias. It is not hard to find cases where emotions have distorted science, such as Nazis rejecting Einstein's theory of relativity because they hated Jews, and researchers fabricating data because they wanted to be famous.

While recognizing that there are many such cases in which emotions have undermined scientific objectivity and truth seeking, I think it would be a mistake to demand that scientists strive to eliminate emotions from the cognitive processes that enable them to do science. First, severing the cognitive from the emotional is probably impossible. As I mentioned in the introduction, there is abundant evidence from psychology and neuroscience that cognitive and emotional processes are intimately intertwined. The ethical principle that *ought implies can* applies: we cannot insist that a person's thinking should be emotion-free when it is biologically impossible for people to think that way. Mitroff (1974, p. 99) found in his study of Apollo moon scientists that they held strongly emotional views, especially of other researchers, but concluded that 'the existence of such strong emotions does not prevent scientists from doing excellent scientific work'.

Second, even if scientists could become free of the sway of emotions, it is likely that science would suffer as a result. The discussion above of emotions in the contexts of investigation and discovery showed that positive emotions such as happiness and even negative emotions such as anxiety play an important role in motivating, energizing and directing science. Given that science is often difficult and frustrating, why would anyone do it if it were not also sometimes exciting and satisfying? According to Mitroff (1974, p. 248): 'To eliminate strong emotions and intense commitments may be to eliminate some of science's most vital sustaining forces.' Moreover, emotions help to focus scientific research on what is important. According to Polanyi (1958, p. 135): 'Any process of inquiry unguided by intellectual passions would inevitably spread out into a desert of trivialities.' Because major discoveries require breaking the rules that govern mundane research, 'originality must be passionate' (1958, p. 143).

Even in the context of justification, there is a useful cognitive contribution of feelings of elegance and beauty, because they are attached to theories that excel in explanatory coherence which, I would argue, is what scientific theory is intended most immediately to maximize. An interesting open question is how the training of scientists might be expanded to include enhancement of the kinds of emotional responses that foster good scientific research. I suspect that most such training occurs implicitly during the interactions between budding researchers and their mentors, through a kind of contagion of taste and enthusiasm. It is possible, however, that scientists individually possess sets of inherited and learned emotional dispositions that are impervious to emotional training.

In conclusion, I would emphatically not urge that scientists aim to become robotically unemotional. Despite dramatic increases in computer speed and intelligent algorithms, computers currently are no threat to replace scientists. A key aspect of intelligence that is rarely recognized is the ability to set new goals, not just as ways of accomplishing other existing goals, but as novel ends in themselves. There is no inkling in the field of artificial intelligence of how computers could acquire the ability to set goals for themselves, which is probably just as well because there is no guarantee that the goals of such artificially ultra-intelligent computers would coincide with the goals of human beings. In humans, the development of non-instrumental goals is intimately tied with emotions such as interest, wonder and curiosity, which are just the goals that spur scientists on. As the eminent neurologist Santiago Ramón y Cajal (1999, p. 7) asserted: 'All outstanding work, in art as well as in science, results from immense zeal applied to a great idea.' A scientist without passion would be at best a mediocre scientist.

13 Emotions and epistemic evaluations

Christopher Hookway

Emotional states make a positive contribution to the epistemic evaluations we use in carrying out everyday inquiries and scientific investigations. After sketching a role for them in regulating our beliefs through expressing unarticulated values, the chapter suggests that this illuminates the roles of doubts in spurring inquiries, and of dogmatism in science (section 3). These emotional evaluations possess a kind of *immediacy* although they are often grounded in a mass of unstated experience and background belief (section 4). One important role for such immediate evaluations is in telling us when to *stop* our investigations and remain satisfied with the evidence we possess (section 5). Section 6 considers de Sousa's (1987) suggestion that emotions can serve this role by rendering items of information salient or relevant, enabling us to treat them as reasons while ignoring much else. The final section considers some special roles that emotional evaluations might have in the conduct of co-operative scientific inquiry.

1 Introduction

Epistemology (the study of epistemic evaluation) generally exploits often unstated assumptions about the kinds of representations and psychological mechanisms involved in cognition: we won't understand the nature of epistemic norms unless we can say something about how they are realized – about the cognitive processes that they regulate and the ways in which they engage with them. Since science is a co-operative activity which exploits the judgements and inquiries of individual agents, it is plausible that the same holds for the philosophy of science: we need an account of the kinds of representations and psychological mechanisms that are exploited in co-operative scientific inquiry. In the past, it may have been taken for granted that we need to attend only to explicit beliefs and hypotheses, and to inferences and the logical principles that

The discussion of my paper at the Hang Seng conference was extremely helpful. Although the finished chapter has benefited from it, I still have much to learn from reflecting upon it. I particularly recall comments from: Richard Boyd, Peter Carruthers, Peter Lipton, David Over, Stephen Stich and Peter Todd. I have also benefited from correspondence with Antonia Barke which introduced the example of obsessive cognitive behaviour, and comments on the final draft from Jennifer Saul.

govern them. Over the last few decades, the complexity and variety of processes involved in epistemic evaluation has become more apparent. The present chapter explores one aspect of this complexity: how far do our emotional responses have a fundamental role in our practices of epistemic evaluation?

It hardly needs argument that emotions and other affective states will often provide an *obstacle* to cognition: wishful thinking, excessive pride or ambition, fear of seeing favoured theories refuted, can all interfere with the rational assessment of evidence. My concern here is with a rather different question: how far can emotions contribute to, or, indeed, be necessary for, *successful* cognitive and scientific activity?

As already suggested, these questions can be raised in two sorts of context. First – a topic for general epistemology – we can ask about the role of emotional responses in ordinary individual inquiries, even those concerned with everyday matters of fact. Alongside their importance for understanding different forms of irrationality (Hookway, 2001), three different strengths of positive answer could be found. Perhaps appropriate emotional responses are required if we are to be able to conduct inquiries into what is the case at all: without such responses, inquiry is simply impossible. Alternatively, we might find that such responses are required if we are to be able to inquire well. And finally, even if we can inquire reasonably well without such responses, we might still conclude that they contribute to our inquiring *better* than we otherwise would.

The second context – scientific inquiry – raises a special problem because it is a public, co-operative kind of activity; one which is bound up with a variety of institutions and is concerned with making progress in the long run rather than with arriving at usable truths here and now. Do these special features introduce additional reasons to take account of the emotional responses of individuals?

It is evident that emotions may enter into these different sorts of cognitive activity in different ways. For example, emotional ties may be required to hold a community of scientific inquirers together, making it possible for them to co-operate, even if emotions do not serve as vehicles of epistemic evaluations. A number of scholars, for example Alison Barnes and Paul Thagard, have provided illuminating discussions of the first of these issues (1997, and Thagard's chapter 12 in this volume). Thagard's observations about the role of 'emotion words' in scientists' description of their activities, and the role of curiosity, wonder, surprise, the avoidance of boredom, fear, anger and frustration in explaining why scientists adopt the projects that they do and pursue their investigations as they do are fascinating and, I hope, complement the present discussion. I am more concerned with how allowing a role for emotions in cognition need not be in tension with more traditional claims about epistemic justification.

Although epistemologists and philosophers of science rarely insist that epistemic evaluations involve our emotional or affective responses, those interested in moral and practical deliberation are open to the suggestion that emotions can make a positive contribution to our rationality. It may be held, for example, that success in such deliberations depends upon the possession of (for example) benevolent or sympathetic sentiments; or it is argued that acting well requires *reverence* for the moral law. The following section makes the suggestion that the cases are parallel: once we take note of the contributions that emotions and sentiments make to practical deliberation, we will be in a position to see that they are required to occupy similar roles in theoretical deliberation. The remainder of the chapter explores some examples of how our emotional or affective responses can contribute to the search for knowledge. The current discussion is programmatic. It offers reasons for taking seriously the idea that our affective responses play an important role in our practice of epistemic evaluation and points to some directions for further research. (Compare Hookway, 1993, 1999, 2001.)

2 Comparisons

Why is it supposed that emotions (and other affective states) are important for practical and moral deliberation? It is plausible that many of the deepest *evaluations* are reflected in our emotional responses: in what we fear, resent, are outraged by, sympathise with, and so on. Indeed, there is an especially close connection between emotions and evaluations. Moreover, these evaluations are often tacit. They involve standards that we have *not articulated*: their influence need not depend upon our using expressions of them as premises in conscious deliberate reasoning. Indeed it can be difficult, or even impossible, to achieve reflective clarity about just what these standards are: doing so will often involve thought experiments that try to reveal patterns in our habitual or emotional evaluations of people and their actions. We can formulate our evaluative standards only as a result of a search for an explanation of our habits of evaluation. Thus the first role for emotional responses in deliberation: they can serve as vehicles for unarticulated evaluations.

To identify a second role, we should recall that an adequate account of practical reasoning must explain how our deliberations can *motivate* us to act and, ultimately, how our evaluations can motivate us to act. Beliefs, it is supposed, will not motivate us to act without an appropriate desire (for example). Emotions, being affective states, have the required effects upon conduct: they are Janus-faced, both embodying evaluations and providing motivational force. Appeal to the affective character of states such as emotions may also help with another problem. It can explain how our evaluations *spread*, of how outrage at someone's actions can lead to further suspicion of her other actions, and so on. This does not always occur through any process of reflective deliberation or inference. If we depended on reflection for this onward transmission of our evaluations, it seems, we would be bound to fail to bring our evaluations into harmony.

To summarize: evaluations which are manifested in our emotional responses can reflect unarticulated standards of evaluation, can explain how our evaluations lead us to act in accordance with them and can explain how the influence of our evaluations can spread through our systems of beliefs and attitudes in appropriate ways, in spite of the fact that their influence is not subject to reflective rational control.

It seems plausible that an account of the norms that govern theoretical inquiry and deliberation should have a similar structure. This, too, is concerned with a study of part of our *evaluative* practice, with when we are justified in our beliefs and with how we ought to conduct our deliberations. In many areas (e.g. induction or inference to the best explanation), these standards are not *articulated*: as Quine has insisted, epistemic reflection is generally very shallow and non-reflective, the goodness of an inductive argument (or an explanation) is more easily *felt* than articulated, we generally cannot *say* what makes such an argument a good one (1960, p. 23). We might *say* that the inference seems simple – but what that amounts to is explained by saying that it is inductively good, rather than being something that can itself be used to explain and justify our inductive practice. Quine claims, plausibly, that a hypothesis is 'simple' because it provides the best explanation of the evidence; we cannot say that it provides a good explanation because it is simple. Epistemic evaluations must have '*motivational*' force too: if I evaluate a proposition as poorly supported by evidence, I must be motivated (when appropriate) to reconsider other beliefs that depend upon it, or to search for more evidence when it becomes important for my inquiries. And epistemic evaluations have to *spread*, too: we become suspicious of the further testimony of someone who has misled us once; and even if we frequently find it difficult to do so, we ought to question beliefs that have been inferred from others that have subsequently been doubted.

In the light of these parallels, why should anyone doubt the relevance of emotions to epistemic evaluation? There is an important difference. Theoretical inquiry and practical deliberation are both problem-solving activities, and the norms we follow in carrying them out are internal to these activities of solving problems – in the one case, problems of what is the case, in the other, problems about what to do. One reason for taking emotions seriously in the practical case is that it could provide a (Humean) account of how we recognize our ethical and practical *ends*. Our fundamental ends, interests and values are reflected in our emotions and sentiments. In the epistemic case, the position is different. Although we may seek a (practical) explanation of which questions interest us on any particular occasion; it can *seem* unnecessary to seek an emotive explanation of the standards we want our conclusions to meet, of the fundamental norms which support our epistemic evaluations of beliefs and inquiries. We want our conclusions to be true; we want them to be justified in the light of the evidence: neither of these properties seems to be grounded in desires or emotions. This means that even if emotional responses serve as the *vehicles* of our evaluations, as the form taken by our evaluation of some belief as justified or as unjustified, they are irrelevant to the *content* and to the *authority* of these evaluations. As someone sympathetic to pragmatism, this does not convince me – nor am I convinced that the point would be decisive even if correct. Indeed, the examples we shall examine below of how emotional responses can have a role in epistemic evaluation will not depend upon rejecting this point.

3 Some examples: doubt and dogmatism

The seventeenth-century common-sense philosopher John Wilkins criticized scepticism thus: '*Doubt* is a kind of *fear* . . . and 'tis the same kind of Madness for Man to *doubt* of any thing, as to *hope for* or *fear* it, upon a mere possibility' (1675, pp. 25–6). When I come to doubt some proposition that I previously accepted – perhaps due to encountering disagreement from others I respect or by finding evidence that conflicts with it – then I evaluate that belief as 'epistemically unsafe', as something it would be rash to believe on the basis of the reasons I currently possess. Wilkins urges that this evaluation takes the form of an emotional reaction: I become anxious about my belief. He also suggests that the appropriateness of emotional response (of fear or hope) provides a suitable model for thinking about the rationality or appropriateness of doubt. Sceptical doubts are dismissed as epistemic timidity. It is a consequence of this that common-sense philosophers emphasize 'real' or 'felt' doubt, over the more speculative claims of philosophers about what *ought* to be doubted. So long as our dispositions to doubt are well trained, our epistemic hopes and fears are probably more trustworthy than our epistemological speculations.

The case of doubt shares the features of emotional evaluations we have noticed above (Hookway, 1999). First, as we have seen, doubt of a proposition is an evaluation of it. Second, doubt often has an holistic character: although I may be able to point to the new information that led me to doubt a favoured belief, the doubt is usually only assessable as appropriate or otherwise by reference to the totality of information that I possess. It is unusual for me to be able to point to the principles and background assumptions that led me to be pushed over into doubt at just this point. Confidence in my doubt rests upon trust that any of my information that did not come to consciousness during the formation of my doubt was not relevant to the evaluation in question. As our quotation from Quine (1960) suggests, the doubt reflects weightings attached to a large number of items of information employing standards which (often) I cannot articulate. And thirdly, the doubt has a 'motivational' force: real doubts motivate me to collect more evidence, to reflect upon the credentials of the erstwhile belief, to worry about other beliefs which depend upon this one, and so on. Philosophical doubts generally lack this motivational force. We might also think that doubt can have a *felt quality* – a phenomenology – analogous to some other strong emotions. There is a 'gut reaction' – a 'felt doubt'.

One question raised by sceptical arguments concerns whether they merely reveal the limits of our habits of doubt, of epistemic fear: do they reveal that there are things we *ought* to doubt what we do not really doubt?; or do our habits of doubt display greater wisdom than our philosophical reflections? Experiments reveal the many ways in which our habits of inference and doubt are irrational, thus indicating the fallibility of emotional evaluations of our beliefs (Nisbett and Ross, 1980). Do sceptical arguments reveal the need to reflect more carefully and extensively than we naturally do? Or do they reveal the likely disastrous consequences of doing so, pointing to the deep need to trust our habitual and emotional evaluations? The suggestion that reflection cannot do justice to the holistic character of our epistemic evaluations, that it always fails to take note of relevant information, provides one reason for preferring the second of these reactions.

One ground for suspicion about the role of emotions in epistemic evaluation is that it can encourage dogmatism: our hopes and fears can encourage us to ignore evidence that challenges our favourite beliefs, or lead us to cut off self-questioning too fast. This is why most have emphasized the role of emotions in undermining epistemic rationality rather than any positive role they may have in supporting it. We evaluate considerations as irrelevant, and we are motivated to ignore their relevance to our cognitive enterprises, when careful reflection would show otherwise. However these observations can be used to support the thesis that emotions can have a fundamental positive role in epistemic evaluations. A number of philosophers of science – most notably Kuhn – have urged that dogmatism is a necessary condition for scientific progress. The capacity to ignore apparent refutations of theoretical frameworks, guided by the hope that further progress will reveal how we can accommodate such 'anomalies', may be required if frameworks are to be developed to their best advantage. When difficulties for beliefs become salient for us, doubt results; often it is best that they lack such salience, that we proceed as if they do not exist. The evaluations that produce this lack of salience seem to be most naturally emotional. Doubt and dogmatism may be related as fear is to hope: effective inquiry requires the effective management of both.

4 Holism and immediacy

These sorts of cases seem to have two important features. On the one hand, my doubt in a proposition, or my sense of the irrelevance of some surprising observation, has a kind of phenomenological *immediacy*. It is not the conclusion of a complex reflective argument. But, on the other hand, the basis of my judgement – the body of information that is relevant to it – is likely to be large and very complex. The two elements are combined in the observation that the evaluation is one whose rational basis I cannot always articulate. If I am right, the emotionally charged evaluation frees us from the need to carry out some enormously complicated assessments of the relevant evidence and information. Moreover, such evaluations can be sensitive to information that is not even available to conscious reflection. In that case, effective epistemic evaluation could turn out to be impossible without – rather then merely burdensome without – appropriate emotional responses (Hookway, 1993).

The above remarks suggest that emotional evaluations provide cognitive shortcuts. Their role reflects the fact that our rationality is reflected in the things we *ignore* as well as in the things we take explicit account of. Reflection involves asking questions about our opinions and about the ways we have arrived at them. Real doubt in some proposition encourages us to inquire into it without first asking about whether and why we were right to move from settled belief into doubt. The holistic character of our system of beliefs means that, even when we do reflect upon our opinions, reflection must come to an end: we must stop asking questions when there are further questions we could, in fact, ask; we must stop looking for evidence when there is more evidence to be had. Emotional judgements ensure that our reflections stop at the right place.

There can be different kinds of reasons that we must stop asking questions. For example, we save time and energy if we stop looking for further evidence when further evidence would make no difference to our conclusions. In such cases, the questions we don't ask we would be answerable if we chose to answer them. In other cases, when I ask what grounds I have for my belief, I may not be able to say or, owing to the holistic character of confirmation, any answer I offer may underestimate the range of considerations that hold it in place. I can retain my confidence in the belief for only so long as the question of what justification it has does not become a salient one.

5 Calling a halt

Some anecdotal examples will help here. The first concerns obsessives who are unable to leave a problem behind, forever checking facts that should be evident to them. Such people, it appears, are of two kinds. There are those who are vividly aware of the absurdity of their behaviour: they know that no further inquiry would be appropriate, but still find themselves forced to carry out tasks they take to be pointless. They feel no *epistemic* anxiety about the matters they keep inquiring into. Another kind does feel epistemic anxiety: they are unable to bring their inquiries to an end, unable to feel that the stage they have reached is an appropriate end point. Comparison of the two cases illustrates two themes. First, inquiry is often, quite properly, driven by a state that we can describe as epistemic anxiety: the 'fear' that one's evidence is insufficient is sufficient to make further inquiry appear rational to the subject. Second, that the end of inquiry must be regulated by an emotional change. This certainly involves the elimination of anxiety about one's epistemic position, and it may also involve an emotional acknowledgement that one's position renders further inquiries inappropriate. Obsessives of the first kind will, presumably, be emotionally sensitive to the absurdity of their continuing inquiries: appropriate reactive emotions will be found. (See Rapoport, 1989.)

The suggestion that emotions have this kind of role in the regulation of belief and inquiry is found in the work of Gigerenzer, Todd and the ABC Research Group on simple heuristics (1999). One role of heuristics is as 'stopping rules', as devices that ensure that we stop the search for further information at an appropriate time and without excessive calculation about the current state of inquiry: 'For instance, falling in love can be seen as a powerful stopping rule that ends the current search for a partner . . . and strengthens commitment to the loved one' (1999, p. 31). They also suggest that feelings of parental love can be 'a means of preventing cost-benefit computations with respect to proximal goals, so that the question of whether it is worthwhile to endure all the sleepless nights and other challenges associated with baby care never arises' (1999, p. 31). Emotions, especially strong ones, 'may facilitate rapid decision making by putting strong limits on the search for information or alternatives' (1999, p. 363). The examples they offer are not crucial to the point I want to make, which is that something similar occurs in scientific and everyday inquiry: one's dogmatic attachment to a theory or paradigm can prevent the question of whether it should be abandoned from arising. That this should not be a matter for constant cost–benefit analysis may easily be a means for cognitive progress. Similarly, that one does not reflect on whether some proposition should be doubted in the absence of felt doubt will often, once again, facilitate effective inquiry.

Is there any evidence in support of these, surely plausible, speculations? Damasio (1994) has presented evidence that damage to the areas of the brain involved in the production of emotions impairs subjects' ability to give weight to remote future goods and prevents their coming to decisions. He explains this by arguing that emotions are involved in making decisions with a holistic basis: the 'gut reaction' resolves uncertainties, leaving us confident that further exploration of a vast mass of relevant evidence may lead to a better decision. Perhaps the relative weights of potential goods and harms that are relevant to the decision are reflected in emotional discomforts and we form decisions that will ease the discomfort. It is hardly plausible that something of this kind is always going on. A more complex story is required. However there are some clear connections between these claims and the conjectures of earlier sections of this chapter. Without a 'felt doubt', we find it hard to take seriously the bearing of remote considerations upon the truth or falsity of our beliefs; and, without some device that will enable us happily to stop surveying vast quantities of potentially relevant information, we may find it very hard to avoid remaining in a state of tentative agnosticism about the issues that concern us. Still the required story will be more complex than is suggested by this talk of 'gut reactions'. I can see two strategies we might follow, and I shall now say something about each. The chief ideas are:

(a) There's more to cognition than assessing the relative weight of bodies of evidence: just as important is that we seek out the right reasons, interrogating nature in the right ways and so forth.
(b) Especially in the case of science: perhaps we could argue that there are advantages in assessments of evidence which are not logically sound.

6 De Sousa's hypothesis

Ronald de Sousa (1987) has offered a hypothesis about one 'function' that emotions can serve which seems to accord with the views just described. As we have noted, the holistic character of decision and of belief formation point to 'the strategic insufficiency of reason': there are always issues left unresolved by principles of 'pure reason'. Searches can always continue; concerns can always arise about how well searches have been conducted. As we have seen, the most threatening questions concern the sort of overload that Gigerenzer and Todd's heuristics are intended to enable us to overcome. According to De Sousa:

> The function of emotions is to fill gaps left by (mere wanting plus) 'pure reason' in the determination of action and belief, by mimicking the encapsulation of perception. (1987, p. 195)

What does this mean? 'For a variable but always limited time, an emotion limits the range of information that the organism will take into account, the inferences actually drawn from a potential infinity, and the set of live options from among which it will choose' (1987, p. 195). How do they do this? De Sousa suggests that these emotions provide 'determinate patterns of salience among objects of attention, lines of inquiry, and inferential strategies. Emotions can serve rationality by: dealing with the insufficiencies of (conscious deliberative) reason by controlling salience' (1987, p. 201). This fits our remarks about doubt: my doubt of a proposition renders particular questions and lines of inquiry salient. It also blocks the danger of a real regress of questions and considerations in dealing with matters of fact (and practical considerations): unless something becomes 'salient', we see no need to take it into consideration.

How, then, are emotions related to rationality? The answer is complex. First they can contribute to the success of our activities including activities of inquiry. Possessing patterns of salience that enable us to dispense with the need consciously to weigh evidence and consider alternatives can, in the contexts in which we find ourselves, contribute to our ability to form beliefs effectively. They may do this even if they impede our desires to conform to what are generally looked upon as paradigm norms of rational deliberation: Kuhnian inquirers may be rational to be dogmatic; and we may often be rational to follow our gut reactions and fail to take account of every piece of available evidence.

A further connection between emotions and rationality is suggested by de Sousa's discussion. He argues that the patterns of salience that are provided by emotions can function as 'a *source* of reasons', as something that comes '*before* reasons'. The reasons we offer, for both actions and beliefs, are offered as considerations that are relevant to what we plan to do or believe. Philosophers tend to think of my 'reasons' as composed of *all* the information I possess (or I am aware that I possess) which has a bearing on the issue. When we pick out some proposition as 'our reason', this is just because it was the most decisive, most surprising consideration; or because it is the one there is most point in explaining to others. We pick out some salient consideration that indicates why belief, doubt or action is appropriate. If the determinants of belief and action have the holistic character we have supposed, then our reasons for any belief and action will include a large set of propositions and desires. This may be wrong. It may be crucial to our epistemic strategies that we identify particular propositions as reasons, as the salient considerations that guide us in belief and action. Their salience, in that case, may be what makes them reasons, not what guides us in choosing to mention these among a much larger set of reasons. It explains how some factors are operative in causing belief or action while others are relegated to the background and ignored. As Gigerenzer, Todd and the ABC Research Group might express the matter: our reliance upon reasons involves using simplifying heuristics that enable us to ignore the non-salient considerations.

This is not merely a verbal matter of how to use the word 'reason'. It concerns how we should think about what goes on when we doubt or form beliefs. Do our emotional responses select particular propositions or considerations to serve a special role in cognition? This seems plausible: rationality requires that some propositions acquire a kind of salience for us: they have a role in our giving of reasons; they provide a focus when we reflect on our beliefs and reasonings. There are others that are, apparently, cut off from such influence, the possibility that I had misread the newspaper and should check it simply does not generally occur to me as a relevant consideration. If Kuhn's account of the role of dogmatism in science is accepted, then anomalies that face an established approach or paradigm may similarly be insulated from having any effects on reasoning and epistemic evaluation. Unless such phenomena occurred, it is suggested, effective and efficient management of our opinions would be impossible. And some emotions facilitate cognition by creating this precondition for effective cognitive management.

7 Science

One nineteenth-century epistemologist, Charles S. Peirce, argued that participation in *science*, and rationality in the use of the scientific method, depended upon the possession of specifically *altruistic* sentiments and emotions. The argument is probably flawed, but interesting for all that.

How do we reconcile confident belief with fallibilism (the idea that there is no metaphysical guarantee concerning any of our beliefs that it won't turn out to be mistaken and that, in the sciences, we can anticipate perhaps quite radical revisions in the current consensus)? One answer is: in part we value our current 'opinions' as of instrumental value – taking them seriously can be a means to our successors eventually reaching beliefs that are actually true. Since it is almost certain that our current scientific opinions contain much falsity, our holding them will lose its instrumental rationality if we ask too many questions, worry too much about a few inconsistencies or experimental failures. Progress may best be served by a slightly irrational attachment to our current views – ignoring known falsifications; paying no attention to the evidence that most theories succumb to falsification eventually; attaching little significance to the thought that all the evidence suggests that these views will eventually be abandoned. Hiding from myself the weakness of my own epistemic position may enable me better to contribute to the growth of *our* knowledge.

This is another side of the possible merits of dogmatism that we mentioned above. Science may be served by the fact that questions don't get raised – or don't receive what we might suppose was their due respect and attention. This ensures that routes of inquiry are fixed in distinctive ways. However, once these questions *are* raised, we cannot but acknowledge their relevance and their implications. We do better not even to think of them: we do better if they have no salience. This means that progress may best be served by groups of people with an epistemically irrational engagement with their current theories and beliefs. Habitual grounded patterns of salience contribute to distortions of what questions arise, of what doubts are treated as significant. Earlier sections of this chapter should help us to see how such evaluations can be carried by, and sustained by, patterns of emotional salience.

Our earlier examples illustrate how individual epistemic rationality is served by a role for emotions in regulating inquiries and managing the ways in which some beliefs emerge as matters of relevant concern and others do not. The observations in this section make a slightly different claim: what could be seen as mismanagement of individual epistemic performance can contribute to the success of a community of inquirers. That some should be careful or cautious and others bold, some sceptical and others dogmatic, may be means to co-operative progress. And, once again, emotions can have a distinctive role in possession of the different epistemic characters that are here manifested.

14 Social psychology and the theory of science

Philip Kitcher

Evolutionary studies of animal behaviour have benefited enormously from the integration of mathematical models and field data. I suggest that the same harmonious combination could transform our theorizing about science. To support this, I try to show how simple models can illuminate various aspects of scientific practice, and also how such models need to rest on empirical psychological and sociological studies. As an illustration of the possibilities and the problems, I consider the issue of public involvement in scientific decision making.

1 An analogy

Between the 1950s and the 1970s, the integration of mathematical models and field research transformed earlier ventures in natural history into a sophisticated branch of ecology. Recognizing that natural selection affects the morphology, distribution and behaviour of organisms, mathematically inclined biologists formulated hypotheses about the expected characteristics of plants and animals by devising rigorous models about the constraints on, and optimal forms of, organismal phenotypes. Their endeavours drew on the observations of generations of field naturalists and, in turn, their results supplied concepts and hypotheses that a new cohort of observers could take into the field. At its best, the joint activity of mathematical theorizing and painstaking observation proved extraordinarily fruitful. So, for example, in Geoffrey Parker's extended study of the behaviour of dung-flies, precise expectations about the times spent foraging, searching for mates and in copula, have been formulated, tested and refined in an ever-more penetrating and detailed understanding of the natural phenomena (see, for example, Parker, 1978, and the discussion of this work in Kitcher, 1985).

I think that contemporary theories about science would benefit from a similar exchange between theoretical model-builders and careful observers. Perhaps that has already occurred in individualistic attempts to provide models of certain types of inference and to acquire psychological and historical evidence about the inferences that are actually made. But what concerns me here is a

I am grateful to the editors for their patience, encouragement and good advice.

complex of issues about the social organization of science which I believe to have been neglected by philosophers since the early seventeenth century and to have surfaced only in the most abortive and botched forms in empirical studies of science. So I want to offer an invitation for a new kind of psychological study of science, one that will dovetail with a rigorous philosophical programme that is still in its infancy.

2 An example

Let me start with an example. It's often been assumed by rationalistically inclined philosophers that if scientists were not terribly high-minded, motivated by some pure and disinterested love of truth, various epistemological disasters would follow – thus, for example, the *angst* produced in philosophical circles by sociological tales from the lab bench and by the earlier historical excavations of Thomas Kuhn (1962) and Paul Feyerabend (1975). No such bogey should terrify us, for the assumption is false. Mundane motivations may sometimes work beautifully in the collective discovery of truth.

Suppose, for example, that there is a problem that some group of scientists agrees to be really worthy of solution, and that there are two or more methods for attacking it. If the available workforce is large enough, it may well be a poor strategy to have everybody working on one approach, even if one is clearly more promising than its rivals. The division of labour we want can readily be produced if the scientists have special incentives to solve the problem – maybe because there's a prestigious prize that will be given to the first solver – since, as the density of people pursuing any single method increases, investigators are going to perceive that their chances of winning the prize are increased if they expend their efforts on one of the rival approaches.

So far, the argument is impressionistic, but we can easily make it rigorous by building a model (Kitcher, 1990, 1993, chapter 8). Suppose that we have N available scientists, that there are two methods, and that the probability of success for method j ($j = 1, 2$) if n scientists work on that method is $p_j(n)$. Then, from the epistemological perspective of the community, the ideal situation would be to maximize:

$$p_1(n) + p_2(N - n) - \text{prob(both methods deliver)}$$

Assume now that $p_j(n) = q_j(1 - e^{-kn})$, that prob(both methods deliver) $= 0$, and that $q_1 > q_2$. (Here the q_i are constants, representing the different rates at which supply of workers is reflected in increased probability of success.) The best situation would then be to have n^* scientists working on Method 1, where:

$$n^* = (kN + \ln q_1 - \ln q_2)/2k$$

(This is the solution to the problem of maximizing the overall probability of success.) If our scientists decide to opt for that method that gives them the best chance of winning the prize, then, when n are working on Method 1 and the rest on Method 2, it will pay for someone in the first group to switch to the second if:

$$p_2(N - n + 1)/(N - n + 1) > p_1(n)/n$$

(For if this condition holds the scientist will increase her chance of winning.) There's now a stable distribution in the vicinity of $n^{\#}$ on Method 1, where:

$$n^{\#} = q_1 N/(q_1 + q_2)$$

It's not hard to verify that, for a large range of values of the parameters, $n^{\#}$ is close to n^* and that the value of $p_1(n^{\#}) + p_2(N - n^{\#})$ is close to the maximum value of $p_1(n) + p_2(N - n)$. This shows that individually selfish motivations, desires to be the first solver and thus to win the prize, can generate a very good epistemological outcome for the community.

What does the mathematical model, with all its apparently arbitrary assumptions, add to the intuitive argument with which we began? Three things. First, it shows us that the intuitive argument doesn't have some hidden flaw. Second, by making all the assumptions fully explicit, it reveals just where the intuitive argument would break down – perhaps in regions of the parameter space that correspond to rather unlikely empirical conditions. Third, it enables us to gauge the distance between the distribution achieved by our mundane scientists and the epistemological ideal, showing how far the value of $n^{\#}$ is from n^* and, perhaps more importantly, what this difference means for the probability of community success; indeed, as we pursue the mathematics of this example, we begin to see that the community can do fairly well, even if it remains rather far from the optimal distribution; there are lots of different ways of dividing the labour that are almost equally good.

3 Advertisement for a theory of science

Let me generalize from this example. The theory of science I envisage treats a wide range of problem contexts, each of which resembles the situation that figures in my story. More specifically, a problem context is defined by a goal that the community is to achieve, a set of individuals and a set of strategies that those individuals can pursue. The first question to ask is how the individuals are best distributed among the strategies to yield the maximal chance of attaining the goal. Secondly, we consider sets of individual preferences and sets of social conditions and relations that may shape or constrain those preferences, and ask, in each instance, what distribution of individuals among strategies would be expected and how this relates to the optimal distribution for attaining the goal. In this way, we can try to discover, *in the particular problem contexts we study*, which types of motivation and which social arrangements are good for promoting various goals that interest us.

I began by suggesting an analogy between a flourishing part of ecology and an embryonic study of science. With a more explicit account of the counterpart of ecological model-building, we should now scrutinize the analogy a little more closely. The first point to note is that models in evolutionary ecology are typically fashioned with confidence about the optimizing power of natural selection: the model-builder supposes that, if the constraints are properly understood, then the phenotypes actually found will correspond to the optimal form. Deviations from the predicted optimum are used to uncover hitherto unsuspected constraints, and thus to increase our understanding about the population, species, or other group that is the target of the study. Plainly, no such assumption should be made in the scientific case. We have no reason to think that the social institutions and the individual preferences they mold are at all apt for promoting the ideal distribution. Indeed, given the haphazard way in which the social arrangements of science have developed from the early modern period to the present, there's every reason to be sceptical about our current ways of organizing inquiry.

Second, although evolutionary ecology typically uses proxies for the notion of reproductive success – asking, for example, how an organism might maximize the rate of intake of some needed nutrient or secure the greatest number of matings, it's always recognized that there's a final currency in which successes and failures are counted, the number of offspring who survive to sexual maturity (or, in some versions, the number of copies of an allele transmitted to the next generation). In advertizing a theory of science, I assumed no such overarching measure of success. Instead, I spoke, rather vaguely, of goals that were shared by a community and constitutive of the various problem contexts. Inspired by the ventures of individualistic epistemology, we might make a grander proposal, asking how the motivations of scientists and the social arrangements among them should be fashioned so as to maximize the chances that they will arrive at truth.

In my judgement, this proposal is doubly unsatisfactory. For, in the first place, the mere adumbration of truth is quite irrelevant: what matters to the sciences is the discovery of significant truth (or maybe approximations to significant truth). Nor can the notion of significance be given some explication that will specify a single notion across epochs and communities. What counts as significant, epistemically as well as practically, evolves in the history of inquiry and of the human institutions that not only frame our cognitive endeavours but all other aspects of our lives. Thus, even were one to consider the sciences only in epistemic terms, the appropriate way of thinking about goals would be to retain the dependence on a particular problem context. But, in the second place, the epistemological focus that has dominated the philosophy of science in our century seems to me too narrow. To undertake the project I've begun to outline we ought to consider the possibility that the goals whose attainment we're considering may form a heterogeneous cluster, only some of which are understood in terms of our standard epistemic notions. Hence my analogy with evolutionary ecology should not mislead us into thinking that there's a universal characterization of the goals of the scientific enterprise against which actual practices can be assessed. We'll return to this point later.

In other respects, however, the analogy is just and indeed serves to guard against overly ambitious versions of the project. What exactly does my story about the division of cognitive labour show? I claimed that it liberated us from fears that we might have had, allowing us to see that admitting the mixed character of actual human motivations doesn't spell doom for the growth of scientific knowledge. That modest claim probably undersells the analysis of the problem context. It would, however, be a mistake to argue that the model reveals that a particular type of institution – dangling prizes in front of ambitious scientists – furthers the work of discovery. Mindful of the debate about the adaptationist programme in evolutionary theory (Gould and Lewontin, 1979), we should consider the possibility that there may be many contexts in which the incentives that work in this instance prove counter-productive. Nor would it be fruitful to envisage some grandiose future enterprise in theoretical science policy, an enterprise that proceeded by taking an inventory of the various problem contexts that are likely to occur in the next stages of the development of the sciences, charting their relative frequencies and trying to fathom which social arrangements offer the best chances of attaining our current goals.[1] That would be like trying to achieve some characterization of the inorganic physical environment and then constructing a theoretical ecology that would identify types of organisms with maximal reproductive success. Just as the actual work of biologists concentrates on local understanding of the traits of particular groups, so too the theory I'm proposing would proceed piecemeal, looking at particular problem contexts that strike us as offering opportunities either for understanding or reform, trying to discover the best ways of achieving our goals and appraising the merits of various social institutions that can be envisaged. Further, in light of my previous qualifications, our judgements are always vulnerable to the possibility that an arrangement fruitful in one context (or set of contexts) may be deleterious elsewhere.

Consider one last aspect of the analogy. So far, I've been lavishing attention on a counterpart for the model-building that theoretical ecologists undertake. Contemporary evolutionary ecology flourishes, however, precisely because the armchair work of the model-builders is informed by dispatches from the field:

[1] Many years ago Stephen Stich forcefully brought home to me the madness of this idea.

when Parker sits down to think about male dung-flies searching for mates, he has data on the distribution of females, on male flying abilities and dietary requirements for expenditure of energy in travel. (In this case, atypically, the field observations come from Parker himself.) But where is the corresponding venture in the study of the sciences? Or, to pose the question more concretely, what supports my claims about the shaping of scientific motivation in the illustrative example about the division of labour?

4 The psychological vacuum

It's time for true confessions. I made up the psycho-social story about scientific motivations. Not completely out of thin air, of course, but by drawing on my everyday understanding of human behaviour in general, and, more specifically, on my impressions of the motivations of the scientists I know. Recall that my intention was to explore the community-wide pattern of behaviour if scientists were to be individually moved by worldly – that is, non-epistemic – concerns. Here are some of the assumptions I made:

1. Worldly scientists who know that a prestigious prize will be offered to the first solver of a problem that they are equipped to tackle will seek the course of action that offers them the best chance of winning that prize.
2. Such scientists will proceed by assessing the probabilities of the available methods.
3. In particular, they will compute the probability of their success through using a given method by dividing the probability that that method will succeed (given the number of people using it) by the number of users.

Any or all of these assumptions could easily be wrong.

There is, of course, evidence that we can cite in favour of each of them. Those who have read *The Double Helix* (Watson, 1969) will surely recognize the first as confirmed by the behaviour of many of the central characters. Researchers of my acquaintance (and I don't think my data are idiosyncratic) routinely explain their decisions about which projects to undertake by mentioning the relative probabilities of success, suggesting that they have at least rough ideas about which approaches are more likely to work. Further, in discussing the differences among research fields, scientists routinely talk of some ventures as 'overcrowded' and others as 'undersubscribed', suggesting that they use something like a lottery model for figuring the chances, the kind of model embodied in my third assumption. But, of course, this evidence is thoroughly anecdotal or 'folkish'. To the best of my knowledge, nobody has done a systematic investigation of the motivations of scientists and, even when we remain in the ambit of folk theory, it's easy to see the limitations of the evidence. Are scientists so worldly that they give no weight at all to intellectual aspects of problems, aspects that are differentially salient for different people? Isn't it likely that their assessment of their own talents may play an overriding role in the decision? Those who are never seen in a modest mood may assume that their chances of winning through pursuing any method are significantly greater than those of any of their rivals.

I began by pointing out that one of the triumphs of contemporary evolutionary ecology lay in the ability of the modelling tradition to supply new questions and concepts to field observers. In just the same way, ventures in analysing the patterns of behaviour that emerge in scientific communities, inspired as they initially are by folk conceptions of motivation, offer new foci for empirical studies of science. What kinds of considerations *do* affect scientists' decisions? Are those decisions responsive to particular incentives which governments, businesses or private foundations might offer? Are the decisions assimilable to the framework of rational decision theory?

The cognitive science of science is a new field, and its brief literature is dominated by discussions of individual reasoning. The most prominent research has been focused on the extent to which actual scientific reasoning conforms to the ideals that philosophers have proposed. A second topic (perhaps, these days, a close second) has been the investigation of kinship between scientific explorations and the learning of young children. Both these ventures strike me as entirely legitimate, and worthy of the attention they've received. But I want to emphasize that they are not the only places where serious psychological research can inform our views about the ways in which scientific inquiry proceeds. As we enlarge our visions from the individual scientist to the scientific community, new issues flood in.

Return one more time to the analogy with evolutionary ecology. The trajectory of fruitful research is typically a zigzag affair. Early model-builders start with some impressions about the group they want to study, sometimes good hunches (as in many instances of the study of foraging), sometimes disastrously bad ideas (as in notorious examples about dominance and aggression in primates). Their models inspire field observations that revise the initial presuppositions, sometimes in quite radical ways. Based on the revised assumptions, a second generation of models is built. These, too, are taken into the field and refined in light of the observed phenomena. Is there any reason why a similar to-and-fro process couldn't transform our understanding of inquiry?

You might think that there is, that the social psychological questions about scientists are simply intractable. In the rest of this section, I want to suggest that there are reasons for greater optimism.

How can one discover what scientists want, or, more exactly, what inclines them to take up one enterprise rather than another? The obvious first thought is that one might ask them. That's immediately vulnerable to the complaint that researchers may not be aware of their own motives, or that they may feel compelled to disguise them for presentation to outsiders. Perhaps one might think that the official ideology of scientific inquiry, with its emphasis on the disinterested curiosity of the investigator, would even foster a particular type of response, whatever the fact of the matter. Although these doubts are reasonable, we might eventually discover that they are unwarranted. To do that, we'd have to have some independent avenue to the fathoming of scientific motivation, one that we could follow to find out that the real springs of scientific decision correspond to those that scientists identify.

Field studies of animal behaviour integrate the findings from two kinds of situations. On the one hand, we can patiently observe the ways in which animals act in complex natural situations; on the other, we can devise an artificial context, designed to highlight the role of some factor we suspect of being influential. The problem with the former is that the complications of the ordinary cases may obscure what's actually going on; the difficulty of the latter is that the causal role of the factor in which we're interested may be importantly modified in the transition to the artificial situation. Although these troubles are genuine, they don't render the investigation of behaviour completely hopeless. Moreover, what we can do for non-human animals (or for human infants) can also be done for fully grown people, even for scientists. Experimental economists have devised clever ways of revealing the preferences of subjects, and I can envisage comparable games that psychologists of science could invite their subjects to play. Combining the resultant data with naturally occurring situations that promise probative power – for example, cases in which one society offers certain kinds of incentives to undertake a particular type of research while another doesn't – we may be able to answer the kinds of questions I've posed.

I offer these suggestions tentatively, for I recognize that experimental psychologists are often far cleverer than philosophers give them credit for being, and that, consequently, there may be more subtle, and more revealing, ways of fathoming the motivations of researchers. But, rather than trying to pronounce from the armchair how the social psychology of science might be done, my aim has only been to address an obvious sceptical concern. Thus if my relatively crude proposals for approaching scientific behaviour (effectively modelled on methods that have helped in the study of non-human animals) can be superseded by more sophisticated ideas, so much the better.

5 A partial agenda

But why should busy people expend ingenuity and effort on trying to address questions about the springs of scientific decision making? Just so that we can resolve a small issue about the division of cognitive labour, thereby putting to rest worries caused by historically minded philosophers and the self-styled sociologists who have followed in their footsteps? No. There are a large number of issues about the collective practice of inquiry that we can't treat in an informed way unless we have some ideas about the motives and attitudes of scientists under a range of background conditions. I'll offer a small sample, not trying to suggest any precise models – although it should be fairly obvious how to begin model-building, given pertinent empirical information.

Case 1: Co-operation. It would be good to understand the kinds of circumstances under which collective inquiry would benefit from the formation of teams of co-operating scientists, and also the conditions under which this isn't so. There are obvious threats to successful co-operation, familiar from the game-theoretic literature (the free-rider problem, Rousseau's stag hunt, and so forth). A fundamental issue about scientific decisions to co-operate is whether the parties can be viewed as rational agents in the style of decision theory (an issue that has already emerged in the case of the division of cognitive labour). Assuming that this issue is resolved positively, we can't offer any serious analysis until we know the kinds of things that scientists expect to gain, either from co-operative or from individual ventures, and until we understand the ways in which the benefits might be divided among partners. Thus, more specifically, if one goal for a researcher is to obtain credit, it's worth understanding whether individual scientists think of themselves as receiving less credit if they operate as part of a team and, if so, what the effective discount rate is. Even more concretely, in instances in which every member of the community can benefit from time-consuming and tedious work, performed by some sub-group, any account of how people are induced to undertake the work must identify the rewards that they might receive; many people have pointed out that describing the Human Genome Project in grandiose terms (the 'quest for the grail') misleads the general public, but it's interesting to speculate that the inflated rhetoric plays a role within the community of inquirers, investing the mundane work of sequencing with attractions that it would otherwise lack.

Case 2: Competition. Scientists clearly do compete with one another. It's an open empirical question whether the kinds and amount of competition in scientific research conduces to goals that we'd reflectively value. Among the obvious worries are the possibility that competition might limit the exchange of information (or even foster the spread of misinformation), that it might provoke people to announce results prematurely, or that significant fields of inquiry might be shut out entirely. Any attempt to model competitive behaviour in science must try to understand if researchers can be regarded as akin to the agents in competitive game theory and, if so, to fathom the structure of the benefits they expect to receive.

Case 3: Consensus. An important sub-case for the exploration of competitive behaviour concerns those occasions on which established authorities are challenged. Many researchers have initially made their names by challenging what they proclaim as 'accepted dogma'. At any given stage of inquiry, we can divide the scientific population, or the population within a field, into two groups – the devout and the iconoclasts. Emulating the reasoning of a once-celebrated economist, we might argue that the percentage of iconoclasts should be somewhere strictly between 0 and 100, for, if all are devout the fundamental ideas that guide inquiry are never challenged, while if all are iconoclasts the field never builds on the achievements of predecessors. It's an interesting normative project to try to build a model to determine just in which roughly defined region the optimal division would lie, and an equally interesting socio-psychological endeavour to try to discover just how the division would respond to various kinds of social arrangements. Approximately how much consensus should we want, and how do we achieve something like that? Neither of these questions can be seriously addressed without increased understandings of the rewards that attract scientists and how those rewards are distributed to those who faithfully follow their elders, those who raise disturbing questions about accepted principles and those who expose mistakes in common assumptions.

A further issue about consensus concerns the conditions under which new findings are integrated into the community-wide lore. Again, there's an argument for the avoidance of extremes. If every member of the community has to check before a result is endorsed, then a vast amount of effort will be spent in replication; conversely, if nobody ever checks anything there'll be an obvious danger that erroneous beliefs will become widely disseminated. I strongly suspect that the normative issue would be far better understood if we had a clearer view of the kinds of distinctions that inquirers make in using the work of others, and that we would then be able to proceed to ask if the kinds of incentives currently offered to scientists produce anything like the replicating behaviour we want.

Case 4: Public dissemination. Within some scientific communities, most notably in astronomy, paleoanthropology and parts of biology, there's now a serious tradition of attempting to explain current research to a broad audience. One obvious question is whether this dissemination of knowledge is best done if there's a special class of individuals who undertake it, or whether it's important that those actively involved in the research participate in the process. Supposing that there's a real public benefit from the engagement of some researchers, even eminent researchers, in the work of 'popularization', we can ask just how much effort should be thus assigned. How ought a scientific community distribute its workforce between addressing the outstanding central problems of the field and explaining existing solutions to a broader audience? Further, if we assume, from our folk knowledge of scientific interactions, that obtaining a public reputation has both advantages and drawbacks (including sneers from more or less envious colleagues), we can inquire about the kinds of distributions that are likely to be reached. Here, as in the other instances, we need much more information about the rewards scientists perceive and how they are generated or held back under various social conditions.

6 Practical goals and democratic ideals

I come now to the most important issue for contemporary thinking about the organization of inquiry. In what ways is it most appropriate to exercise democratic control of scientific research? The answers embodied in American institutions (and it's unlikely that the situation is different elsewhere) rest on guesswork. At the end of the Second World War, Vannevar Bush and his collaborators wrote a landmark document (Bush, 1945) that promised scientific research in the service of the nation, without a clear analysis of the intended goal or any empirical research about how to attain it. Half a century later, the perception that NIH is insufficiently responsive to public input inspired a report that offered well-intentioned, but vague, proposals (Rosenberg, 1998). In this case, too, there was no clear specification of a goal, nor any empirical study of how to move towards it.

As I noted earlier, when we raise our vision of the goals of science beyond the epistemic, we face serious questions about precisely what it might mean to say that scientific research advances the goals of the broader society. Pluralistic democracies contain a wide variety of groups with very different preferences, and if scientific research is to represent 'the society' then it must respond to this diversity by finding ways to compromise disagreements. Vulgar democracy identifies the goal with the course of action that would be chosen by majority vote, but this is plainly unsatisfactory. How, then, are we to understand the collective interest?

Here's the outline of a proposal. Start with a homely analogy. Imagine a family with a free evening and a strong shared wish to spend it together in some form of entertainment. The family members begin with a number of different proposals, explaining to one another their preferences, the strength of the preferences, the considerations that move them. Each learns new things about the character of the various options, and each learns how the others view the possibilities. Nobody wants to do anything that any of the others regards as too unattractive, and they end up with a plan that reflects their collective wishes. Those collective wishes can't be conceived as the wishes that would emerge from a simple vote in the initial state of mutual ignorance; rather, they are the wishes that would be produced by a more intricate procedure.

Let's try to mimic this procedure for decisions about the research agenda for the sciences, decisions to commit resources, such as investigators and equipment, in particular amounts to particular projects. I envisage ideal representatives of members of the society, individuals with different initial preferences, coming together, like the family, to discuss the available courses for inquiry to pursue. The first thing to recognize is that, unlike the family, they are likely to begin from a very partial understanding of the possibilities. An obvious remedy for their ignorance is to insist on transmission of information so that each deliberator becomes aware of the significance, epistemic and practical, that attaches to potential lines of inquiry. Once this has been accomplished, each deliberator revises the initial preferences to accommodate the new information. Specifically, I imagine that each considers how possible inquiries might bear on goals that were antecedently adopted. The product of that consideration is a list of outcomes that the deliberator would like scientific inquiry to promote, coupled with some index that measures how intensely those outcomes are desired. Personal preferences have given way to *tutored* personal preferences.

The next step is for ideal deliberators to imitate the imaginary discussion of the family. They exchange their tutored personal preferences, offering their explanations of why they want particular outcomes to particular degrees and listening to the explanations given by others. In this process, I assume that each is moved by respect for the preferences of others, and aims to arrive at a consensus list in which none of the others is substantially underrepresented. The deliberators are committed to seeing the others as having, like themselves, a claim to realize their aspirations, and thus to take seriously the others' descriptions of their preferences and predicaments, and the rationales they provide for choosing as they do. Ideal deliberators thus recognize that they are engaged in a long-term sequence of interactions with people whose situations and fundamental wishes may be quite different from their own, and that such people cannot be expected to sacrifice their desires to the preferences of others (Gutmann and Thompson, 1996: chapter 2).

At the end of this exchange, the preferences of each ideal deliberator are again modified, this time to absorb their recognition of the needs of others. The next step is for them to attempt to draw up a list that represents their priorities concerning the outcomes to which inquiry might prove relevant. One possibility is that there is consensus. After coming to understand both the sources of scientific significance and the tutored preferences of other deliberators, each party formulates the same list, assigning exactly the same value to each outcome. If this is so, then the resulting list expresses the collective preferences, and no further accommodation is needed. A second possibility is that some deliberators favour different lists but each is prepared to accept a set of lists as fair and the intersection of the sets is non-empty. Under these circumstances, if the intersection contains a unique member, then that expresses the collective preferences. If not, then the ideal deliberators must decide by vote which of the lists in the intersection is to be preferred. Finally, if the intersection of the sets of lists that deliberators accept as fair turns out to be empty, collective preferences are determined by vote on all candidates drawn from the union of these sets of lists.[2]

At this point, our deliberators have formulated the issues they'd like inquiry to address, and have indicated the relative weight to be given to these issues. Their formulation, tutored by a clear understanding of the sources of significance for scientific endeavours already completed as well as those that might now be undertaken, can be expected to recognize possibilities for satisfying curiosity as well as opportunities for practical intervention, long-term benefits as well as immediate pay-offs. The next step is to assess the possibilities that particular scientific ventures might deliver what the ideal deliberators collectively want. Given a potential line of inquiry that might bear on some items on the collective list, we require an estimate of the chances that the desired outcomes will be delivered, and it's appropriate to turn at this point to groups of experts. How are the experts to be identified? I suppose that the ideal deliberators can pick out a group of people to whom they defer on scientific matters generally, that this group defers to a particular sub-group with respect to questions in a particular field, that that sub-group defers to a particular sub-sub-group with respect to questions in a particular sub-field, and so forth. Further, it's assumed that the experts identified are disinterested – or that any members of a group whose personal preferences would be affected by the project under scrutiny are disqualified from participating in the process. If matters are completely straightforward, there'll be consensus (or virtual consensus) at each stage concerning the appropriate people to consult, and these people will agree on exact probabilities with respect to the outcomes of research. In that case, the output of the search for probabilities is just the collection of chances assigned by the groups singled out at the ends of the various chains of deference.

Complications can arise in any of three ways. First, there may be disagreement on those to whom deference is warranted. Second, the experts may be unable to do more than assign a range of probabilities, possibly even a wide range. Third, the experts may be divided on which probabilities (or ranges of probabilities) should be assigned. I deal with all these complications in the same general way,

[2] It's clear from the extensive literature on social choice theory stemming from Kenneth Arrow's famous impossibility theorem that any procedure like the one described here may lead to counter-intuitive conclusions under certain hypothetical circumstances. I don't assume that my suggestion is immune to such problems, but I hope they arise only in sufficiently recherché situations to make the standard I attempt to explicate appropriate for the purposes of characterizing well-ordered science. For lucid discussions of Arrow's theorem and its significance, see Sen (1985).

namely through being inclusive. If any of them arises, the output of the search for probabilities is no longer a single set of values, but an explicit record of the verdicts offered by different groups, coupled with the extent to which those groups are supported by deliberators who have full information on the current state of inquiry and the past performances of the groups in question. Hence, instead of a simple judgement that the probability that a scientific project will yield a particular desired outcome is a certain definite value, we may have a more complex report to the effect that there are several groups of people viewed as experts, that deliberators in full command of the track records of the various groups select particular groups with particular frequencies, that the groups divide in their judgements in specified proportions, and that these judgements assign specified ranges of probability values.[3]

At the next stage, we suppose that a disinterested arbitrator uses the information about probabilities just derived, together with the collective wish list, to draw up possible agendas for inquiry. The arbitrator begins by identifying potential levels of investment in inquiry (possibly an infinite number of them). With respect to each level, the task is to pick out either a single assignment of resources to scientific projects that would be best suited to the advancement of the deliberators' collective wishes, given the information about probabilities, or a set of such assignments that represent rival ways of proceeding that cannot be decisively ranked with respect to one another. In the simplest case, when the arbitrator is given point probability values, the decision procedure can be specified precisely: with respect to each budgetary level, one identifies the set of possible distributions of resources among scientific projects, and picks from this set that option (or set of options) yielding maximal expected utility, where the utilities are generated from the collective wish list and the probabilities obtained from the experts. If there are disagreements about who the experts are, disagreements among the experts, or latitude in the responsible assignment of probabilities, then the arbitrator must proceed by considering the distributions of resources that would meet both budgetary and moral constraints, subject to different choices for probability values, picking that set of distributions that best fits the views of the majority of those who are most often considered experts.[4]

The last stage of the process consists in a judgement by the ideal deliberators of the appropriate budgetary level and the research agenda to be followed at that budgetary level. Perhaps there is consensus among the ideal deliberators about which level of support for inquiry should be preferred, and perhaps the

[3] I am indebted to Stephanie Ruphy for a discussion that brought home to me the potential complexities of the appeal to experts in this context.

[4] There are several ways of treating this problem formally. For present purposes, I suppose that the arbitrator proceeds in the way we take disinterested, intelligent and reasonable people to make their judgements when confronting divergent 'expert' opinions.

arbitrator assigns a single distribution of resources among lines of inquiry at that particular level. If that is not so, then the final resolution must be reached by majority vote. The result (whether it comes from consensus or voting) is the course of inquiry that best reflects the wishes of the community that the ideal deliberators represent. We want to organize science so it proceeds towards the lines of inquiry that would have been picked out in the ideal procedure I have sketched.

At this point we can start to ask serious psycho-social questions about the character of the institutions we've inherited. Consider two polar dangers, first that the conduct of research systematically fails to respond to the needs and interests of a particular segment of the wider population, and second that the practice of inquiry is distorted and limited by the ignorance of outsiders who insist on some projects and refuse to pay for others. Looking at the arrangements that are currently in place, from features of the training of scientists through the dissemination of knowledge to the methods of funding research, we can, and should, ask how these ameliorate or exacerbate the problems I've noted. Do systems of education and funding effectively exclude the preferences of some social groups? Do they give rise to a tyranny of the ignorant? Many friends of science believe that the tyranny of the ignorant is a larger danger; critics of science, by contrast, often argue that socially disadvantaged groups are under-represented in scientific research. In my view, both positions are matters of guesswork. Until we've done some serious analysis and informed it with some empirical studies of scientific motivation, there's just no basis for claims about whether or not the practice of the sciences promotes societal goals.

7 Elitism and democracy: first steps towards an analysis

So, let's formulate the problem a bit more carefully (and a bit less realistically). Imagine that the community with which we're concerned faces a sequence of pairwise decision situations $\{D_1, \ldots, D_m\}$. In each such situation D_i the options will be S_{i1} and S_{i2} where, in every case, the members of the elite group of inquirers uniformly prefer S_{i1} to S_{i2} – that is, elite utilities are measured by a function V^E such that $V^E{}_{i1} > V^E{}_{i2}$.

For a sub-set of the decision situations, the outsiders who, like the elite, are uniform in their desires and aspirations, have reversed preferences. Without any loss of generality, we can re-order the sequence of decisions so that these situations come first. So for some n such that $m \geq n$ and for every i ($n \geq i \geq 1$), $V^O{}_{i1} < V^O{}_{i2}$; but for $i > n$, $V^O{}_{i1} \geq V^O{}_{i2}$.

Let's make a further simplifying assumption, supposing that everyone – elite or outsider – is very good at seeing the value of whichever option is more attractive. Conflict comes about only because the elite sometimes ascribe too low a value to S_{i2} or because outsiders sometimes underrate the importance of S_{i1}. More exactly, I'll take it that the value assigned to the preferred option would always be the value that would be assigned as the output of ideal deliberation: $V^E{}_{i1} = V^C{}_{i1}$ and $V^O{}_{i2} = V^C{}_{i2}$.

The next task will be to try to understand the mistakes that occur when the decision is simply left to the elite. In the conflict situations, $D_1, \ldots, D_n$, elite decision making goes awry when $V^C{}_{i1} < V^C{}_{i2}$. If there are any such situations, we can re-order the D_I so that they come first. So there's a number k between 1 and n such that this inequality holds when $k \geq i \geq 1$. The total cost of the elite errors is thus:

$$\Sigma^k{}_1\left(V^C{}_{i2} - V^E{}_{i1}\right) = \Sigma^k{}_1\left(V^C{}_{i2} - V^C{}_{i1}\right)$$

If we make the very strong (and implausible) assumption that the absolute value of $(V^C{}_{i2} - V^C{}_{i1})$ is always the same, L, then the cost of elite decision making in conflict situations is just kL.

Do the elite make mistakes outside conflict situations? Not in $D_{k+1} \ldots D_n$, for in those cases, *ex hypothesi*, $V^C{}_{i1} \geq V^C{}_{i2}$. In $D_{n+1} \ldots D_m$, we know that $V^E{}_{i1} > V^E{}_{i2}$ and $V^O{}_{i1} > V^O{}_{i2}$. By the assumption about the relations between actual preferences and those that would result from ideal deliberation:

$$V^E{}_{i1} = V^C{}_{i1} \quad \text{and} \quad V^C{}_{i1} \geq V^E{}_{i2} \quad \text{and} \quad V^O{}_{i2} = V^C{}_{i2} \quad \text{and} \quad V^C{}_{i1} \geq V^O{}_{i1}$$

An error would occur only if $V^C{}_{i2} > V^C{}_{i1}$; that is only if $V^O{}_{i2} > V^C{}_{i1}$. But, in the cases under consideration, $V^O{}_{i1} > V^O{}_{i2}$, which implies $V^C{}_{i1} > V^O{}_{i2}$. It follows that elite decision making can deviate from ideal decision making in $D_{n+1} \ldots D_m$ only if:

$$V^O{}_{i2} > V^O{}_{i2}$$

which is impossible. So the previous analysis identified all the costs of elite decision making.

Compare elite decision making with three rival scenarios. In the first, there's no tutoring of preferences, and the decision is made by majority vote among representatives of elite and outsiders; since the outsiders outnumber the elite, we've effectively replaced elite decision making with outsider decision making. For exactly the reasons just given, the shortcomings of this method – that is, the cases in which it departs from what would have been favoured by ideal deliberation – must be confined to the conflict situations $D_{k+1} \ldots D_n$. Let's suppose, then, that outsider decision making goes astray in $D_{k+1} \ldots D_{k+k^*}$, where k^* lies between 0 and $n - k$. It's not hard to show that $k^* = n - k$. For if it were less, then there'd be some decision situation – e.g. D_n – in which neither the elite nor the outsiders deviated from the ideal. But since the elite and the outsiders disagree about this situation, one or the other party must deviate.

Hence if we continue to make the very strong assumption that all errors incur the same costs, the crucial inequality for elite decision making to be preferable is:

$$k < n - k \quad \text{or} \quad k < n/2$$

The formalism thus captures the intuitive grounds on which we resist vulgar democracy about science. We think that the elite will see the relations among the values more frequently than the outsiders. Notice, however, that this diagnosis rests on the strong assumption that all error costs are the same. It's not hard to show that, when that assumption is relaxed, it's theoretically possible for outsider decision making to be preferable.

The real interest of the model doesn't lie with this first scenario, but rather with two other possibilities. In either case, decision making proceeds in two stages. At the first phase, there's exchange of information about the significance and prospects of various lines of inquiry and about the needs of the outsiders, and that discussion issues in a set of transformed preferences. At the second stage, there's a voting procedure of one of two types: in the elitist version the outsiders are purely advisory and democracy figures only to the extent that outsiders have the chance to tutor the preferences of the elite; in the fully democratic version, the final decision is taken by majority vote, and since the outsiders greatly outnumber the elite, their preferences (possibly modified because of the first phase) hold sway. Call these *tutored elite decision* and *tutored outsider decision*, respectively.

To keep matters simple, let's suppose that the discussion doesn't modify the parties' evaluations of the options that bear most directly on their own concerns. The elite don't change their assessment of S_{i1} and the outsiders don't alter their evaluation of S_{i2}. There's a chance, however, that members of the elite will change their assessment of S_{i2} to agree with that given by the outsiders, and a chance that the outsiders will modify their evaluation of S_{i1} to agree with that given by the elite. Further (unrealistically) I assume that if one of the elite changes then all change, and similarly for the outsiders. At the end both classes are homogeneous, just as they were at the beginning.

Suppose, then, that the initial discussion modifies the original preferences, $V \rightarrow V^T$, where:

$$V^{ET}{}_{i1} = V^{E}{}_{i1}, V^{ET}{}_{i2} = V^{O}{}_{i2} \text{ with prob } p,$$
$$V^{ET}{}_{i2} = V^{E}{}_{i2} \text{ with probability } 1 - p$$

$$V^{OT}{}_{i2} = V^{O}{}_{i2}, V^{OT}{}_{i1} = V^{E}{}_{i1} \text{ with prob } q,$$
$$V^{OT}{}_{i1} = V^{O}{}_{i1} \text{ with probability } 1 - q$$

Discussion occurs only in those situations where there's controversy, so that $D_{n+1} \ldots D_m$ are decided as before. Assuming (probably unrealistically) that modifications of elite opinion and outsider opinion are independent, the old class of conflict situations is now partitioned into four sub-classes with probabilities as follows:

α	β	γ	δ
$V^{ET}{}_{i1} = V^{OT}{}_{i1}$	$V^{ET}{}_{i1} \neq V^{OT}{}_{i1}$	$V^{ET}{}_{i1} = V^{OT}{}_{i1}$	$V^{ET}{}_{i1} \neq V^{OT}{}_{i1}$
$V^{ET}{}_{i2} = V^{OT}{}_{i2}$	$V^{ET}{}_{i2} = V^{OT}{}_{i2}$	$V^{ET}{}_{i2} \neq V^{OT}{}_{i2}$	$V^{ET}{}_{i2} \neq V^{OT}{}_{i2}$
pq	$p(1-q)$	$(1-p)q$	$(1-p)(1-q)$

The situations in α show complete agreement between elite and outsiders and, indeed, both groups reach the assessments they'd have achieved after ideal deliberation. In all the other cases, one or other of the groups fails to assign the value that would have resulted from ideal deliberation. The task is to compare the expected losses from elite decision making without discussion, elite decision making after discussion and outsider decision making after discussion.

As we saw earlier, the decision situations divide into two classes $\{D_1, \ldots, D_k\}$, $\{D_{k+1}, \ldots, D_n\}$. In the former, the untutored elite deviate from the ideal outcome, and in the latter the untutored outsiders deviate. In any situation in which an untutored group agrees with the ideal outcome, the tutored group does, too (tutoring never hurts). But tutored groups might sometimes do better.

Assume that the outcomes of types $\alpha, \beta, \gamma, \delta$ are randomly distributed across the sequence of controversial cases, so that the probability of modifying preferences in a decision situation is independent of which of the sub-classes that situation belongs to. So for the outsiders, deviations in $\{D_{k+1}, \ldots, D_n\}$ are corrected with probability q, and for the elite, deviations in $\{D_1, \ldots, D_k\}$ are corrected with probability p. Hence, assuming, as before, that each deviation brings a standard loss L, the expected losses from untutored elite, tutored outsider and tutored elite decision making are as follows:

Untutored elite:	kL
Tutored outsiders:	$(1-q)(n-k)L$
Tutored elite:	$(1-p)kL$

If $p > 0$, and there are no costs of tutoring, then, plainly, tutored elite is preferable to untutored elite. But we should surely recognize such costs. Let's suppose that each occasion of deliberative exchange of information is associated with a fixed cost, C. We now obtain the following conditions:

(1) Tutored outsider is preferable to untutored elite iff:
$(1-q)(n-k)L + nC < kL$ – i.e. $nC < [k(2-q) - n(1-q)]L$

(2) Tutored outsider is preferable to untutored elite iff:
$(1-p)kL + nC < kL$ – i.e. $nC < pkL$.

To defend the idea that current ways of resolving the research agenda – that is, untutored elite decision making – are preferable, one would have to contend that realistic values of the parameters satisfy these inequalities. This seems to me implausible.

Focus first on (2). L is surely likely to be large in relation to C, representing the obvious thought that costs of exchanging information are small in comparison with failure to reach the ideal outcome. Hence if (2) is not to hold, the value of pk/n will have to be a small fraction. Now k/n is just the proportion of cases in which the elite fail to appreciate the ideal ordering of the options, and p is the probability that the exchange of information will enlighten the elite. Notice that it's rather hard to defend the idea that *both* quantities are small, since the picture of a wise elite (small k/n) is in tension with the idea of an uneducable elite (small p). The best strategy for the defender of contemporary versions of elitism is probably to argue that k/n is tiny, effectively expressing confidence that those who decide how inquiry should be pursued almost always provide the public with what they would choose if they were reflective and informed. That looks like an article of blind faith.

At first sight, (2) seems easier to support than does (1). If, however, we believe that it's possible to design an effective programme of education for outsiders, giving them full insight into the promises of various research projects, then we can bring the value of q close to 1. If we're optimistic about the potential for this type of education, then condition (1) reduces to:

$$(1^*) \quad C < kL/n$$

Tutored outsider decision making will thus be preferable unless the proportion of cases in which the untutored elite deviate is truly tiny. If this is correct, then the same article of faith lies at the foundation of resistance to the democratization of science.

8 The psychological vacuum again

These are *first steps* towards an analysis of an important problem. They do not take us very far, as is evident from the number of places in which I've made highly unrealistic (even ludicrous) assumptions. What the model does reveal is the sort of information we'd like to have. How frequently would experts and outsiders shift their preferences after exchange of information? How often do experts diverge from the outcomes that would have been preferred by fully enlightened outsiders? These are empirical questions that ought to be undertaken by psychologically informed social studies of science. If they were pursued, then the answers to them might enable us to formulate more adequate models, and to emulate the strategy that has been so fruitful in behavioural biology. I hope to have indicated why I think that this is an enterprise worth pursuing.

Part four

Science and the social

15 Scientific cognition as distributed cognition

Ronald Giere

After introducing several different approaches to distributed cognition, I consider the application of these ideas to modern science, especially the role of instrumentation and visual representations in science. I then examine several apparent difficulties with taking distributed cognition seriously. After arguing that these difficulties are only apparent, I note the ease with which distributed cognition accommodates normative concerns. I also present an example showing that understanding cognition as distributed bridges the often perceived gap between cognitive and social theories of science. The chapter concludes by suggesting some implications for the history of science and for the cognitive study of science in general.

1 Introduction

Pursuing the philosophy of science within the context of philosophy has long been justified on the grounds that scientific knowledge is the best example of knowledge there is. It is KNOWLEDGE WRIT LARGE, as the saying goes. And epistemology is one of the main areas of philosophical inquiry. Philosophers of science have tended to regard the relationship as asymmetrical. Philosophy of science illuminates the problems of epistemology, but not much the other way around.

One can now say something similar about cognition in general. Science provides arguably the best example of a higher cognitive activity. Since the emergence of cognitive science as a recognizable discipline, however, the relationship has definitely been symmetrical, if not somewhat asymmetrical in favour of the cognitive sciences. A number of philosophers of science have explicitly drawn on developments in the cognitive sciences in attempting to illuminate the practice of science (Nersessian, 1984b, 1992a, 1999; Giere, 1988, 1999a; Thagard, 1988, 1992, 1999; Churchland, 1989; Darden, 1991; Bechtel,

I thank Linnda Caporael and Nancy Nersessian for introducing me to the work of Ed Hutchins and others developing the concept of distributed cognition in the cognitive science community. Nersessian, in particular, shared with me her grant proposal on cognitive and social understandings of science, and even dragged me off to hear Ed Hutchins at a Cognitive Science Meeting.

1996a). In this chapter I wish to continue this tradition by suggesting that some new developments within the cognitive sciences provide a useful framework for thinking about cognition in the sciences generally. These developments may be classified under the title of *distributed cognition.*

2 Distributed cognition

The idea of distributed *processing* has long been a staple in computer science. A dramatic contemporary example is the project *SETI at Home*, in which currently a million and a half participants contribute spare time on their personal computers to analyse data from the Arecibo Radio Telescope in Puerto Rico as part of a Search for Extra-Terrestrial Intelligence. Distributed *cognition* is an extension of the basic idea of distributed processing. I will focus on two of several contemporary sources of the notion of distributed cognition within the cognitive sciences.

2.1 Hutchins' cognition in the wild

One source for the concept of distributed cognition within the cognitive sciences is Ed Hutchins' study of navigation in his 1995 book, *Cognition in the Wild.* This is an ethnographic study of traditional 'pilotage', that is, navigation near land as when coming into port. Hutchins demonstrates that individual humans may be merely components in a complex cognitive system. No one human could physically do all the things that must be done to fulfil the cognitive task, in this case repeatedly determining the relative location of a traditional navy ship as it nears port. For example, there are sailors on each side of the ship who telescopically record angular locations of landmarks relative to the ship's gyrocompass. These readings are then passed on, e.g. by the ship's telephone, to the pilot-house where they are combined by the navigator on a specially designed chart to plot the location of the ship. In this system, no one person could possibly perform all these tasks in the required time interval. And only the navigator, and perhaps his assistant, knows the outcome of the task until it is communicated to others in the pilot-house.

In Hutchins' detailed analysis, the social structure aboard the ship, and even the culture of the US Navy, play a central role in the operation of this cognitive system. For example, it is important for the smooth operation of the system that the navigator holds a higher rank than those making the sightings. The navigator must be in a position to give orders to the others. The navigator, in turn, is responsible to the ship's pilot and captain for producing locations and bearings when they are needed. So the social system relating the human components is as much a part of the whole cognitive system as the physical arrangement of the required instruments. One might say that the social system is part of the overall cognitive system.

Now one might treat Hutchins' case as an example of 'socially shared cognition' (Resnick, Levine and Teasley, 1991) or, more simply, *collective cognition*. The cognitive task – determining the location of the ship – is performed by a collective, an organized group and, moreover, in the circumstances, could not physically be carried out by a single individual. In this sense, collective cognition is ubiquitous in modern societies. In many workplaces there are some tasks that are clearly cognitive and, in the circumstances, could not be carried out by a single individual acting alone. Completing the task requires co-ordinated action by several different people. So Hutchins is inviting us to think differently about common situations. Rather than simply assuming that all cognition is restricted to individuals, we are invited to think of some actual cognition as being distributed among several individuals.

How does this differ from the simple 'pooling' of knowledge possessed by several individuals, an already well-known phenomenon? Why introduce the new notion of 'distributed cognition'? Part of the answer is that categorizing some activities as a type of cognition is a more fruitful way of thinking about them. Among other things, it brings them within the scope of cognitive science. Another consequence of thinking in terms of distributed cognition is a focus on the *process* of acquiring knowledge rather than the static possession of knowledge. The knowledge acquired by the members of the crew is not 'pooled' in the sense of their simply bringing their individual bits of knowledge together. Rather, the individual bits of knowledge are acquired and co-ordinated in a carefully organized system operating through real time.

Hutchins' conception of distributed cognition, however, goes beyond collective cognition. He includes not only persons but also instruments and other artefacts as parts of the cognitive system. Thus, among the components of the cognitive system determining the ship's position are the alidade used to observe the bearings of landmarks and the navigational chart on which bearings are drawn with a ruler-like device called a 'hoey'. The ship's position is determined by the intersection of two lines drawn using bearings from two sightings on opposite sides of the ship. So parts of the cognitive process take place not in anyone's head but in an instrument or on a chart. The cognitive process is distributed among humans and material artefacts. (See also Latour, 1986.)

The standard view, of course, has been that things such as instruments and charts are 'aids' to human cognition which takes place only in someone's head. But the concept of an 'aid to cognition' had remained vague. By expanding the concept of cognition to include these artefacts, Hutchins provides a clearer account of what things so different as instruments and charts have in common. They are parts of a distributed cognitive process.

2.2 The PDP research group

Hutchins comes to distributed cognition through anthropology and ethnography. Another source comes through the disciplines usually regarded within the core of cognitive science: computer science, neuroscience and psychology. This is the massive, two-volume 'Exploration in the Microstructure of Cognition', titled simply *Parallel Distributed Processing* produced by James McClelland, David Rumelhart and the PDP Research Group based mainly in San Diego during the early 1980s (McClelland and Rumelhart, 1986). Among many other things, this group explored the capabilities of networks of simple processors thought to be at least somewhat similar to neural structures in the human brain. It was discovered that what such networks do best is recognize and complete *patterns* in input provided by the environment. The generalization to human brains is that humans recognize patterns through the activation of prototypes embodied as changes in the activity of groups of neurons induced by sensory experience. But if something like this is correct, how do humans do the kind of *linear* symbol processing apparently required for such fundamental cognitive activities as using language and doing mathematics?

Their suggestion was that humans do the kind of cognitive processing required for these linear activities by creating and manipulating *external representations*. These latter tasks *can* be done well by a complex pattern matcher. Consider the following simple example (1986, vol. 2, pp. 44–8). Try to multiply two three-digit numbers, say 456 × 789, in your head. Few people can perform even this very simple arithmetical task. Figure 1 shows how many of us learned to do it.

This process involves an *external representation* consisting of written symbols. These symbols are manipulated, literally, by hand. The process involves eye–hand motor co-ordination and is not simply going on in the head of the person doing the multiplying. The person's contribution is (1) constructing the external representation, (2) doing the correct manipulations in the right order and (3) supplying the products for any two integers, which can be done easily from memory.

```
    4 5 6
    7 8 9
    -----
  4 1 0 4
3 6 4 8
3 1 9 2
-----
3 5 9 7 8 4
```

Fig. 1 A common method for multiplying two three-digit numbers

Notice again that this example focuses on the *process* of multiplication; the task, not the product, and not knowledge of the answer. Of course, if the task is done correctly, one does come to know the right answer, but the focus is on the *process* rather than the *product*. The emphasis is on the *cognitive system* instantiating the process rather than cognition *simpliciter*.

Now, what is the cognitive system that performs this task? Their answer was that it is not merely the mind–brain of the person doing the multiplication, nor even the whole *person* doing the multiplication, but the *system* consisting of the person *plus* the external physical representation. It is this whole system that performs the cognitive task, that is, the multiplication. The cognitive process is distributed between a person and an external representation.

Here is a more complex example making a similar point. The diagram in figure 2 embodies a famous proof of the Pythagorean Theorem. The area of the large square is $(a + b)^2$. That of the small square is c^2. Remembering that $(a + b)^2 = a^2 + b^2 + 2ab$, and that the area of a triangle is $\frac{1}{2}ab$, we see that $(a + b)^2$ is also equal to $c^2 + 2ab$, from which it follows that $a^2 + b^2 = c^2$.

The claim is that this reasoning is perfectly valid and that the diagram is essential to the reasoning. One literally *sees* that $(a + b)^2 = c^2 + 2ab$. So, here again, we seem to have a case where the cognitive task involves an external representation as an essential part of completing the task successfully. This is an example of what is now often called 'diagrammatic reasoning' (Chandrasekaran, Glasgow and Narayanan, 1995). It is also a case of what the logicians Jon Barwise and John Etchemendy (1996) call 'heterogeneous inference' because it involves *both* linguistic and visual representations.

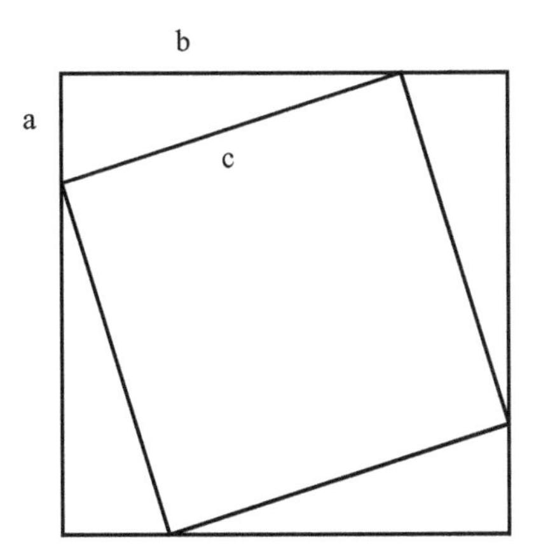

Fig. 2 Diagram for proving the Pythagorean Theorem

2.3 Other sources

Although I have focused on just two sources of recent interest in distributed cognition, the idea is much more widespread, as one can learn from Andy Clark's *Being There: Putting Brain, Body, and World Together Again* (1997). Lucy Suchman's *Plans and Situated Actions* (1987) and Varela, Thompson and Rosch's *The Embodied Mind* (1993) are among earlier influential works. A common theme here is that the human brain evolved primarily to co-ordinate movements of the *body*, thereby increasing effectiveness in activities such as hunting, mating and rearing the young. Evolution favoured cognition for effective action, not for contemplation. *Cognition is embodied.* This point of view argues against there being a single central processor controlling all activities and for there being many more specialized processors. Central processing is just too cumbersome. Thus, an emphasis on distributed cognition goes along with a rejection of strongly computational approaches to cognition. In fact, some recent advocates of Dynamic Systems Theory have gone so far as to argue against there being any need at all for computation as the manipulation of internal representations (Thelen and Smith, 1994).

Clark, in particular, invokes the notion of 'scaffolding' to describe the role of such things as diagrams and arithmetical schemas. They provide support for human capabilities. So the above examples involving multiplication and the Pythagorean Theorem are instances of scaffolded cognition. Such external structures make it possible for a person with a pattern matching and pattern completing brain to perform cognitive tasks it could not otherwise accomplish.

The most ambitious claim made in behalf of distributed cognition is that *language* itself is an elaborate external scaffold supporting not only communication, but thinking as well (Clark, 1997, chapter 10). During childhood, the scaffolding is maintained by adults who provide instruction and oral examples. This is seen as analogous to parents supporting an infant as it learns to walk. Inner speech develops as the child learns to repeat instructions and examples to itself. Later, thinking and talking to oneself (often silently) make it seem as though language is fundamentally the external expression of inner thought, whereas, in origin, just the reverse is true. The capacity for *inner* thought expressed in language results from an internalization of the originally *external* forms of representation. There is no 'language of thought'. Rather, thinking in language is a manifestation of a pattern matching brain trained on external linguistic structures (see also Bechtel, 1996b).

This distributed view of language implies that cognition is not only embodied, but also *embedded* in a society and in a historically developed culture. The scaffolding that supports language is a cultural product (Clark, 1997). Interestingly, an externalized and socialized view of language was advocated in the 1920s by the Soviet psychologist, Lev Vygotsky (Vygotsky, 1962, 1978;

Wertsch 1985). A product of the intellectual ferment inspired by the Russian Revolution, Vygotsky explicitly appealed to Marxist thought for the idea that the key to understanding how language developed lay not so much in the mind as in society. Nevertheless, his major 1934 book (translated in 1962 as *Thought and Language*) was suppressed within the Soviet Union from 1936 to 1956, and has only since the 1960s received notice in the English-speaking world.

A view similar to Vygotsky's, with no apparent connection, has recently been developed under the title of 'cognitive linguistics' or 'functional linguistics' (Velichkovsky and Rumbaugh, 1996). Like Vygotsky, some advocates of cognitive linguistics emphasize comparative studies of apes and humans (Tomasello, 1996). The most striking claim is that the major difference between apes and human children is *socialization*. Of course there are some differences in genotype and anatomy, but these are surprisingly small. The bonobo, Kanzi, raised somewhat like a human child by Duane Rumbaugh and Sue Savage-Rumbaugh (Savage-Rumbaugh *et al.*, 1993), is said to have reached the linguistic level of a two-year-old human child. Tomasello argues that what makes the difference between a natural and an enculturated chimpanzee is developing a sense of oneself and others as *intentional* agents. Natural chimpanzees do not achieve this, but enculturated ones can.

From the standpoint of computational linguistics, there has always been a question of how the neural machinery to support the necessary computations could possibly have evolved. From the standpoint of cognitive linguistics, this problem simply disappears. Language is not fundamentally computational at all, but the product of a pattern matching neural structure, which biologically could evolve, supported by an elaborate scaffolding of social interaction within an established culture.

3 Distributed cognition in scientific research

I will now consider several examples for which the concept of distributed cognition provides resources for understanding aspects of scientific research (see also Bechtel, 1996a). I take these examples from my own earlier research partly because I am familiar with them and partly because I wish to avoid the risk of re-interpreting someone else's work in ways they might not appreciate.

3.1 Instrumentation

In the mid-1980s I spent considerable time observing the operations of the Indiana University Cyclotron Facility. This facility was designed for the investigation of nuclear structures using light ions such as hydrogen and helium to bombard heavy nuclei. At the time I was very impressed by the role of technological artefacts in scientific research, but, never having heard of distributed cognition, I did not know what to say about it. The best I could do, in just three pages of *Explaining Science* (1988, pp. 137–40), was to suggest we think of technology as 'embodied knowledge'. Thus, what philosophers typically regard as 'background knowledge' or 'auxiliary assumptions' may not best be understood in terms of symbolic representations but as being physically encapsulated in experimental apparatus.

I took many photographs around the laboratory, many of them showing the instrumentation being employed. In fact, I sometimes took great pains to picture specimens of a type of instrument in isolation both from other instruments and from the human researchers. I now think that this orientation was completely mistaken. On the contrary, it is particularly enlightening to think of the whole facility as one big cognitive system comprised, in part, of lots of smaller cognitive systems. To understand the workings of the big cognitive system one has to consider the human–machine interactions as well as the human–human interactions.

In thinking about this facility, one might be tempted to ask, *Who* is gathering the data? From the standpoint of distributed cognition, that is a poorly framed question. A better description of the situation is to say that the data is being gathered by a complex cognitive system consisting of the accelerator, detectors, computers and all the people actively working on the experiment.

Understanding such a complex cognitive system requires more than just enumerating the components. It requires also understanding the organization of the components. And, as with Hutchins' ship, this includes the *social* organization. It is not irrelevant to the operation of this cognitive system that the people monitoring the data acquisition are most likely the PhD faculty members who participated in designing the experiment. Those tending the detectors may have PhDs, but may well be full-time laboratory employees rather than faculty. Those keeping the accelerator in tune are probably technicians without PhDs. One cannot adequately understand the operation of the whole cognitive system without understanding these differences in roles and status among the human participants.

3.2 Visual representations

Another result of my time in the laboratory was an appreciation for how big a role *visual* representations play in the actual doing of science. Here again I did not at the time know what to say about this. I devoted just one page of *Explaining Science* (1988, p. 190) to describing how physicists judge goodness of fit *qualitatively* by visually comparing two-dimensional curves representing, respectively, data and theoretical predictions. In fact, the laboratory had a light-table in a corner of the control room so one could visually compare curves and data points by superimposing one sheet of graph paper over another.

In a later publication on visual models and scientific judgement (Giere, 1996a), I speculated that one of the roles of visual representations is to help scientists organize their knowledge of the situation so as to be able to judge the fit between abstract models and the world. I now think that this suggestion erred in putting all of the cognitive activity in the heads of scientists. It is more enlightening to think of the scientist plus the external visual representation as an extended cognitive system that produces a judgement about the fit between an abstract model and the world. The visual representation is not merely an aid to human cognition; it is part of the system engaged in cognition.

3.3 Models and theories

For those philosophers of science who understand scientific theories to be axiom systems ideally reconstructed in a formalized language, there seems little new to learn from at least the computational parts of cognitive science. For those philosophers of science who, like myself, think of theories more in terms of *families of models*, this connection is not nearly so strong. But there is other work in cognitive science that is relevant, namely, that on concepts and categorization. This is because a model is a lot like a concept. Like concepts, models themselves make no claims about the world. But they can be *used* to make such claims. That is, both concepts and models can be applied to things in the world.

There is, however, a conflict between how most philosophers of science think of models and how most cognitive psychologists think of concepts. Philosophers typically hold a classical view of models in which a model can be defined in terms of a set of necessary and sufficient conditions. Cognitive psychologists typically hold a graded view of concepts in which there are no necessary and sufficient conditions for application, only more and less central cases (Smith and Medin, 1981).

My own resolution of this conflict is to use the classical account for individual models, but to realize that a family of classical models radiating out from a central model can produce a graded structure among the things to which various members of the family of models can be applied (Giere, 1994, 1999b).

There is another difference between the notion of model employed in the philosophy of science and the account of concepts in cognitive psychology. The idea of distributed cognition provides a way of resolving this difference. Almost all studies of concepts in psychology deal only with concepts like 'bird' or 'chair' which are simple enough to be deployed by normal humans without relying on external representations. Most models in science, even in classical mechanics, are too complex to be fully realized as mental models. Not even authors of science textbooks can have in their heads all the details of the models presented in their texts. Rather, the details of these models are *reconstructed* as external representations when needed. These reconstructions typically take the form of equations or diagrams. What scientists have inside their skins are representations of a few general principles together with bits and pieces of prototypical models. They also possess the *skills* necessary to use these internal representations to construct the required *external* representations that are then part of an extended cognitive system.

4 Objections

I will now consider two objections to taking seriously the notion of distributed cognition for understanding scientific research.

4.1 Individuation

Once the boundaries of a cognitive system are allowed to extend beyond the skin of a human being, where do they stop? I earlier claimed that the cyclotron facility should be regarded as a large cognitive system. Now this facility requires much electricity to operate. The electricity comes over wires from a generating plant, which is fired by coal, which is mined in the western USA. Are we to suppose that the boundary of this cognitive system thus extends all the way from Indiana to Montana? Hardly.

Here I think we can distinguish those features of the system that *differentially* influence the output of the system in scientifically relevant ways from those features that merely make it possible for the system to generate any output at all. The person at the control panel maintaining the coherence of the beam contributes to the quality of the output. The electricity makes it possible for the machine to run, and thus for there to be output, but it does not differentially influence the particular output obtained. Coal mines in Montana are thus not part of this cognitive system. On the other hand, if the data were to be transmitted directly to a computer at the University of Montana and returned after appropriate processing to the facility in Indiana, then the boundary of the system in Indiana would indeed extend to Montana.

This response can be generalized. Anything, anywhere that is *designed* to contribute to the specific quality of the output of a cognitive system, and functions as designed, is surely part of that system. I am inclined to go further and affirm that anything that *actually* contributes to the specific quality of the output of a cognitive system, whether or not it was designed to do so, and even if it is completely unknown to anyone, is part of that system. However, I do not think it is worth arguing over such niceties. The main point is that, for scientific purposes, one can *identify* particular cognitive systems in space and time. They are not unmanageably amorphous.

4.2 *Mind*

In countenancing extended cognition, are we not thereby committed to the existence of something like 'group minds'? And has not that notion long been discredited?

Now it is true that some advocates of distributed cognition are tempted by the notion of *distributed minds* (Clark, 1997, chapter 9). The presupposition behind this move seems to be that cognition is necessarily a property of minds. So, where there is cognition, there is also a mind. But I do not think that one is forced to take this route. I do not see why one cannot maintain an *ordinary* conception of mind, which restricts it to creatures with brains, that is, humans and perhaps some animals. So, much cognition does involve minds. Perhaps we can also agree that all of our currently existing cognitive systems have a human somewhere in the system. So every cognitive system has a mind somewhere. But as a concept in cognitive science, cognition need not be restricted exclusively to places where there are also minds. There is no reason why we cannot extend the *scientific* concept of cognition to places where there are no minds.

I do not think that the question of whether there are extended minds is one for which there is now a determinate answer. Rather, the question will be settled, if it is ever settled at all, by many individual decisions about the best way to develop a science of cognition.

5 Normative considerations

Thinking in terms of cognitive systems provides a natural way to incorporate *normative* considerations into a cognitive theory of science. Any individual cognitive system is designed to do a particular job. It can be evaluated as to how well it does that job. And evaluation may require consideration of tradeoffs among several desirable characteristics. A low energy cyclotron, for example, cannot tell us much about elementary particles, but it can provide high-resolution determinations of various characteristics of heavy nuclei.

The main point I want to make here is that the evaluation of a cognitive system is largely an *empirical* matter. It is a matter for empirical investigation to determine how well a particular cognitive system does what it is supposed to do. Determining the desirable characteristics of any system is a matter of experienced professional judgement. There are no perfect cognitive systems and thus no absolute standards.

6 Between cognitivism and constructivism

The main competitor for a cognitive philosophy of science is some form of social constructivism. Constructivism is currently the dominant meta-theory of science in the sociology of science and constitutes a widespread orientation in the history of science. Theories of science emphasizing cognitive operations have been marginalized in the broader science studies community because so much of what seems important to understanding the workings of modern science takes place in a public arena, and not only in the heads of scientists. This includes particularly the technology employed in theoretical as well as in experimental contexts.

Thinking of science in terms of systems of distributed cognition enlarges the domain of the cognitive in our understanding of science. It is typically assumed that there is a sharp divide between the cognitive and the social. From the perspective of distributed cognition, what many regard as purely social determinants of scientific belief can be seen as part of a cognitive system, and thus within the purview of a cognitive understanding of science. There is no longer a sharp divide. The cognitive and the social overlap.

6.1 Epistemic cultures

That the concept of distributed cognition potentially provides a bridge across the gap between the cognitive and the social is illustrated in a book by one of the founders of social constructivist sociology of science, Karin Knorr-Cetina. In *Epistemic Cultures: How the Sciences Make Knowledge* (1999), Knorr moves beyond promoting social constructivism to examine once again just how scientific work produces knowledge. She answers that it is done partly through the creation of *epistemic cultures*, cultures that may be different in different sciences. Her primary example is the culture of High Energy Physics (HEP) at CERN. She argues that the scale of these experiments in time (several years) and numbers of participants (several hundreds) has produced a communitarian structure in which the individual is effectively replaced by 'the experiment' as the *epistemic subject*.

She even evokes the concept of distributed cognition, although tentatively and without any elaboration. For example, in the introduction, she writes (1999, p. 25): In HEP, 'the subjectivity of participants is . . . successfully replaced by something like distributed cognition.' Much later, when discussing the management of large experiments, she remarks:

> Discourse channels individual knowledge into the experiment, providing it with a sort of *distributed cognition* or a stream of (collective) *self-knowledge*, which flows from the astonishingly intricate webs of communication pathways. (1999, p. 173, emphasis in original)

Again, commenting on how experiments are named, she makes the point emphasized by Hutchins:

> Naming, then, has shifted to the experiment, and so has epistemic agency – the capacity to produce knowledge. The point is that no single individual or small group of individuals can, by themselves, produce the kind of results these experiments are after . . . It is this impossibility which the authorship conventions of experimental HEP exhibit. They signify that the individual has been turned into an element of a much larger unit that functions as a collective epistemic subject. (1999, pp. 167–8)

In spite of the tentativeness with which she refers to distributed cognition, I think she has got it exactly right. In HEP, new knowledge is produced by distributed cognitive systems consisting of both humans and machines.

Knorr clearly assumes that if knowledge is being produced, there must be an epistemic subject, the thing that knows what is known. The standard assumption that permeates modern philosophy is that the subject is an individual human being. But Knorr's deep understanding of the organization of experiments in HEP makes this assumption problematic in that setting. Feeling herself forced to find another epistemic subject, she settles on the experiment itself. But could we not think of scientific knowledge more impersonally, say in the form: 'It has been scientifically established that . . . ' or: 'These scientific experiments indicate that . . . ' These forms of expression free us from the need to find a special sort of epistemic subject. Individuals cannot *produce* the knowledge in question, but they can in some ordinary sense come to *know* the final result.

A similar problem arises for Knorr because the traditional epistemic subject is a *conscious* subject. But if we take the extended experiment to be the epistemic subject, do we have to introduce an extended form of consciousness as well? Knorr is tempted by this implication. Speaking of stories scientists tell among themselves, she writes:

> The stories articulated in formal and informal reports provide the experiments with a sort of consciousness: an uninterrupted hum of self-knowledge in which all efforts are anchored and from which new lines of work will follow. (1999, p. 178)

And on the following page, she continues:

> Collective consciousness distinguishes itself from individual consciousness in that it is public: the discourse which runs through an experiment provides for the extended 'publicity' of technical objects and activities and, as a consequence, for everyone having the possibility to know and assess for themselves what needs to be done. (1999, p. 179)

Again, Knorr is not alone in making such connections. Such speculations could be avoided, however, if we adopt a more impersonal attitude toward scientific knowledge.

7 Implications

7.1 Distributed cognition in the history of science

It is often claimed that the scientific revolution introduced a new way of thinking about the world, but there is less agreement as to what constituted the 'new way'. The historiography of the scientific revolution has long included both theoretical and experimental bents. Those on the theoretical side emphasize the role of mathematics, Platonic idealization and thought experiments. The experimentalists emphasize the role of experimental methods and new instruments such as the telescope and microscope. Everyone acknowledges, of course, that both theory and experiment were crucial, but these remain a happy conjunction.

The concept of distributed cognition provides a unified way of understanding what was new in the way of thinking. It was the creation of new distributed cognitive systems. Cartesian co-ordinates and the calculus, for example, provided a wealth of new external representations that could be manipulated to good advantage. And the new instruments such as the telescope and microscope made possible the creation of extended cognitive systems for acquiring new empirical knowledge of the material world. From this perspective, what powered the scientific revolution was an explosion of new forms of distributed cognitive systems. There remains, of course, the historical question of how all these new forms of cognitive systems happened to come together when they did, but understanding the source of their power should now be much easier.

7.2 The cognitive study of science

Taking seriously the role of distributed cognition in science obviously has implications for the cognitive study of science as a whole. Here I will only note several of the most prominent. In general, insisting on a role for distributed cognition does not automatically invalidate other approaches. It does, however, call into question their often implicit claims to be the whole, or most of, the story about scientific cognition.

In a burst of enthusiasm for parallel distributed processing, Paul Churchland (1989, chapter 9, p. 188) suggests that we 'construe a global theory as a global configuration of synaptic weights' in a very high dimensional vector space. He seems to be claiming, for example, that the general theory of relativity can be represented in such a vector space. Now that might be a theoretical possibility, but it is almost certainly not how it is represented by real scientists. Rather, given our current understanding of distributed cognition, it is much more likely that the structure of scientific theories is a feature of *external* representations, such as systems of equations, diagrams and prototypical models. These are manipulated using the kind of pattern matching brain Churchland describes. All that need be explicitly represented in the brain is bits and pieces of the external representations necessary to facilitate external manipulations. The idea of parallel distributed processing, however, retains its relevance for understanding how science works.

A decade ago, Paul Thagard (1992) attempted to explain scientific revolutions as due to the greater explanatory coherence of a new theory, where explanatory coherence depends on logical relations among statements of theories. He suggested that these relationships might even be instantiated in the minds of scientists, leading them to choose the new theory over the old. Missing from this picture, of course, is any place for the influence of a new form of external representation or new instrumentation. The whole dynamics of theory choice is reduced to logical relationships among statements.

One of the most active areas of the cognitive study of science is cognitive development. The underlying assumption, often explicit as in Gopnik (1996a) and Gopnik and Meltzoff (1997), is that what we learn about conceptual development in infants and children can be applied to conceptual development in science. If it is true, however, that a major factor in the development of modern science has been the invention of elaborate means both for creating external representations and for observing the world, then the cognitive development of children in less scaffolded conditions loses much of its relevance. There may be much to learn about human cognition in general by studying cognitive development in children, but not much to learn specifically about the practice of science (Giere, 1996b).

16 The science of childhood

Michael Siegal

> One has all the goodness and the other all the appearance of it.
>
> (Jane Austen, *Pride and Prejudice*)

A fundamental aspect of children's scientific understanding concerns their ability to distinguish between reality and the phenomenal world of appearances. If children are to understand the nature of scientific phenomena, they need to recognize the workings of invisible causal mechanisms. In this chapter, I address research on the issue of conceptual change in children's scientific understanding, notably in the case of their understanding of biological causality.

1 The Piagetian view of cognitive development and scientific understanding – and beyond

The traditional Piagetian position is based on the notion that the thinking of young children is egocentric and limited by a realism in which they cannot accurately consider the underlying causes of effects. As children under the age of about seven years are restricted to focusing on only one visual aspect of a problem, they cannot see how processes that are invisible or microscopic create events. For example, Piaget (1929) asked children questions such as, 'Can a bench feel anything if someone burnt it?' or, 'Are bicycles alive?' He interpreted their answers to indicate that young children assign internal states and motives to inanimate objects. Children believe that any object may possess consciousness at a particular moment, especially a thing that moves like a bicycle. In other words, they do not easily distinguish between reality and the phenomenal world of appearances.

Other examples come from children's explanations for the appearance of shadows and for the origin of dreams. For Piaget (1930), young children typically believe that a shadow is a substance emanating from the object but occurring outside in the night. They realistically attribute animistic qualities to leaves in projecting shadows. Only with increasing age do children know how

I am grateful to commentators at the Cognitive Basis of Science Conferences at Rutgers in November 1999 and Sheffield in July 2000 for their many constructive suggestions.

shadows are projected and deny that objects cast shadows at night. Similarly, children are said to believe that the origin and location of dreams is outside the head of the dreamer, and that others can see them while the dreamer is asleep (Piaget, 1962). A fundamental conceptual change is required for children to relinquish these naïve views in embracing the scientific theories of effects that can be traced to invisible causal forces.

However, more recent research has shown that children as young as four years and, in some instances, as young as two do have knowledge of the distinction between appearance and reality. In a series of intriguing articles, Flavell and his colleagues (Flavell, Flavell and Green, 1983; Flavell, Green and Flavell, 1986; Flavell, 1993, 1999) have arrived at the conclusion that by the age of four years children have substantial knowledge of the appearance–reality distinction. Flavell devised tests to determine whether children can distinguish between the true nature of a substance and its appearance under, for example, coloured filters or masks and costumes. For example, in one task, children are shown milk in a glass with a red filter wrapped around it. The children were asked, 'What colour is the milk really and truly? Is it really and truly red or really and truly white? Now here is the second question. When you look at the milk with your eyes right now, does it look white or does it look red?' Whereas most four-year-olds succeeded on this task, less than half of three-year-olds tested by Flavell correctly identified the milk to *look red* but to *be white* really and truly.

In another of Flavell's studies, children saw an adult named Ellie put on a disguise as a bear. The children were asked, 'When you look at her with your eyes right now, does she look like Ellie or does she look like a bear? Here is a different question. Who is over there really and truly? Is she really and truly Ellie, or is she really and truly a bear?' On this task, the performance of three-year-olds was no better than what would be expected by chance. Two types of errors were equally common. Children would either say that Ellie both looked like and truly was a bear, or she looked like and truly was Ellie. In this sense, they may give 'phenomenalist' answers and report the *appearance* of an object when asked about its real properties or they may give 'realist' answers and report the *reality* of an object when asked about its appearance. By contrast with three-year-olds, children aged four and older are quite proficient at avoiding such errors.

Building on Flavell's work, it has been demonstrated that three-year-olds display an understanding of the appearance–reality distinction when asked to respond in the course of natural conversation to an experimenter's request (Sapp, Lee and Muir, 2000). In these studies, children were given deceptive objects such as a blue birthday candle covered by a blue crayon wrapper so that it looked like a crayon though in reality it was a candle, and a grey, irregular piece of sponge that looked like a rock. They were also provided with non-deceptive objects: a red crayon, a white birthday candle, a granite rock and a yellow sponge. Children's responses were in keeping with the findings of other studies in which an early inability to respond correctly on cognitive developmental measures can be attributed to children's conversational awareness of the purpose and relevance of the questioning rather than to a conceptual deficit (Siegal, 1997, 1999). At a nearly perfect success rate, they could pick out the correct object when the experimenter asked questions about appearance ('I want you to take a picture of Teddy with something that looks like a crayon. Can you help me?') and reality ('I want a sponge to clean up some spilled water. Can you help me?'). By contrast, children often responded incorrectly when questioned using Flavell's procedure. Therefore in this instance, children may have interpreted the purpose and relevance of appearance–reality questioning as a request to evaluate the effectiveness of a deception rather than as a task to determine whether they can distinguish between appearance and reality. These results are consistent with reports that children as young as two years can describe the appearance and display the real function of objects in the context of a show-and-tell game (Gauvain and Greene, 1994).

The early ability to distinguish between appearance and reality is compatible with the ability to use representations that are not confined to the world of observable appearances in drawing conclusions to problems that require scientific inferences. For example, in a seminal investigation, Shultz (1982, experiment 3) gave children aged three, five, seven and nine years three problems. The children viewed: (1) an electric blower and a lamp directed toward a spot of light on the wall; (2) an electric blower and tuning fork directed toward a candle light which was then extinguished; and (3) a lamp and tuning fork directed toward the opening of a wooden box which began to resonate audibly. Even many of the three-year-olds were able to determine the cause of the effect. They told the experimenter that the lamp created the spot of light on the wall, the electric blower extinguished the candlelight and the tuning fork made the sound come from the box. The children were able to identify the blower as irrelevant to the spot of light, the tuning fork as irrelevant to the candle and the lamp as irrelevant to the sound from the box. Shultz concluded that young children can understand cause and effect relations in a sophisticated 'generative' sense rather than as the simple co-variation of two events at the same point in time.

In the case of Shultz's studies, it could be said that the cause was a visible event and so that success on the problems did not require a stringent test of children's ability to distinguish appearance from reality. However, children have been shown capable of responding on logical reasoning tasks in which the premises are unfamiliar or run contrary to their own experience in the context of make-believe play situations. For example, Gelman, Collman and Maccoby (1986) told four-year-olds that boys have 'andro' in their blood and girls have 'estro' in theirs. Then they were shown pictures that conflicted with these properties, such as a boy who looked like a girl. The pre-school children were asked to infer whether the boy or girl had andro or estro. The children often correctly ignored perceptual information in specifying the properties of the children in the pictures.

In other studies, children aged four to six years have been presented with syllogisms such as, 'All fish live in trees. Tot is a fish. Does Tot live in a tree?' In such instances, they can use their make-believe abilities to overcome the bias in their beliefs that fish live in water and respond correctly in drawing the logical inference that the fish Tot does indeed live in a tree (Dias and Harris, 1988). Given appropriate contextual formats for questioning, even two- and three-year-olds can demonstrate such logic in their reasoning (Richards and Sanderson, 1999). In this sense, Flavell's research on the development of the appearance–reality distinction has generated research demonstrating that there is an emergent cognitive basis for scientific understanding in young children. Since appearance and reality usually do correspond, it is entirely rational to use appearance as a first step toward to predicting the reality of causal mechanisms – a strategy that in some domains may decline with age and experience (Schlottman, 1999).

2 Conceptual change in childhood: Carey's approach

Carey has made several strong claims about the nature of children and science. For one thing, she has maintained that 'young children are notoriously appearance bound' (Carey, 1985, p. 11). She has subscribed, for example, to the view (Kohlberg, 1966) that young children regard gender to be tied to appearances rather than a matter of biology; they believe that boys and girls can change into the opposite sex if they wear opposite sex clothes or play opposite sex games (Carey, 1985, p. 54). However, as Carey recognizes, concepts can change in many ways in that these can be differentiated, integrated, or re-analysed in the course of cognitive development. In this regard, her approach differs from that of Piaget. Her proposal (2000a, pp. 13–14) is that the 'main barrier to learning the curricular materials we so painstakingly developed is not what the student lacks, but what the student *has,* namely, alternative conceptual frameworks for understanding the phenomena covered by the theories we are trying to teach'. Children's initial core concepts in domains of knowledge can act as obstacles toward a mature understanding (Carey and Spelke, 1994, 1996; Carey, 1999).

Carey's position on conceptual change in childhood understanding of biology can be seen in relation to the work of Thagard (1996, 1999) who has provided a detailed analysis of the nature of conceptual change in how scientists reason about disease. Most notably, Thagard has made a distinction between conservative and non-conservative conceptual changes. Conservative conceptual changes involve extensions to existing concepts and beliefs – a form of 'knowledge enrichment'. The expansion during the twentieth century of the causes of diseases to include nutritional, immunological and metabolic factors does not challenge the standing of germ theory and thus involves conservative conceptual change. However, the humoral theory of disease causation held by the ancient Greeks has made way for germ theory in a process of non-conservative conceptual change. The Greeks, led by Hippocrates, proposed that imbalance in the humours or fluids of blood, phlegm, yellow bile and black bile associated with heredity, diet, or climate result in illness. It was not until Fracastoro in 1546 published his treatise on contagion that infection through the transmission of invisible particles was seen to be the cause of some diseases, although this proposal co-existed with the notion of humoral causation. As Thagard observes, in the nineteenth century, the eventual transition from humoral theory to germ theory involved a conceptual change that may be deemed non-conservative in substantial respects as new causal concepts and rules replaced the old ones.

For Carey (1995), young children's ideas about biology are said to progress through two phases of development. In the first phase, from the pre-school years to approximately age six, children learn facts about the biological world. For instance, pre-school children know that animals are alive, that babies come from inside their mothers and look like their parents, that people can get sick from dirty food or from playing with a sick friend, and that medicine makes people better. Though this knowledge is impressive, it is quite different from having a 'framework theory' that involves the connecting of facts to create a coherent, unified conceptual structure (Wellman and Gelman, 1992; Keil, 1994). Not until the age of seven years or so are children said to begin to construct a coherent framework theory of biology through a process of 'conceptual change'.

Carey maintains that such change in the domain of biology is liable to be of the non-conservative, strong variety. Using insights from the history and philosophy of science (especially Kuhn, 1962) and comparing shifts in the knowledge of adults who are novices in physics as opposed to experts (Larkin, 1983), she is committed to the view that: 'the child's knowledge of animals and living things is restructured in the years from four to ten. The kind of restructuring involved is at least of the weaker sort needed to characterize novice–expert shifts that do not involve theory change and most probably is of the stronger sort needed to characterize theory changes that involve conceptual change' (Carey, 1985, p. 190). Children's early concepts are undifferentiated in that these embrace notions that later appear to the child to be incommensurable and demand cognitive restructuring in the strong sense. Change involves differentiation and re-analysis to the extent that it involves strong restructuring and children acquire new causal concepts.

According to this analysis, young children cling to appearances rather than underlying realities that involve true causal mechanisms. In their biological understanding, they think that irritants such as pepper as well as germs transmit appearances that correspond to colds, that traits such as eye colour are the result of environmental influence rather than biological inheritance and that a dead corpse retains certain biological attributes of life. Change does not proceed very rapidly as children have presuppositions of the nature of the world and cling to their initial theories in the face of conflicting evidence. Thus it can be very hard for them to concede that death involves the breakdown of the organization and functions of body organs. They think that the buried still require bodily functions such as eating, breathing and excretion (Slaughter, Jaakola and Carey, 1999).

Carey's approach is a novel, provocative and powerful account of conceptual change in childhood. It therefore requires careful scrutiny. Below is an examination of research on children's biological understanding in three areas: germs and the biological basis for illness, inheritance of physical traits and life and the permanence of death.

3 Children's understanding of germs and the biological basis of illness

A biological conception of disease involves the notion that germs are living organisms that reproduce. However, results from several recent studies have suggested that children's early understanding falls short of this conception. For example, instead of identifying germs as living organisms that multiply, children aged four to seven years appear to maintain (1) that germs are not alive and do not eat or die and (2) that colds are equally likely to be transmitted by poisons or by irritants such as pepper as by germs (Solomon and Cassimatis, 1999). They may also express the view that illness transmission through germs occurs immediately on contact rather than after a period of incubation (Kalish, 1997) and that germs grow like tumours but do not reproduce inside the body (Au, Romo and DeWitt, 1999). Solomon and Cassimatis (1999) interpret this pattern of results to support Carey's analysis of young children's biological understanding. On this basis, it has been concluded that young children may make judgements about the role of germs in illness without considering biological causality at all. They regard illness as simply due to contact with noxious substances rather than as the outcome of microscopic infection by germs. They may simply be reproducing a learned fact ('colds come from being close to someone else who is germy with a cold') rather than embedding their knowledge of illness within a biological theory.

However, Hatano and Inagaki (1994) note that children are often involved in caring for other things and so learn rules for maintaining the health of animals and plants, as well as for themselves. This understanding is adaptive in that it enables them to form predictions about the behaviour of familiar natural kinds such as mammals regarding food procurement and reproduction. Therefore they should be capable of an early understanding in the domain of biology to the extent that they may in some respects be credited with an incipient framework theory. In this respect, the distinction between germs and poison may be a marginal case. As Kalish (1999, p. 108) has observed:

> Poisons and other chemical/physical entities can be viewed as mechanisms of contagion and contamination. Poison is clearly a contaminant. If one contacts poison one may become ill. Poison may also be a vehicle for contagion. For example, if someone gets a particularly virulent poison on his hand and then touches someone else, that second person may come to show the effects of the poison as well. This transfer of materials (and the effects of the materials) represents a coherent model of physical infection . . . Germs function like poisons, as physical agents of contamination.

Thus it is unsurprising that children may not be differentiating between germs and poisons; they may assume simply that a germ is a type of poison since it is an agent that is harmful to health. This assumption is itself adaptive as a conservative bias in judging a potential agent of contamination and contagion to transmit illness acts as a safeguard against illness, and is employed by adults on logical reasoning problems (Occhipinti and Siegal, 1994; Tooby and Cosmides, 1992). Such processes may extend to deontic reasoning in general (Manktelow *et al.*, 2000).

Moreover, even though children may be overinclusive in labelling irritants and poisons as agents of illness transmission and some may appear to believe that germs do not function as living organisms, their responses on the tasks used to test their knowledge do not necessarily reflect the depth of their understanding. In particular, with reference to children's early contamination sensitivity, Siegal and Share (1990) presented three-year-olds with a naturalistic situation in which they viewed some juice in the process of being contaminated by a cockroach. Contrary to the notion that young children lack the capacity to comprehend the nature of contamination, 77% consistently responded that the juice was not good to drink, 83% accurately evaluated the correct and incorrect responses of other children to the incident in both stories, 67% correctly inferred that, while a child may want a chocolate drink, he or she should be given an alternative, and 75% chose a sad face without prompting as a warning against the drink. Several results have been reported that are broadly consistent with these findings (Springer and Keil, 1991; Au, Sidle and Rollins, 1993; Rosen and Rozin, 1993; Springer and Belk, 1994; Kalish, 1996, 1997).

How, then, do children represent germs? If given limited instruction, children may be inhibited in applying their ability to distinguish between appearance and reality and may not be encouraged to represent germs as the biological organisms that they are. In absence of such knowledge enrichment, they may seem limited to a 'physical theory' of transmission. However, in practice, adults may not be so good at distinguishing between appearance and reality either. Rather, they can conceive of germs as nasty or 'cute' depending on their origin, and may become convinced that their loved ones are too pure to carry harmful germs

Fig. 1 A child's representation of a germ

and that others are impure and so are prone to transmit germs and infections (Nemeroff and Rozin, 1994).

Given that adults themselves often have inadequate representations of germs, that children may also not display these is hardly surprising. Yet if asked to draw a germ, this is what may emerge (see figure 1). Learning about germs, for example, is adaptive for survival and well-being. However, children may be deprived of a potentially enriched environment. Instead, they are often told that germs are to be avoided by keeping one's distance and that is all. Despite the now well-documented ability of children to distinguish reality from appearance, few children are asked to draw their ideas of a germ and to discuss how germs infect the body in the form of harmful live organisms. Germs in the form of a tiny menacing human-like organism can be used as a heuristic toward improving children's understanding about germs by unleashing their mental representation and pretence abilities, particularly as pre-school children can discriminate the basis for movement of animate and inanimate objects (Massey and Gelman, 1988). These representations can serve by analogy as a foundation for learning that germs are like other living things that eat, multiply and die though they cannot be seen. In this regard, explicit discussion about the microscopic basis of infection can lead to quite successful training programs (Au, Romo and DeWitt, 1999; Solomon and Johnson, 2000).

Thus there is a need to re-examine the extent to which children can (1) distinguish between the role of germs and specific instances of irritants and poisonous substances in the transmission of illness, (2) identify that germs are alive, feed and can be eradicated and (3) understand the process of germ incubation and prevention of germ infection. To ensure that children understand the purpose and relevance of the experimenter's questions, this investigation could take the form of experiments in which children are asked to recognize the correct answer from alternatives provided by other children (e.g. Siegal, 1996, 1997, 1999; Siegal and Peterson, 1999). As Koslowski (1996) demonstrates in her research on hypothesis testing, tasks in many studies aimed to measure scientific thinking have been inadvertently designed so that the plausibility of alternatives does not arise. Research in a variety of related areas – children's vulnerability to misleading information (Newcombe and Siegal, 1996, 1997), their understanding of lies (Siegal and Peterson, 1996, 1998) and their knowledge of the relation between knowledge and pretend states (German and Leslie, 2001) – converges to demonstrate the importance of presenting salient, relevant alternatives to children. This is necessary in order to rule out the possibility that their responses reflect performance limitations inherent in the experimental setting rather than a genuine conceptual deficit. Altogether, work using a variety of questioning techniques should provide new evidence on the extent to which children can be deemed to have a coherent framework theory and can possess an implicit biological knowledge about the role of germs in contamination and contagion.

4 Children's understanding of biological inheritance

Of related concern to the issue of disease understanding is whether or not children have an understanding of properties that are transmitted through biological inheritance and those that are transmitted by cultural influences such as through non-biological, adoptive parentage. According to Solomon *et al.* (1996, p. 152), 'to be credited with a biological concept of inheritance, children need not understand anything like a genetic mechanism, but they must have some sense that the processes resulting in "Resemblance to Parents" differ from learning or other environmental mechanisms'.

Solomon *et al.* (1996) investigated this issue by carrying out a series of four studies with children aged four to seven years. In their initial study 1, the children were told a story about a little boy who, depending on the counterbalanced version of the story, was born to a shepherd but grew up in the home of a king or vice versa. Before proceeding with the testing, the children were asked two control questions to ensure their comprehension, in the sequence, 'Where was the little boy born? Where did he grow up?' They were then asked test questions. These concerned, for example, pairs of physical traits and beliefs such as, 'When the boy grows up, will he have green eyes like the king or brown eyes like the shepherd?' and, 'When the boy grows up, will he think that skunks can see in the dark like the shepherd or that skunks cannot see in the dark like the king?' Many of the four-year-olds answered that both physical traits and beliefs are determined environmentally. Not until seven years of age did children often report that physical traits are associated with the biological parent and beliefs with the adoptive parent. The results of study 2 indicated that pre-school children recognize that physical traits cannot change whereas beliefs can change. However, critically, their judgements of whether beliefs can change were dependent upon whether this change was desirable or not. Study 3 replicated the results of study 1 using female story characters as did study 4 in which an attempt was made to lessen the environmental focus of the stories by showing the children only schematic pictures of the adoptive mothers rather than pictures of their homes. Thus the outcome of this set of studies permitted Solomon *et al.* to conclude that children undergo conceptual change in their biological understanding in that they start to differentiate biological from environmental influences within a framework theory only after the age of six years.

However, this position seems incompatible with other findings. For example, Hickling and Gelman (1995) report that children aged as young as four and a half years were generally able to identify that same-species plants are the sole originator of seeds for new plants of that species. Springer (1995) has shown that four- and five-year-olds who understand that human babies grow inside their mothers (77% of the total number of 56 children in his first experiment) possess a 'naïve theory of kinship', in that they can use this knowledge to predict the properties of offspring. They can say that a baby that is physically dissimilar to the mother will likely share her stable internal properties (e.g. 'grey bones inside her fingers') and lack transitory properties (e.g. 'scrapes on her legs though running through some bushes'). More recently, Springer (1999) has studied the effects of adoption on children's responses to phenotypic 'surprises'. He gave adopted and non-adopted children aged four to seven years situations such as, 'There's a red-haired baby. His hair is bright red. But his mom and dad both have black hair. How could that happen? How could two black-haired parents have a red-haired baby?' In contrast to the non-adopted group, most of the younger adopted children generated biological kinship explanations in referring to grandparents or properties running in families. To some extent, adopted children's concepts of kinship may simply be more 'accessible' than are those of non-adopted children. If asked simply to select outcomes even young non-adopted children may distinguish appearance as a trait that is biologically rather socially determined. Hirschfeld (1995, experiment 5) gave children aged three to five years two simple situations. In one, they were asked to indicate whether the baby of a black couple who grew up with a white couple would be black or white. The other situation involved the inverse in which the child of the white couple grew up with the black couple. Both the four- and five-year-olds clearly favoured nature over nurture and were able to give justifications to this effect.

What is the basis for the discrepancy between these various results? One possibility has to do with differences in the method of asking children questions. In this connection, as discussed elsewhere (Siegal and Peterson, 1999, p. 10), several potential concerns arise.

First, in the case of Solomon *et al.*'s (1996) method, pre-school children may have not easily followed the narratives that were used to evaluate their understanding of biological and adoptive influences, since the adoptive mother (who was present at the birth and brought the baby home from hospital) was depicted as loving her very much and immediately calling her 'daughter'. Pre-school children may have imported their own relevance to incidental features of the narrative and assumed that some relationship between the birth and adoptive mothers must have existed to make these behaviours possible (e.g. they were sisters or sisters-in-law). If so, heritable traits could conceivably be linked in the baby and the adoptive mother, especially as they were pictured as of the same race and no information about the fathers was provided.

Second, children's answers to questions about whether offspring physically resemble a biological or adoptive parent may be influenced by the description of the loving relationship between the biological or adoptive mother and the son or daughter. For example, children may interpret the question to imply that the relationship between the child and the adoptive parents is so loving that their answers should convey the strength of this relationship by indicating that it could even transform the appearance of the adopted child to resemble that of his or her adoptive parents. In this regard, the children were asked to make many more judgements about traits that are environmentally as opposed to biologically transmitted. According to Hirschfeld (1995, p. 239), this procedure may have prompted them to respond that even biological traits such as eye colour are the result of adoptive parentage.

Third, the two control questions in Solomon *et al.*'s studies were apparently asked in the same order (e.g. 'Where was the little boy born? Where did he grow up?') with the question about the biological parent first and the one about the adoptive parent second. Thus children might have answered the adoption question correctly simply because it was the more recent and salient.

Solomon *et al.*'s (1996) studies do provide valuable data relevant to children's biological knowledge. However, as is the case for children's understanding of germs and the biological basis for illness, the research to date falls short of providing decisive evidence on what children can and do know about biological influences on appearance, beliefs and behaviour. Further research is again needed on the nature of the conceptual change that takes place in development.

5 Children's understanding of life and the permanence of death

At an early age, children are adept at distinguishing between processes that are controlled voluntarily by the mind as distinct from involuntary processes controlled by the body. Inagaki and Hatano (1993) found that most four- and five-year-olds distinguish between bodily characteristics such as eye colour that are not modifiable in contrast to the modifiability of other bodily characteristics such as the speed of running and mental characteristics such as memory. Almost all can say that they could not stop their heartbeat or stop their breathing for a couple of days. However, this does not necessarily mean that they have an understanding of the biological mechanisms underlying life and death. Carey (1985) has proposed that the pre-school child's concept of death is incommensurable with that of older children and that it likely undergoes strong structuring in the course of conceptual change. For Carey, the pre-school child's has an undifferentiated core concept of 'life' in that life is a concept that embraces things as alive, active, real and existent. By contrast, 'death' refers to things are not alive, inactive, unreal and absent. This undifferentiated concept becomes incoherent and is replaced once children construct a vitalistic biology involving an understanding of biological processes such as eating and breathing. Further, Carey maintains that the pre-school child's concept of 'animal' is incommensurable with a biological understanding of animals. The core of this concept derives from its role in an early intuitive theory of psychology that is in place by age four, and is essentially a psychological concept involving an entity that behaves. Conceptual change occurs with the construction of a vitalistic biology. The concepts of animal and plant now coalesce into a single category, with *living being* at the core. The concept of death undergoes a re-analysis with death becoming the opposite of a living thing. A dead animal and plant is one that no longer eats and breathes.

According to this analysis, the pre-school child's concept of 'death' is essentially not biological. Young children should not recognize that death entails the cessation of the biological signs of life. As evidence, Slaughter, Jaakola and Carey (1999) report a study in which children, having been classified as life-theorizers or non-life-theorizers, were interviewed about death. Children

Table 1. *Questions in the death interview (Slaughter, Jaakola and Carey, 1999).*

1. Do you know what it means for something to die?
2. Can you name some things that die? If people aren't mentioned: Do people sometimes die?
   2a. Does every person die?
   2b. What happens to a person's body when they die?
3. When a person is dead . . .
   a. do they need food?
   b. do they need to pee and poop?
   c. do they need air?
   d. can they move around?
   e. do they have dreams?
   f. do they need water?
   g. If a dead person had a cut on their hand, would it heal?
4. How can you tell if a person is dead?
5. Can you name some things that don't die?
6. Can you think of something that might cause something to die?
7. Once something dies, is there anything anyone can do to make it live again?
8. Could a doctor make a dead person live again?

aged four to six years were classified on the basis of their answers to questions about the heart, brain, eyes, lungs, blood, stomach and hands. They were asked three questions about each body part: (1) 'Where is X?' (2) 'What is X for?' and (3) 'What would happen if somebody didn't have an X?' They were also asked questions about the bodily processes of eating food and breathing air: 'Why do we eat food/breathe air? What happens to the food we eat/air we breathe?' Those classified as life-theorizers mentioned staying alive, or not dying, at least twice. All children were then given a 'death interview' designed to measure children's understanding of various sub-components of the concept of death: recognizing only living things can die, knowing that death is a necessary part of the life cycle and how death ultimately comes about and understanding that death is a permanent state and involves a cessation of all bodily needs (see table 1).

As predicted, many of the non-life-theorizers indicated that the dead would retain biological functions. For example, of the 18 non-life-theorizers, the numbers who claimed that dead people need food, need to excrete, need air, move, need water, dream, and have cuts heal were 8, 5, 8, 1, 5, 6 and 10, respectively. Comparable figures for the 20 life-theorists were 1, 1, 0, 0, 2, 2 and 8.

Such results are certainly impressive. However, they represent a single set of data limited mostly to an open-ended interview methodology, and require confirmation through converging evidence using a variety of methodologies. At the present time, it is not difficult to conceive of other explanations for children's responses that do not appeal to strong restructuring in conceptual change.

On the one hand, responses that indicate a belief in the dead possessing the biological attributes of life may reflect aspects of a particular questioning methodology as children are notoriously likely to make up responses to well-meaning questions that are interpreted in ways other than those intended. For example, the position has been challenged that pre-school children lack gender constancy in that they believe gender corresponds to appearance. Rather, children attribute responses that boys and girls can change into the opposite sex if they wear opposite sex clothes or play opposite sex games to be due to pretence rather than to genuine conviction that appearance and reality are one and the same (Siegal and Robinson, 1987; Bussey and Bandura, 2000). Similarly, asking children to evaluate whether the dead possess biological properties of life in terms of real and pretend outcomes may give more accurate indication of concepts of life and death.

On the other hand, children's responses may indicate that they genuinely do believe the dead possess biological life attributes. However, in this respect, it is unclear how far children depart from adults in the rationality of their beliefs for it is apparent that children are quite susceptible to normative cultural influences from an early age. For example, in India, their judgements of whether contaminated food can be made edible are influenced by whether the mother has tasted it first (Hejmadi, Rozin and Siegal, 2002). In Italy, whether young children maintain that the perpetrator of a deception has lied or make a mistake can rest upon whether the culprit has previously been blessed in church and thus can be regarded in this sense as 'holy' (Siegal *et al.*, 2001). In the same way, the expressed beliefs of American and Australian children in Slaughter *et al.*'s research that the dead possess the properties of life may reflect beliefs that are held in the wider culture. These involve a normative, anthropomorphic view of the outcomes for persons who have died but live on in a form of heaven (or hell) that is held by many adults (Barrett and Keil, 1996). Generally, as noted above, young children's judgements that beliefs can change are influenced by whether they regard change to be desirable (Solomon *et al.*, 1996). At an early age, children are sensitive not only to the importance of individual desires in predictions of behaviour but also what is regarded as desirable in their culture (Rieff *et al.*, 2001).

As Botterill and Carruthers (1999) point out, that humans have limited cognitive abilities requires a consideration of both formal logical and cultural norms for rationality. Therefore in an analysis of what constitutes human rationality, reasoning in line with the dictates of culture can be rational in that it very often provides good problem-solving solutions when faced with cognitive limitations (Evans and Over, 1996; Occhipinti and Siegal, 1996). A key concept in this instance may that be of 'vitalism' – involving the transmission of energy by an organism. By this account, vitalism is not necessarily beyond the grasp of young children. It can be seen as a rational metaphor used to some unidentified causal mechanism that is distinct from an organ employing psychological attributes of intentionality to maintain biological functions. Thus both young children and adults are more likely to ascribe to the conception of the stomach as taking in energy from food than of the stomach wanting or thinking about taking in food (Morris, Taplin and Gelman, 2000). In this sense, vitalism serves as a form of 'placeholder' that dissolves when the causal mechanism is identified.

6 Conclusion: conservative vs. non-conservative conceptual change

In the cases of germs and the biological basis for illness, inheritance of physical traits, and life and the permanence of death, children can be capable at an early age of recognizing the existence of invisible causal processes though they are not yet able to specify their nature. Intelligence in many aspects of biology can be seen as an adaptive specialization that involves preparedness for acquiring knowledge (Rozin and Srull, 1988; Siegal, 1995). Thus development in the domain of biology may often involve knowledge enrichment and conservative conceptual change. Until alternative explanations are eliminated, there is reason to be cautious in embracing the position that children's biological understanding necessarily undergoes a non-conservative conceptual change that requires a strong restructuring in which previous causal concepts that are incommensurable with the culturally received scientific view are substantially abandoned.

The extent to which non-conservative conceptual change does take place depends on the domain in question. As researchers such as Carey (1985, pp. 191–2) and Rozin (1976) point out, cognitive development is probably better characterized in terms of change that is specific to domains rather than as following a general stage sequence. The pattern of performance varies so greatly across different tasks that an explanation in terms of a Piagetian notion of décalage is inadequate. Clearly, each domain of knowledge has its own rules and criteria for understanding. As such, conceptual development in each domain must be considered on its own merits. Here it is instructive to contrast knowledge in biology with knowledge in the domains of astronomy and number.

It has been maintained in the domain of astronomy, for example, that children's early concept of the shape of the earth reflects a synthesis of the spontaneous observation that the earth is flat with the culturally received notion that the earth is a round sphere. Thus their entrenched pre-conception of the earth as flat precludes a mature grasp of a spherical earth concept and children may indicate that the earth consists of a hemisphere topped by a platform on which people live (Vosniadou and Brewer, 1992). Yet Australian children are likely to display an early understanding that one cannot fall off the bottom of a spherical earth and that the earth's revolution rather than the sun's creates the day–night cycle. They recognize that it is daytime in Australia when it is night for family and friends in England and vice versa. It is perhaps the early knowledge conveyed by the culture that Australia is a distinctive landmass remotely located from major population centres that prompts an advanced geographical and astronomical understanding, pre-empting the need for restructuring in this aspect of the conceptual development of Australian children (Butterworth *et al.*, 2002). Therefore a universal, non-conservative conceptual change in children's scientific reasoning may not be needed in this domain. Rather, the process may be better characterized in terms of knowledge enrichment that involves an accumulation of scientific facts backed up by powers of mental representation. But with regard to children's astronomical understanding, more research again needs to be carried out in examining relevant theoretical and methodological issues.

By contrast, a strong restructuring in conceptual development of the sort proposed by Carey may apply to other domains. In particular, number is a domain that may be demanding of a non-conservative conceptual change (Gelman, 1991, 2000). Though counting appears to occur spontaneously and even pre-verbally in young children, multiplication and division need to be taught formally (Geary, 1995). In particular, children's core concept of number seems to be limited to the natural integers used for counting. They display considerable difficulty in accommodating their theory that number consists of the counting integers to account for fractions as numbers that fill in the gaps between the integers and to represent both fractions and integers in terms of infinity. Indeed, it may be the understanding in this domain that sweeps across others such as biology and astronomy and allows science to be systematized and explicitly accessible in cognition. Mathematical representations of this sort may allow children to think of an infinite number of germs in reproduction and of the infinity of the solar system and the celestial bodies within it.

In conclusion, the knowledge that children can have about science is often implicit and skeletal but is characterized by an awareness that permits them to separate reality from appearance. The analysis that young children have little or no understanding of causal processes involving knowledge of the distinction between appearance and reality is not in accordance with the developmental evidence. In the domain of biology, to date there is no decisive data that children normally undergo a process of strong restructuring in conceptual change. Astronomy is a different domain from biology and is liable to entail a different pattern of conceptual change, as is the domain of number. This complexity warrants caution in the scientific investigation of conceptual change in its many varieties.

17 What do children learn from testimony?

Paul L. Harris

Contemporary accounts of cognitive development have emphasized that children build up a conception of the world on the basis of their own personal observation. The extent to which children also make use of testimony provided by other people, especially adults, has rarely been systematically examined. I argue that children are well equipped to make use of such information. Evidence for children's early use of testimony emerges from research on their understanding of the shape of the earth, the relationship between mind and body, and the origin of species. By implication, adult testimony helps children to construct a more objective or 'scientific' conception of the world. However, further evidence also illustrates that children also use adult testimony in developing various metaphysical ideas, for example concerning the special powers of God.

Two views of children's use of testimony are then considered. On the 'amplification' view, adult testimony is a useful supplement to children's personal observation but does not fundamentally alter the nature of the information that children gather. On the 'transcendence' view, adult testimony is not just a supplement to personal observation. It can provide children with a new type of information – information that is not grounded in direct observation. Children's receptivity to such information raises the question of when and how they begin to differentiate between different bodies of belief, notably between science and religion.

1 What do children learn from testimony?

Many contemporary accounts of cognitive development assume that children revise their conception of the world in the wake of incoming evidence. The exact way that revision is triggered by evidence is not well understood but in this respect accounts of conceptual change in the course of children's cognitive development are probably not much worse off than accounts of theory change in the history of science. There too explanatory, as opposed to descriptive, accounts of how conceptual change is brought about are rare.

In this chapter, I shall not tackle the question of the way that concepts or theories are revised in the light of evidence. I focus instead on a relatively neglected issue: the nature of the evidence gathered by the child. In particular, I consider the sources of evidence that the child uses. At first sight, there seems little to discuss. Surely, children – like scientists – make observations, and try to make sense of those observations in the light of their existing theories. Admittedly, children do not deliberately set out to produce critical observations, as do experimental scientists. Rather, in the course of their everyday commerce with the world, they are bombarded by a range of observable events. Yet so long as they can observe and encode those events, the fact that children make no deliberate effort to gather evidence need not prevent that evidence from appropriately undermining or reinforcing their existing concepts. On this account, then, children differ from scientists in that they do not actively seek out decisive data. Nevertheless, they resemble scientists in that they are 'brought to book' by a mixed set of observations, some consistent with their existing expectations, some not.

I shall argue that this conception of the child as observer–scientist ignores an important issue – namely the extent to which children make use of verbal testimony[1] as a source of information about the world. As a result, cognitive–developmental theory has adopted an overly individualistic account of conceptual change in childhood.

2 The possible role of testimony

The eighteenth-century Scottish philosopher, Thomas Reid, made the striking proposal that human beings are innately equipped to process two distinct sources of information: information about the world that is provided directly by their own senses, and information about the world that is provided by their fellow human beings in the context of verbal testimony (Reid, 1764/1997). We are naturally disposed, he argues, to treat these two sources of evidence in an equivalent fashion. Moreover, we are inclined to trust or believe in the evidence provided by either source. Admittedly, there are occasions when we come to treat the evidence of our senses as misleading or illusory – we distinguish between the real identity and properties of an object on the one hand and its apparent identity or properties on the other. Yet our natural inclination is to ignore the reality–appearance distinction. Indeed, it is a developmental achievement of some magnitude to make the distinction between what really is the case and what appears to be the case in the absence of full perceptual access – a distinction that is understood somewhere between three and five years of age, according to the latest and best evidence (Flavell, 1986; Wellman, Cross and Watson, 2001).

[1] I follow philosophical convention in using 'testimony' in the broad sense of 'assertions of purported truth or fact' rather than in the more narrow sense associated with the evidence provided by a witness in a court of law.

Similarly, it is plausible to suppose that the natural or default stance is to ignore the distinction between veridical and misleading testimony and to treat all testimony as accurate. Again, there will be occasions when we come to regard a given piece of testimony as wrong or misleading. We distinguish, for example, between the actual location of an object and where someone asserts it to be. Yet our natural tendency is to accept information that is provided by testimony. As in the case of the reality–appearance distinction, the ability to distinguish between what someone claims to be the case and what really is the case is not available to the very young child. Although the developmental evidence is less clear-cut, the ability to make that distinction appears to emerge in the pre-school period.[2]

Thus, in line with Reid's arguments, we may propose that children enjoy two sources of evidence – the evidence provided by their senses and the evidence provided by the testimony of their fellow human beings. In neither case are children initially inclined to doubt that evidence. Moreover, even when they are capable of such doubt, it is likely that the default position remains one of trust.

However, the claim that children are disposed to use and trust both sources of information is not one that is easy to square with current descriptions of early cognitive development. These descriptions typically present the young child as a stubborn autodidact – a creature who is prepared to revise his or her ideas in the light of personal observation but resists, or is slow to assimilate, information that is provided by the testimony of other people. For example, if we examine Piaget's account of cognitive development, the child is portrayed as someone who acts upon the world – making objects disappear and reappear or bunching and spreading a group of objects – and learns from such personal observation (Piaget, 1952, 1954). Testimony by other people is not considered to play a major role in the child's eventual understanding of object permanence or number conservation. These concepts are primarily discovered – or constructed – by the child.

Moreover, contemporary portraits of the child as scientist (Gopnik, Meltzoff and Kuhl, 1999b) do not propose any fundamental change in the learning procedures of the pre-verbal infant as compared to the older child. In each case, domain-specific observations are made and used to modify existing concepts. This presumed continuity between infancy and childhood implies that the arrival of a completely new source of evidence toward the end of the infancy period (namely, the testimony of other people) has no major impact on the course of cognitive development. If testimony did play a role, we might expect either a change in the rate of progress, granted that children would have two distinct sources of information rather than one, or, alternatively, we might expect a

[2] By the age of approximately three–four years, children start to deliberately produce false utterances in order to mislead other people (Polak and Harris, 1999).

change in the nature of that progress, insofar as testimony is likely to offer children information that they could not normally obtain via personal observation.

In summary, cognitive developmental theory has emphasized what children learn from their own investigation. The potential contribution offered by the testimony of others has not been systematically investigated. There are, of course, caveats to this broad generalization. For example, Piaget (1929) recognized that, in response to questioning, young children would offer both their own spontaneous ideas and also tutored beliefs. Similarly, contemporary research on the way that children acquire various scientific concepts has underlined that children's spontaneous ideas may not marry up with the concepts that are presented to them in the classroom (Chinn and Brewer, 2000). Nonetheless, the emphasis has been on the fact that children initially construct their own ideas and are relatively resistant to information offered via testimony. The possibility that young children might be readily disposed to construct or change their ideas about the world on the basis of testimony has rarely been seriously examined.

In the sections that follow, I review evidence suggesting that young children do benefit from information acquired by testimony. My strategy is to focus on those claims for which children would have difficulty in obtaining relevant, first-hand evidence. I first discuss children's understanding of various scientific conclusions – regarding the shape of the earth, the relationship between the mind and the brain and the origin of species. With respect to each of these conclusions, children are presumably not in a position to make the kind of empirical observations that have lead to such conclusions – but they are routinely presented with the conclusions themselves. I then focus on children's understanding of various theological claims regarding the powers of God. With respect to these latter claims, it is not clear that there are any relevant empirical observations to make. Again, however, children are routinely presented with the claims themselves.

To the extent that children understand and adopt any of these scientific claims, it is plausible to draw the inference that they relied on the testimony of other people. There are certain risks inherent in my strategy of focusing on claims for which children could not normally make the relevant observations. In the first place, this strategy can only establish the role of testimony with respect to a restricted class of conclusions – precisely those where first-hand observation by the child is unlikely to have played any role. Hence, my strategy may fail to identify cases where testimony is actually used by the child even when the alternative possibility of gathering evidence first-hand is available to the child but not actually deployed. However, for the time being, this means that my strategy is a conservative one: it is likely to underestimate rather than overestimate the cases in which children rely on testimony.

A second risk is that my strategy rests on an informal assessment of the extent to which relevant first-hand observations can be made and such a casual assessment might overlook ways in which children can gather information first-hand. For example, in learning about the shape of the earth, children might learn from models or pictures of the earth. However, without the assistance of others children can rarely interpret such representational devices. Some labelling or explanation is typically required, and these aids to interpretation constitute a form of testimony.

Below I present various pieces of evidence indicating that children do indeed benefit from adult testimony. I then step back to ask how we should construe that benefit. Should we regard children's use of testimony simply as an amplification of their powers of personal observation, or should we see it as something more radical – a source of information that is likely to re-set the child's cognitive agenda?

3 The shape of the earth

Vosniadou and Brewer (1992) emphasize that young children subscribe to two presuppositions about the earth that are rooted in their own personal observation and experience: they believe that the ground is flat and that unsupported things fall. They argue that both of these presuppositions stand in the way of children's acceptance of the claim that the earth is a sphere with people inhabiting both its upper and lower surface. Nevertheless, children do come to accept that claim. When Vosniadou and Brewer (1992) asked children ranging from six to eleven years where you would end up if you walked for many days in a straight line, 63% denied that you would fall off the edge of the earth. In a replication study that included younger children and introduced more child-friendly methods,[3] Butterworth *et al.* (2002) posed a similar question: 'If you walked for many days in a straight line, would you fall off the edge of the world?' They found that denials were widespread even among pre-school children as shown in figure 1. Thus, whatever young children's early observations and presuppositions about the world, and despite its comparative recency – at least in relation to the overall cultural history of *homo sapiens* – children come to accept the claim that the

[3] Vosniadou and Brewer (1992) asked children: 'What is below the earth?' – a question that most children answered (understandably) by referring to what is below the ground (e.g. 'dirt') rather than below the planet earth. Although Vosniadou and Brewer appropriately ignored children's replies to this confusing question, it may have contaminated children's interpretation of what the experimenter meant by the term 'earth' in subsequent questions. For example, after children had drawn a picture of the earth (the large majority drew a circle), Vosniadou and Brewer introduced a picture of a house and asked them to explain how in that picture 'the earth is flat'. Again, children might reasonably have construed the experimenter as referring to the ground rather than the planet. The overall effect of such misinterpretations would have been to mask children's appreciation of the spherical nature of the earth. Nonetheless, despite the possibility that such misinterpretations may have occurred, it is worth emphasizing that among those children whose mental model of the earth could be assigned to a distinct category, the model that was most frequently adopted was that of a sphere. Consistent adoption of a 'flat earth' model was rare. This overall conclusion is consistent with the findings of Butterworth *et al.* (2001).

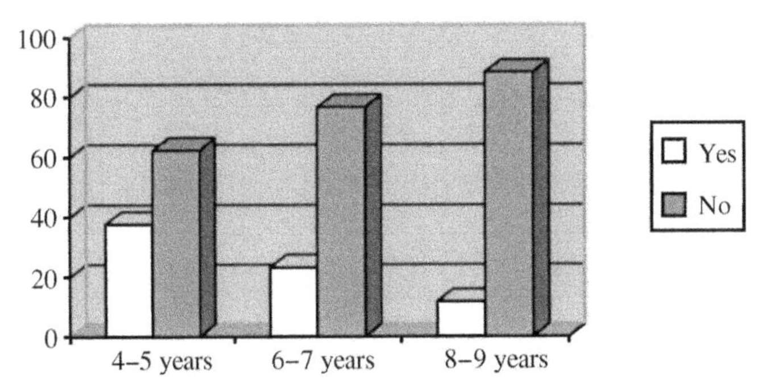

Fig. 1 Number of children in each of three age groups offering a 'yes' or 'no' reply
Source: Based on Butterworth *et al.* (2002).

earth is a sphere. Moreover, it is unlikely that children are simply parroting an isolated, factual claim that they have heard adults make. After all, children are not usually told what happens if you walk for several days in a straight line. To answer this question, children must presumably work out the implications of the claim that the earth has a spherical shape – namely that you cannot reach an edge even if you keep on walking.

4 The relationship between mind and body

A large body of work now attests to young children's developing conception of mental states, notably beliefs and desires. Arguably, children's understanding in this domain is largely based on their own observations – whether of other people or of themselves. However, children's understanding of the neural correlates of mental states, and more specifically their understanding of the link between their own mental states and their own brain is not likely to be based on personal observation. By implication, any appreciation of the connection between mind and brain must depend on testimony, and we may ask at what point in development such an appreciation emerges.

To examine this issue, Johnson and Wellman (1982) probed young children's conception of the brain. They found that even four- and five-year-olds recognize that the brain is needed to engage in cognitive activities such as thinking or remembering but they fail to understand that it is also needed for non-cognitive activities. Older children increasingly understand the brain's role in non-cognitive activities as well – for example, motor movements or feelings. Subsequent research established that children appreciate that it is particularly the brain and not just any internal organ that underpins mental processes. Johnson (1990, study 2) asked children about the consequences of a brain transplant for the ability to

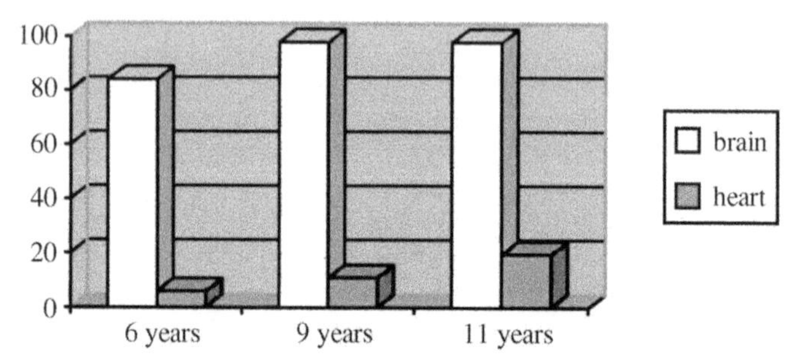

Fig. 2 Percentage of children who claimed that a brain transplant and a heart transplant would affect the performance of cognitive activities
Source: Based on Johnson (1990, study 1).

engage in various cognitive activities – counting, remembering, and so forth. In a control set of questions, children were asked about the consequences of a heart transplant. Even six-year-olds distinguished between the effects of the two organ transplants, typically claiming that a brain transplant would affect performance whereas a heart transplant would not. This clear-cut pattern of results is illustrated in figure 2.[4]

Finally, Johnson (1990, study 1) posed five- to nine-year-olds a more indirect and taxing question about the mind–brain connection. Children were asked about the sense of personal identity that a creature would have following a brain transplant. Specifically, they were asked to say how a pig would respond to the question: 'Who are you?' if its own brain were replaced by that of a child – indeed by the brain of the child being questioned. Five- and six-year-olds mostly ignored the hypothetical transplant and said that the pig would still claim to be a pig. Older children, by contrast, said that the pig would now claim their own identity. The pattern of results is illustrated in figure 3. The replies of the older children show that they have not just grasped the general connection between mind and brain – they have also come to recognize the role of the brain in the maintenance of personal identity.

Taken together these results indicate that children build an important and coherent connection between a domain that they can effectively encounter first-hand, through their own experience, namely the domain of mental activities, and an organ that they presumably learn about exclusively from other people's reports. Note, however, that children's emerging grasp of this connection is not likely to be based on explicit teaching even if it does depend on testimony.

[4] Johnson (1990; study 2) asked about two directions of transplant: other to self and self to other. The results shown in figure 2 are averaged across this manipulation, which had little effect on the overall pattern.

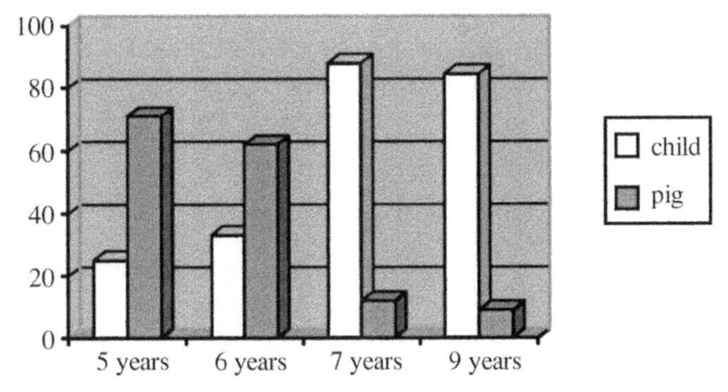

Fig. 3 Percentage of children in four age groups saying that pig with child's brain would claim to be a pig or a child
Source: Based on Johnson (1990, study 1).

First, even nursery children have some understanding of the connection. Second, Johnson and Wellman (1982) found that children's replies were relatively similar both before and after their exposure to a school science unit about the brain. By implication, the pattern of results reported by Johnson and Wellman reflects what children learn from 'informal' testimony – everyday dialogue that arises outside of the classroom.

Finally, it is worth noting that Piaget (1929, chapter 1), in the context of his pioneering investigation of young children's conception of mental phenomena, also recorded references by young children to the brain. However, he noted such references among children aged seven or eight years and upward and when they occurred he was inclined to regard them either as empty phrases or as an index of the way that children distorted ideas they had assimilated from adults.

5 The origin of species

It would not be surprising if children gave some thought to the origin of life in general and certain species in particular. Once they have grasped the cycle of life and death, and the succession of generations, the way is open for them to ponder how life came into being. Evans (2001) probed children's conception of the origins of life by asking them how the very first member of a given species got here on earth. In one task, for example, children were asked to say how much they agreed with different types of explanation. I focus on three key explanations, namely *Creationist* explanations (e.g. 'God made it and put it on earth'), *Evolutionist* explanations (e.g. 'It changed from a different kind of animal that used to live on earth') and *Spontaneous generation* explanations

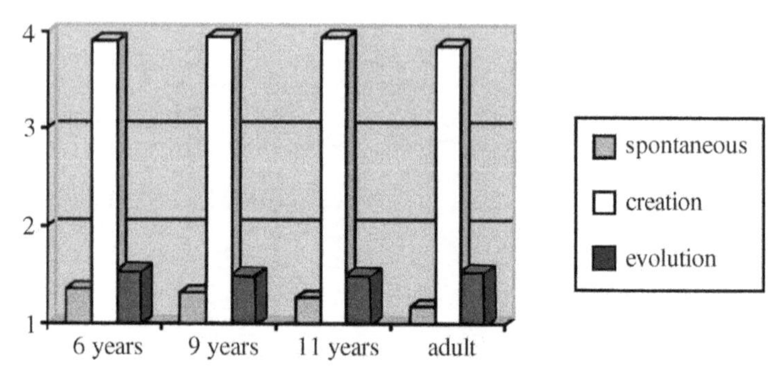

Fig. 4 Degree to which three different types of explanation were accepted by children and adults in a fundamentalist sample
Source: Based on Evans (2001).

(e.g. 'It just appeared; it came out of the ground'). Figures 4 and 5 show the replies of two different North American samples – fundamentalist children (plus their parents) attending Christian schools and non-fundamentalist children (plus their parents) mostly attending public (i.e. community-maintained) schools. The figures show that the pattern of development is different in the two samples. Creationist explanations are firmly endorsed throughout the course of development in the fundamentalist group, whereas evolutionist or spontaneous generation explanations are rarely accepted.

In the non-fundamentalist community, by contrast, there is a less marked endorsement of creationist explanations. Moreover, there is a developmental shift in the relative frequency with which creationist and evolutionist explanations are accepted, with the former yielding ground to the latter in the course of development.

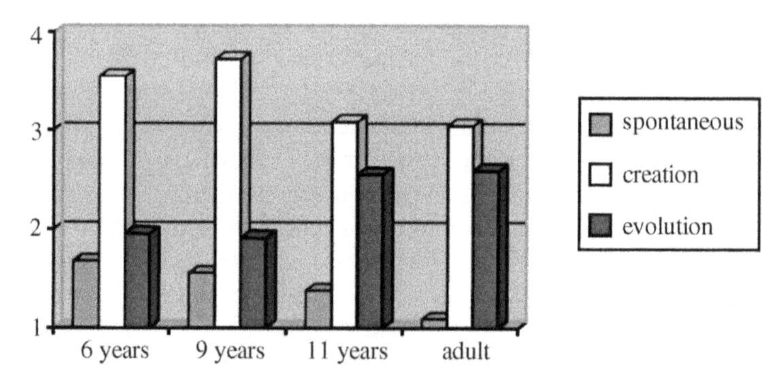

Fig. 5 Degree to which three different types of explanation were accepted by children and adults in a non-fundamentalist sample
Source: Based on Evans (2001).

Assuming that children in these two communities have roughly the same relatively meagre data-base of pertinent first-hand observations at their disposal, the most plausible way to interpret the different pattern of development in each community is to suppose that the type of adult testimony that is on offer – whether in the context of the home or the school – is different. Evans (2001) went on to explore various factors that might be associated with the acceptance of a given type of explanation. There were indeed several indications that children's endorsement of a particular explanation varied depending on the type of testimony that they had assimilated. Thus, among the six-year-olds endorsement of creationist explanations was associated with attendance at a fundamentalist school. Endorsement of evolutionist explanations, by contrast, was associated with being well informed about fossils.[5]

6 Interim review

The evidence presented so far shows that young children are influenced by the testimony of adults in drawing conclusions about the shape of the earth, the relationship between mind and brain and the origin of species. With respect to each of these issues, they are disposed to accept – depending on the type of testimony to which they have been exposed – claims that have been established quite recently in human history. Moreover, acceptance is not something that children slowly display in the course of explicit teaching in the classroom. Instead, it appears that children come to school having already assimilated fundamental physical, biological and psychological conclusions. By implication, the world picture that young children construct today is different from the world picture that they constructed in earlier millennia. To the extent that children cannot establish for themselves whether the earth is round or flat, whether they think with their brain or their heart, or whether dinosaurs owe their existence to God or evolution, we are led to the conclusion that children give credence to adult testimony on these matters.

This is not to say that there is no developmental change in children's beliefs. In all three domains, younger children were likely to reply less systematically and sometimes differently as compared to older children. Arguably, such developmental shifts indicate an early overall predilection for conclusions rooted

[5] If we focus, not on the differences between the two communities, but on their similarities, then the relatively widespread endorsement of creationist explanations comes into focus, especially among six- and nine-year-olds. One potential interpretation of that widespread endorsement is that it reflects some natural or easily available mode of causal explanation, namely the invocation of a quasi-human agent. However, it may also reflect the fact that the invocation of such an agent is frequently encountered in the testimony to which children are exposed, whether they are growing up in a fundamentalist or non-fundamentalist community. Stated bluntly, few elementary school children are likely to live in communities where most adult testimony offers a ringing endorsement of Darwinian theory.

in personal observation – a predilection that gives way, in the course of development, to credence in adult authority. However, this interpretation of the developmental evidence is not very convincing. So far as the shape of the earth is concerned there is no indication even among four- to five-year-olds of a majority belief in a flat earth. With respect to the link between mind and brain, it is not obvious what personal observation could lead children to any particular commitment. Finally, if we examine children's responses concerning the origin of species, we have evidence of at least some susceptibility to adult testimony even among six-year-olds. A comparison of figures 4 and 5 shows that the endorsement of creationist explanations is more common at this age among children growing up in a fundamentalist community as opposed to a non-fundamentalist community.

In short, two conclusions emerge from this admittedly brief review. First, children draw on adult testimony concerning the nature of the physical, biological and psychological world. Second, acceptance of such testimony is not a late-emerging phenomenon that occurs when children are taught about science in the classroom. It is found among children who are just starting their school career as well as those who are more advanced.

7 The 'amplification' view

One interpretation of the findings reported so far is that adult testimony serves to amplify the range of mundane phenomena that children can 'observe'. It offers them, vicariously, a superior vantage-point. On this view, testimony serves as a kind of telescope, or X-ray device, or time-machine. It provides them with information about phenomena that would otherwise lie beyond their powers of observation. Human beings – and young children in particular – will frequently not be in the right place at the right time to make particular observations. The testimony of others offers a valid surrogate. By implication, testimony is reliable in that it does not typically offer false or misleading information. Such a view of adult testimony has been set out by Gopnik, Meltzoff and Kuhl (1999b):

> Babies depend on other people for much of their information about the world. But that dependence makes babies more in tune with the real world around them, not less. It gives children more and better information than they could find out for themselves. As children we depend on other people to pass on the information that hundreds of previous generations have accumulated. Together the children and the grown-ups (and other children) who take care of them form a kind of system for getting at the truth.

However, there are two problems with this conception of testimony. First, it implies that, irrespective of culture, children make cognitive progress in the same direction, namely toward the truth. Yet, as we have already seen, children adopt different conclusions about the origin of species depending on the community in which they are growing up. Children growing up in a fundamentalist community rarely accept evolutionary explanations whereas children growing up in a non-fundamentalist community increasingly do so. Of course, it might be argued that children in these two communities differ simply in the amount of veridical testimony that they receive. This would allow us to retain the notion that children and adults alike constitute a system for 'getting at the truth'. By implication, some communities form a better system for getting at the truth than others, but they are all engaged in the same fundamental enterprise. However, this somewhat Panglossian defence is not very plausible. It seems more reasonable to conclude that children in the two communities are given mutually inconsistent accounts of what happened in the course of pre-history rather than accounts that differ only in the amount of true information that gets included.

A second difficulty with the amplification view is its tacit assumption that testimony is simply a matter of empirical report, and stands in for direct observation. The recipient – having been unable to observe a particular phenomenon first-hand – is given an opportunity to contemplate it by proxy because his or her informant was better placed to observe the phenomenon in question. In this fashion, for example, young children might learn from their parents some key facts about their own birth, or indeed about birth in general. Yet parental testimony is not confined to this kind of testimony. Children will often be given information that does not amount to a straightforward empirical report. In particular, depending on their parents' convictions, children will be presented with metaphysical assertions that are not open to empirical checks in any ordinary sense. For example, they will be told about the nature and existence of God or the life hereafter. The data presented earlier on children's belief in the creation of species offer another suggestive example. If children attend to and remember such claims, their conceptual universe will scarcely be based on evidence gathered by their own powers of observation amplified by the testimony of previous generations. It will incorporate various metaphysical claims that go beyond mundane observation. Of course, it might be argued that children typically winnow out such claims, especially when they fly in the face of common sense. At most, they accept them when they are neutral or agnostic about the issue in question. On this view, children would be relatively abstemious in their reliance on testimony. This possibility is discussed in the next section.

8 Omniscience and immortality

Most religions include beliefs about extraordinary powers or possibilities. The Christian God, for example, is claimed to be immortal and omniscient. Do young children reject such claims because they run counter to their notions of what is ordinarily possible? A considerable body of evidence has shown that children of four to five years understand that a person's knowledge of events or objects is constrained by his or her perceptual access – the opportunity to observe the events or objects in question. Any acknowledgement of omniscience would presumably run counter to this assumed psychological constraint. Similarly, children of five to six years understand the nature of the life-cycle: they appreciate that growth is uni-directional and that death is its end-point. Hence an acknowledgement of immortality would run counter to this assumed biological constraint. Granted these considerations, one might expect young children to conceive of God in more or less human terms, denying Him any special status. On this argument, children who are confronted with Christian claims about the omniscience or immortality of God would be expected to ignore or reject such claims. In short, they would display 'testimonial abstinence'.

However, it can also be argued that the very ubiquity and continuity of such claims in a variety of religious ontologies is an index of their cognitive salience and longevity. Anthropologists of religion have made the important point that when we consider the epidemiology of belief systems – and notably their transmission from one generation to the next – only attention-grabbing, comprehensible and memorable principles are likely to survive cross-generational transmission (Boyer and Walker, 2000). This proposal implies that the paradoxical nature of religious beliefs is a key component of their continuity across generations. Far from rejecting or resisting such claims, young children would heed their implications early and easily. More generally, this argument implies that children would latch onto certain kinds of testimony even when that testimony is not an empirical report but a metaphysical claim that flies in the face of ordinary empirical observation.

Extending a provocative series of experiments by Barrett, Richert and Dviesenga (2001), Harris and Giménez (2001) tested between these two proposals by giving young children an 'omniscience' and an 'immortality' interview. For each interview, children were asked a parallel set of questions about an ordinary human being on the one hand, namely their best friend, and an extraordinary being on the other, namely God.

The omniscience interview focused on the everyday principle mentioned earlier, namely that what we know is ordinarily constrained by our perceptual access. If we are shown a closed box, we cannot know what is inside the box until we have looked inside it. Between three and five years of age, children show a sharp increase in their ability to understand this constraint on knowledge (Wellman, Cross and Watson, 2001). To assess whether children believe that all beings are subject to that same constraint Harris and Giménez (2001) asked three-, four- and five-year-olds a series of questions about what a friend would know and what God would know in the absence of relevant perceptual access. For example, one of the interview questions was as follows: 'Let's pretend we gave this present to (your friend/God) and we asked them to tell us what is inside the box without taking the paper off. Would (your friend/God) know for sure what is in the box or would they have to guess?' Children were asked to provide a justification of whichever answer they chose.

The immortality interview focused on children's realization that under ordinary circumstances the life-cycle is uni-directional and finite. To assess whether children believe that all beings are subject to that constraint they were asked whether a friend and God were once babies, are currently getting older and will eventually die. For example, one question was as follows: 'What's going to happen to (your friend/God) next year and the year after that? Will (s/he) get older and older or will (s/he) stay just the same?' As in the omniscience interview, children were asked to justify their choice of answer.

Two different patterns of findings might be expected. First, suppose that children simply ignore claims to the effect that God has a special status. In that case, we might expect the pattern of replies to be similar for each of the two targets: the friend and God. In particular, to the extent that children come to acknowledge that there are constraints on what human beings know and how long they live, they should extend those same constraints to God. Effectively, they should construe God in a fully anthropomorphic fashion and any development in their conception of the constraints that human beings are subject to should be extended to God. Second, suppose that children do increasingly acknowledge the aforementioned constraints, but also come to acknowledge that God violates those constraints. In that case, the pattern of replies for the two targets ought to diverge in the course of development – as children come to appreciate on the one hand that there are constraints on mere mortals and on the other hand that such constraints do not apply to God.

Our findings – for both the omniscience and immortality view – correspond best to this second pattern. This is illustrated in figures 6 and 7 which indicate, respectively, the mean number of justifications in which children referred to

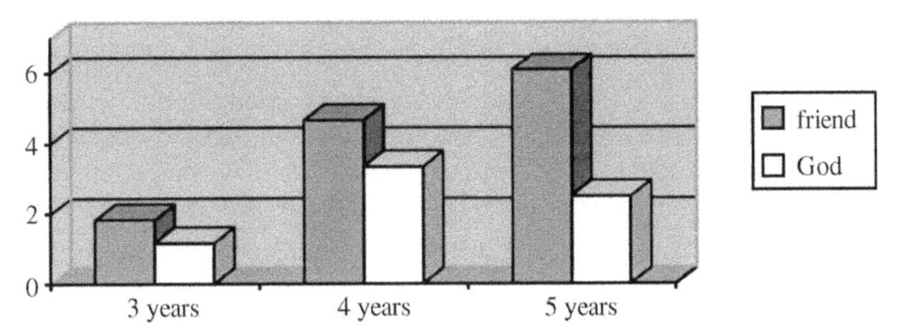

Fig. 6 Mean number of justifications referring to ordinary constraints on knowledge and the life cycle as a function of age
Source: Based on Giménez and Harris (2001).

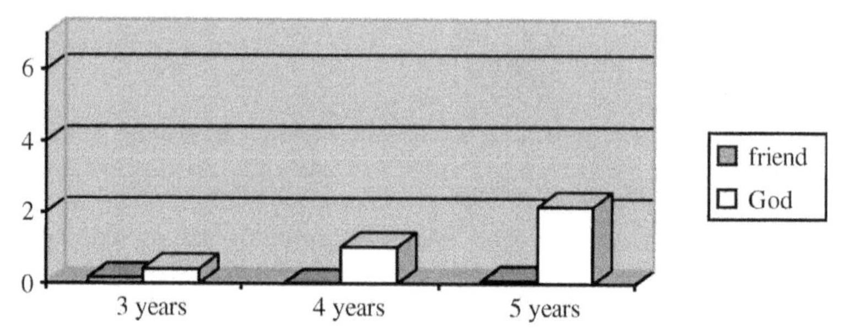

Fig. 7 Mean number of justifications referring to extraordinary violations of constraints on knowledge and the life cycle as a function of age
Source: Based on Giménez and Harris (2001).

ordinary human constraints (e.g. 'He will have to guess because he can't see it') and the mean number of justifications in which they referred to the possibility of violating those constraints (e.g. 'He will know for sure because from Heaven he can see everything'). (In the case of both figures 6 and 7, means were calculated by summing across the omniscience and immortality interviews).

Inspection of figure 6 shows that references to ordinary constraints rise in frequency with age when children discuss their friend. This is in line with standard findings on children's developing appreciation of psychological and biological constraints. The pattern for God, on the other hand, is different. There is no consistent increase with age in the number of references to ordinary constraints. Indeed, by five years of age children differentiate fairly sharply between God and a friend. Thus, five-year-olds often refer to ordinary constraints in talking about their friend but mention them less frequently in talking about God. Inspection of figure 7 shows the converse pattern. Thus, references to extraordinary violations increase with age but only in connection with God. Five-year-olds often refer to such violations in talking about God but rarely do so in talking about their friend.

Taken together, figures 6 and 7 indicate that five-year-olds resist the tendency to anthropomorphize God – they recognize that His knowledge is not restricted by the availability of perceptual access and that His existence is not constrained by the ordinary human life-cycle. Admittedly, children do show some tendency – especially at four years of age – to overextend ordinary human constraints to God but by five years of age the differentiation between God and an ordinary human being is relatively firm. This differentiation is strong evidence against the proposal that young children are prone to testimonial abstinence. If anything, they are surprisingly receptive to testimony. Their receptivity is highlighted by the fact that they accept claims about God that run directly counter to the conclusions that they have reached about human beings.

Making the same point differently, the implication of these findings is that information derived from testimony is not just a supplement to children's first-hand observation, and nor is it to be simply construed as a fast track to a deeper or more adequate 'theory' of the empirical world. Testimony also introduces the lure of the transcendent: it sometimes invites children to set aside causal constraints that they have gradually erected in the course of cognitive development. Children's willingness to accept that invitation calls for some re-thinking of their cognitive agenda.

9 Rethinking the child's cognitive agenda

An orthodox view of cognitive development is that children move toward a more veridical conception of the world. Whether it is in the domain of physics, biology, or psychology, they recognize that certain 'magical' outcomes are impossible. For example, they come to acknowledge that a solid object cannot suddenly materialize out of thin air, that children cannot remain forever young and that simply wishing for something does not make it happen. A plausible generalization from this line of research is that children eventually accept the existence of various constraints on what can ordinarily happen – as discussed in the previous section. Based on that generalization, one might further conclude that children gradually reject the possibility that 'magic' is genuinely possible. It is increasingly recognized as something that might happen in fairy tales but cannot happen in real life. On this view, magical outcomes are eventually relegated to a cognitive backwater, as children proceed to a more objective conception of what can actually happen. In much the same way, research on alchemy or on Lamarckian inheritance was relegated to a scientific backwater as it became increasingly clear that no such phenomena could be reliably produced.

I believe that such analogies with the history of chemistry or biology are misleading (Harris, 2000b, chapter 8). It is true that scientific research depends on the possibility of being able to observe and reproduce a given phenomenon. If that is impossible, investigation gradually ceases because there is literally nothing to investigate. However, the same process does not appear to operate in the sphere of cognitive development. Although children almost certainly do acknowledge an increasing number of constraints on what can ordinarily happen, the realization that a given outcome is ordinarily impossible need not mean that those outcomes are removed from the cognitive agenda. Children are not part of a scientific community whose primary goal is the production and interpretation of reliably observable phenomena. Normally impossible phenomena may not be observable but they can be contemplated in the imagination, they can be talked about and they can be the subject of sincere testimony. As the findings in the previous section show, children accept that testimony even when it concerns phenomena that they would otherwise deem impossible.

Surveying the various findings that have been presented so far, we may reasonably conclude that young children are engaged in the pursuit of three concurrent agendas. First, they are engaged, just as orthodox developmental theory implies, in the construction of various theories or sub-theories, largely based on their own observation and experimentation. The toddler's grasp of naïve physics or the pre-school child's grasp of folk psychology are plausible examples of this type of universal and largely endogenous conceptual development. Second, aided by the testimony of others, children gradually come to understand aspects of the world that are normally hidden from their own first-hand, direct observation. Thus, as we have seen, alongside their acquisition of a naïve physics they also come to understand that the earth is a sphere, and alongside their understanding of folk psychology they come to understand that mental processes depend upon brain processes. Third, and again aided by the testimony of others, children understand and even endorse various metaphysical claims even when those claims run counter to what they know about the ordinary world. They acknowledge, for example, that God is not constrained by the biological and psychological principles that apply to ordinary human beings.

To the extent that children tackle all three of these agendas concurrently, we may ask how far they treat them as one indivisible whole and how far they differentiate among them. In particular, we may ask whether children recognize that some claims are better established, and enjoy a larger consensus, than others. For example, do they recognize that whereas there is little doubt about the existence and permanence of the natural kinds that occupy their everyday world, the existence of an omniscient creator is more debatable? In the next section, I consider three possible answers to this question.

10 Distinguishing among beliefs

As noted in section 2, Reid argued that we are innately equipped to take in and trust information provided via testimony as well as information provided by our senses. A plausible extension of this view is that although children may come to doubt a particular piece of information – insofar as they realize that perceptual appearances or testimony may be misleading – they draw no broad or principled distinction between the two major sources of information at their disposal. Thus, the objects or processes that they can learn about only via testimony – whether they concern matters of scientific investigation or matters of religious belief – are just as real in their eyes as the objects or processes that they learn about through their own personal observation. As a result, children end up with a motley set of conceptual entries, some respectable from the point of view of contemporary science but scarcely entertained in earlier centuries, some derived from the distinctive doctrine of a particular reference group or sect and some consistent with everyday common sense throughout human history. In the child's mind, these various conceptual entries would be on a par with one another. To state the matter baldly, this view implies that children are prone to undifferentiated credulity: they might think of germs, angels and giraffes as all being on a similar ontological footing – all equally and uncontroversially real.

A second possibility is that – contrary to the thrust of Reid's proposals – children make a deep and dichotomous cut in the epistemological landscape. They distinguish between entities that they have learned about via their own observation and experimentation and entities for which they have only testimonial evidence. Moreover, they treat the claims about the existence of the former as more reliable or trustworthy than claims about the existence of the latter. By implication, they regard vast tracts of relatively mundane but objective information – about the historical past, the remote present and the microscopic – as being less trustworthy than information gained from first-hand experience, and indeed they treat such testimonial evidence as being no more trustworthy than assertions about God's existence, omniscience or immortality. Returning to the example introduced above, they would put giraffes into one ontological category and germs and angels into another, less secure category. Note that this would lead children to a decidedly eccentric (yet tenable) ontological stance. As more recent philosophical analyses have shown (Coady, 1992), adults ordinarily navigate in a more or less boundless ocean of testimony, with no disquieting thoughts about its ultimate lack of objectivity. Still, it is possible, albeit unlikely, that young children are sceptical or cautious, in ways that adults are not.

Finally, we may consider a third possibility. When children listen to older children or adults, the conversation is likely to range over each of the three domains that I have sketched out. References will be made to easily perceptible entities in the mundane world. References will also be made to entities whose existence has been established by expert investigation but can scarcely be checked by individual children. Finally, references will be made to transcendent or metaphysical entities that are targets of faith rather than empirical investigation. In each of these domains, the discourse may include cues to the ontological standing of what is being referred to. In the course of development, children might become alert to those cues and use them to draw up an ontological map. For example, in conversation about the ordinary world, the existence of various entities is taken for granted. There may be no explicit assertion of their existence – but neither is the reality of their existence qualified or placed in doubt. Thus, adult speakers will refer to pigs, giraffes, dinosaurs and viruses as if their existence were all equally uncontroversial. By contrast, references to other entities will be mainly confined to distinctive types of discourse. References to giants or monsters are typically embedded in discourse about a fictional world that is discontinuous both in space and time with the ordinary world. Similarly, references to God, to heaven, or to angels, are more likely to occur in the context of distinctive modes of speech – sermons, prayers, blessings, or readings from the Bible – than in the context of an everyday conversation.[6]

On this third view, we might also expect children to make a dichotomous cut but not a cut in terms of their source of knowledge about a given entity. Instead, it would be based on the type of discourse or testimony that they receive. More specifically, children would tend to place entities accessible to their own observation and experimentation in the same ontological category as entities established via expert investigation because both types of entity would be mentioned without any special qualification in routine discourse about the everyday world. Each of these two types of entity would be distinguished from fictional or metaphysical entities in so far as references to either are embedded in special modes of discourse. In short, children would be expected put giraffes and germs in one category and angels in another.

11 Conclusions

I have argued that we radically underestimate children's cognitive agenda if we see them primarily as lone scientists, gathering data first-hand via observation and experimentation. By the time young children start school, they have spent thousands of hours talking with older children and adults. From such wide-ranging testimony, they will draw conclusions about entities that they cannot investigate for themselves – and indeed about entities that are not open to empirical investigation by anyone. This capacity to gather information via testimony opens up an important question about young children's ontological commitments. Are they prone to undifferentiated trust? Do they place greater trust in their own first-hand observation? Alternatively, do they navigate primarily by means of social cues, noting the ways in which other people refer to particular entities? Once we shake off the image of the child as a stubborn autodidact, this last possibility comes into view and can be tested.

[6] It ought to be acknowledged that references to 'special' beings may not always be embedded in a special mode of discourse. For example, Evans-Pritchard's (1976) classic ethnography stressed the extent to which a belief in witches and witchcraft infused everyday practice, and presumably everyday speech, among the Azande.

18 The baby in the lab-coat: why child development is not an adequate model for understanding the development of science

Luc Faucher, Ron Mallon, Daniel Nazer, Shaun Nichols, Aaron Ruby, Stephen Stich and Jonathan Weinberg

Alison Gopnik and her collaborators have recently proposed a novel account of the relationship between scientific cognition and cognitive development in childhood. According to this view, the processes underlying cognitive development in infants and children and the processes underlying scientific cognition are *identical*. We argue that Gopnik's bold hypothesis is untenable because it, along with much of cognitive science, neglects the many important ways in which human minds are designed to operate within a social environment. This leads to a neglect of *norms* and the processes of *social transmission* which have an important effect on scientific cognition and cognition more generally.

1 Introduction

In two recent books and a number of articles, Alison Gopnik and her collaborators have proposed a bold and intriguing hypothesis about the relationship between scientific cognition and cognitive development in early childhood.[1] In this chapter we will argue that Gopnik's bold hypothesis is untenable. More specifically, we will argue that even if Gopnik and her collaborators are right about cognitive development in early childhood they are wrong about science.[2] The minds of normal adults and of older children are more complex than the minds of young children, as Gopnik portrays them, and some of the mechanisms

We are grateful to Peter Carruthers, Paul Harris and other participants in the Hang Seng Conference on the Cognitive Basis of Science for their helpful comments on earlier versions of this chapter. The authors are listed alphabetically.

[1] The books are Gopnik and Meltzoff (1997) and Gopnik, Meltzoff and Kuhl (1999a). The articles include Gopnik and Wellman (1992, 1994) and Gopnik (1996a, 1996b). For brevity, we will often refer to the views we are criticizing as Gopnik's, but it should be borne in mind that her collaborators share many or all of these views and certainly deserve part of the credit (or blame!) for developing them.

[2] For some objections to Gopnik's views on cognitive development in early childhood, see Stich and Nichols (1998).

that play no role in Gopnik's account of cognitive development in early childhood play an essential role in scientific cognition. A central theme in our critique of Gopnik's account will be that it ignores the many important ways in which human minds are designed to operate within a social environment – a phenomenon that we shall sometimes refer to as *the interpenetration of minds and culture*. One aspect of the interpenetration of mind and culture that will loom large in our argument is the human capacity to identify and internalize the *norms* of the surrounding culture. We will argue that the cultural transmission of norms, which has been largely neglected in the cognitive sciences, has a major impact on theoretical reasoning and thus has an essential role to play in explaining the emergence of science. Cultural transmission also plays a central role in learning science and, remarkably, in shaping some quite basic cognitive processes that make science possible. These phenomena have been given little attention in the cognitive sciences and they play no role in Gopnik's account.

2 Gopnik's bold hypothesis

Gopnik reports, with obvious relish, that her theory often provokes 'shocked incredulity' (Gopnik 1996a, p. 486). In this section, we'll set out some of the central components of that shocking theory.

According to Gopnik and her colleagues, infants are born with a rich endowment of theoretical information (Gopnik, 1996a, p. 510). They often describe this innate endowment as 'knowledge', though it is clear that they are using that term in a way that is importantly different from the way that philosophers use it, since a fair amount of the innate theory turns out not to be true, and is ultimately replaced as the child develops. It is the thesis that infants are born with lots of theoretical knowledge, Gopnik suggests, that generates much of the shocked incredulity. Gopnik and her colleagues have quite a lot to say about the nature of this innate knowledge and why they believe it should be regarded as 'theoretical'. For our purposes, the crucial point is that theories are 'defeasible' – they can be and often are replaced when they are not well supported by the evidence that a cognitive agent (child or scientist) encounters (Gopnik and Meltzoff, 1997, p. 39).

Cognitive development in infancy and early childhood, Gopnik maintains, consists in the adoption (and subsequent abandoning) of a sequence of theories. Subserving this process of theory revision is a 'powerful and flexible set of cognitive devices' which are also innate. These devices, unlike the infant's innate theories, remain largely unchanged throughout childhood and beyond. The reason that Gopnik and her colleagues think that there are important connections between scientific cognition and cognitive development in infancy is that, on their view, the psychological processes subserving cognitive development in children, from early infancy onward, are *identical* to those that underlie theory change in science (Gopnik and Meltzoff, 1997, p. 3; Gopnik, 1996a, p. 486). We will call this claim the *Continuity Thesis*, and much of what follows is devoted to arguing that it is false.

On Gopnik's account, scientific cognition is just a continuation of the very same processes of theory revision that children have been engaged in from earliest infancy:

> [T]he moral of [the] story is not that children are little scientists but that scientists are big children. (Gopnik, 1996a, p. 486)

> [E]veryday cognition, on this view, is simply the theory that most of us most of the time have arrived at when we get too old and stupid to do more theorizing . . . We might think of our enterprise as scientists as the further revision of the theory by the fortunate, or possibly just childish, few who are given leisure to collect evidence and think about it. (Gopnik and Meltzoff, 1997, p. 214)

Indeed, Gopnik goes on to suggest that the greatness of many important figures in the history of science can be traced to their 'childishness' (Gopnik, 1996b, p. 561). One of the attractions of Gopnik's bold hypothesis is that, if it is correct, it will unify two fields of investigation – the study of early cognitive development and the study of scientific cognition – that have hitherto been thought quite distinct, with the result that advances in either domain will further our understanding of the other.

Figures A and B are our attempt to capture the fundamental aspects of Gopnik's theory. Figure A depicts the basic components in Gopnik's account of theory change. Figure B – our attempt to capture the Continuity Thesis – shows that, for Gopnik, the same theory revision process is at work in early childhood, later childhood, and science.

3 Why do people have the capacity to do science? An old puzzle 'solved' and a new one created

One virtue that Gopnik and her colleagues claim for their theory is that it solves 'an interesting evolutionary puzzle'. Everyone agrees that at least some humans have the capacity to do science. And few would challenge the claim that in doing science people use a flexible and powerful set of cognitive abilities. But, Gopnik and Meltzoff ask, 'Where did the particularly powerful and flexible devices of science come from? After all, we have only been doing science in an organized way for the last 500 years or so; presumably they didn't evolve so that we could do that' (Gopnik and Meltzoff, 1997, p. 18; Gopnik, 1996a, p. 489). The answer they suggest is that many of the cognitive devices that subserve scientific reasoning and theory change evolved because they facilitate 'the staggering amount of learning that goes on in infancy and early childhood' (Gopnik and Meltzoff, 1997, p. 18; Gopnik, 1996a, p. 489). So, according

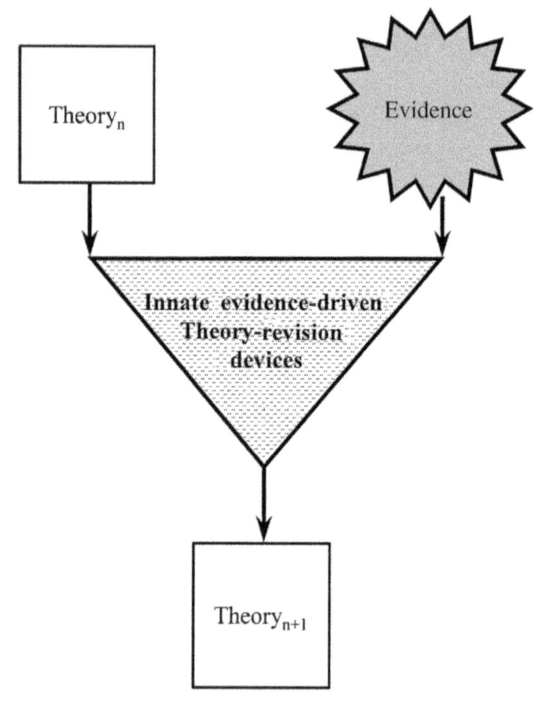

Fig. A The basic components in Gopnik's account of theory revision

to Gopnik, '*science is a kind of spandrel, an epiphenomenon of childhood*' (Gopnik 1996a, p. 490; emphasis added).

This proposed solution immediately suggests another problem, which Giere (1996b, p. 539) has dubbed 'the 1492 problem'. 'Science as we know it,' Giere notes, 'did not exist in 1492.' But if Gopnik and her colleagues are right, then the cognitive devices that give rise to science have been part of our heritage since the Pleistocene. *So why have humans been doing science only for the last 500 years*? The solution that Gopnik proposes turns on the availability of relevant evidence.

> My guess is that children, as well as ordinary adults, do not . . . systematically search for evidence that falsifies their hypotheses, though . . . they do revise their theories when a sufficient amount of falsifying evidence is presented to them. In a very evidentially rich situation, the sort of situation in which children find themselves, there is no point in such a search; falsifying evidence will batter you over the head soon enough. (Gopnik, 1996b, p. 554)

Now what happened about 500 years ago, Gopnik maintains, is that as a result of various historical and social factors a few thinkers found themselves confronted

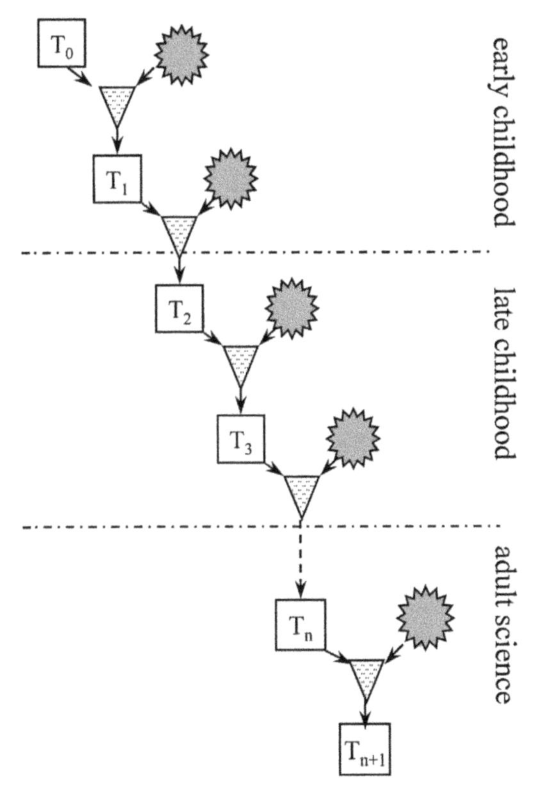

Fig. B Gopnik's Continuity Thesis which maintains that the same theory revision process is at work in early childhood, in later childhood and in science

with unprecedented amounts of new evidence, some of which falsified venerable answers to questions like: Why do the stars and planets move as they do? New technology was one reason for the availability of new evidence; telescopes, microscopes and devices like the air pump were invented. Other technological and social changes greatly facilitated communication allowing 'a mathematician in Italy to know what an astronomer has seen in Denmark' (Gopnik, 1996b, p. 554). Greater leisure (at least for a few) was yet another factor. All of this, and perhaps other factors as well, Gopnik suggests, created an environment in which the theory revision mechanisms that natural selection had designed to enable children to cope with 'the staggering amount of learning that goes on in infancy and childhood' might begin functioning actively in adulthood, long past the stage in life in which they would have made their principal contribution to fitness in the environment in which our ancestors evolved.

Though in an earlier paper two of the current authors expressed some enthusiasm for this solution (Stich and Nichols, 1998), the work of others in the current group of authors has made us all quite sceptical. The problem, as we see it, is that just about all the factors cited in Gopnik's explanation of the emergence of science in the West were present at that time – and indeed much earlier – in China. Well before science emerged in the West, the Chinese had much sophisticated technology that generated vast amounts of information, which they collected systematically for various practical purposes. So there was *lots* of data. They also had a tradition of strong unified government and a centralized bureaucracy that built and maintained systems of transportation and communication far superior to those in the West. Compared with the situation in the West, it was relatively easy for scholars in one part of the country to learn about information generated in another part of the country. Moreover, there was no shortage of scholars with ample leisure. A substantial number of astronomers and other collectors of useful information were employed by rich aristocrats, and there was a substantial class of wealthy individuals who had access to the highly developed system of communication and who were well enough educated to be aware of the large amounts of data being generated in their society by sophisticated equipment and instruments. Despite all this, however, *science as we know it did not emerge in China* (Needham, 1954; Guantao, Hongye and Qingfeng, 1996). So something in addition to the availability of new data is needed to explain why science emerged in the West. In sections 6–9 we will offer an account of some of the additional factors that played an important role.

4 The very limited role that culture plays in Gopnik's account

One feature of Gopnik's account that will play a central role in our critique is that it is fundamentally individualistic or a-social. According to Gopnik, as children develop they proceed from one theory to another and, as indicated in figure A, their trajectory – the path they follow through this space of theories – is determined by only three factors: (i) the nature of the innate theory revision devices; (ii) their current theory in the domain in question; (iii) the evidence that they have available. Since, on Gopnik's view, theory change in science is entirely parallel to theory change in infancy, the same three factors, and *only* these, determine the process of theory revision in science.

Gopnik would, of course, allow that the culture in which a cognitive agent (whether she is a child or a scientist) is embedded can be a source of evidence for *theories about that culture*. And, for both older children and scientists, the culture can also be an indirect source of evidence about the world, when one person reports his observations to another, either in person or in writing. (Recall the example of the astronomer in Denmark who wrote a letter to a mathematician in Italy describing what he had seen.) But apart from using other people as an indirect means of gathering evidence, the social surround is largely irrelevant.

The fundamental *un*importance of culture in Gopnik's account of theory change can be seen particularly clearly in a passage in which Gopnik predicts that *all* cognitive agents – *regardless of the culture in which they are embedded* – would end up with 'precisely' the same theory, provided they start with the same initial theory and get the same evidence:

> [Our] theory proposes that there are powerful cognitive processes that revise existing theories in response to evidence. If cognitive agents began with the same initial theory, tried to solve the same problems, and were presented with similar patterns of evidence over the same period of time they should, precisely, converge on the same theories at about the same time. (Gopnik, 1996a, p. 494)

Gopnik and her colleagues are hardly alone in offering an individualistic and fundamentally asocial account of the processes subserving cognitive development in childhood. As Paul Harris notes in chapter 17 in this volume, most of the leading figures in developmental psychology from Piaget onward have viewed the child as 'a stubborn autodidact'. Arguably, this tradition of 'intellectual individualism' can be traced all the way back to Galileo, Descartes and other central figures in the Scientific Revolution (Shapin, 1996, chapter 2). As we will see in Section 6, there is a certain irony in the fact that Gopnik treats intellectual individualism as a descriptive claim, since for the founding fathers of modern science the individualistic approach to belief formation and theory revision was intended not as a description but as a *pre*scription – a *norm* that specified how inquiry *ought* to proceed.

5 The interpenetration of minds and cultures

What we propose to argue in this section is that Gopnik's account of the cognitive mechanisms subserving theory revision, sketched in figure A, provides an importantly incomplete picture of cognition in adults. The inadequacy on which we will focus is that this highly individualistic picture neglects the important ways in which cultural or social phenomena affect cognition. Of course, this *by itself* is hardly a criticism of Gopnik and her colleagues, since they do not claim to be offering a *complete* account of the processes involved in adult cognition. However, in sections 6–9 we will argue that some of the social aspects of cognition that Gopnik and most other cognitive scientists neglect have had (and continue to have) a profound effect on scientific cognition.

A central theme underlying our critique is a cluster of interrelated theses which we'll refer to collectively as the *interpenetration of minds and culture*. Among the more important claims included in this cluster are the following:

(1) The minds of contemporary humans were designed by natural selection to operate within a cultural environment.
(2) Humans have mental mechanisms that are designed to exploit culturally local beliefs and theories as well as culturally local information about norms,

social roles, local practices and prevailing conditions. In some cases, these mechanisms cannot work *at all* unless this culturally local information has been provided. In other cases, when the culturally local information is not provided the mechanisms will do only a part (often a small part) of what they are supposed to do, and an individual who lacks the culturally local information will be significantly impaired.

(3) In addition to acquiring culturally local *content* (like beliefs, norms and information about social roles), the cultures in which people are embedded also have a profound effect on many cognitive *processes* including perception, attention, categorization and reasoning.

(4) The cultural transmission of both content and cognitive processes is subserved by a variety of mechanisms, most of which are quite different from the sorts of theory revision processes emphasized by Gopnik and her colleagues.

(5) Minds contain a variety of mechanisms and interconnected caches of information, some innate and some acquired from the surrounding culture, that impose important constraints on the ways in which cultures can develop. However, these constraints do not fully determine the ways in which cultural systems evolve. The evolution of culture is *partially autonomous.*

5.1 An example of the interpenetration of minds and culture: the emotions

One of the clearest examples of a system that exhibits the interpenetration of mind and culture is to be found in the work on the psychological mechanisms underlying the emotions done over the last 25 years by Paul Ekman, Richard Lazarus, Robert Levenson and a number of other researchers. On the account that has been emerging from that research, the emotions are produced by a complex system which includes the following elements:

A set of '*affect programs*' (one for each basic emotion). These can be thought of as universal and largely automated or involuntary suites of co-ordinated emotional responses that are subserved by evolved, psychological and physiological mechanisms present in all normal members of the species.

Several sets of *triggering conditions.* Associated with each affect program is a set of abstractly characterized conditions specifying the circumstances under which it is appropriate to have the emotion. Like the affect programs, these triggering conditions (which Ekman calls 'appraisal mechanisms' and Levenson calls 'emotion prototypes') are innate and present in all normal members of the species. Lazarus (1994) offers the following examples of triggering conditions:

For anger: A demeaning offence against me and mine
For fright: An immediate, concrete and overwhelming physical danger

For our purposes, what is important about these innate triggering conditions is that they are designed to work in conjunction with a substantial cache of information about culturally local norms, values, beliefs and circumstances. It is the job of this culturally local information to specify what counts as a demeaning offence, for example, or what sorts of situations pose overwhelming physical danger. And without information of this sort, the emotion triggering system cannot function.

Display rules and other downstream processes. On the 'downstream' side of the affect program, the theory maintains that there is another set of mechanisms that serve to filter and fine-tune emotional responses. Perhaps the most famous example of these are the culturally local 'display rules' which, Ekman demonstrated, lead Japanese subjects, but not Americans, to repress certain emotional responses after they have begun when (but apparently only when) the subjects are in the presence of an authority figure (Ekman 1972). In this case, the culturally local display rules must interact with a body of information about prevailing norms and social roles that enables the subject to assess who counts as a person of sufficient authority. Local norms also often play an important role in determining the downstream reactions after an affect program has been triggered. So, for example, Catherine Lutz (1988) reports that the Ifaluk people (inhabitants of a Micronesian atoll) have a term, *song*, for a special sort of justified anger. In a dispute, only one party can have a legitimate claim to feeling *song*, and according to the norms of the Ifaluk both the aggrieved party and other members of the community can apply sanctions against the person who has provoked *song*. Similarly, Daly and Wilson (1988) note that in many cultures the prevailing norms specify that if a man is angry because another man has had sex with his wife, the aggrieved party has a right to call on others in the community to aid him in punishing the offender.

Figure C is Robert Levenson's (1994) sketch of some of the main elements of this 'bio-cultural' model of the emotions.

5.2 *The importance of norms*

The psychological processes underlying the emotions are, of course, importantly different from those underlying scientific cognition. For our purposes, the example of the emotions is important for two reasons. First, it illustrates one well-studied domain in which the interpenetration of minds and culture has been explored in some detail. Second, it highlights the fact that the *norms* that obtain in a culture can have an important effect on cognition. In a full account of the role of norms in cognition, there are many important questions that would need to be addressed, including questions about the nature of norms and about the psychological mechanisms that subserve them, questions about how norms are transmitted, how they change and what their evolutionary function

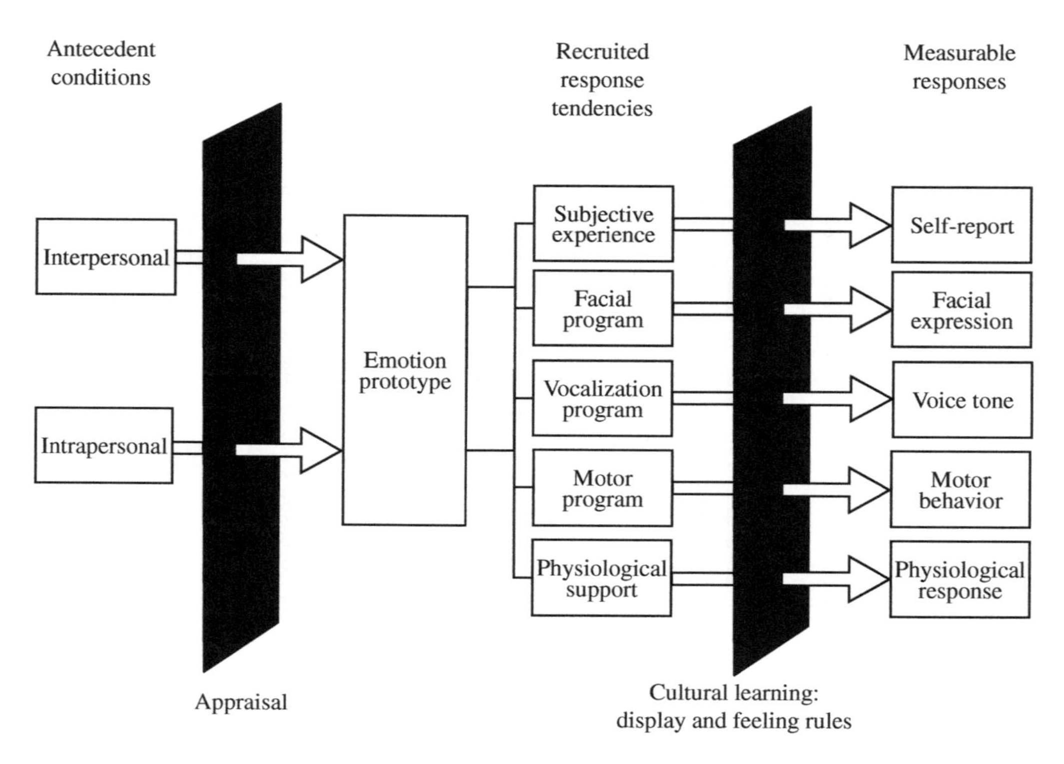

Fig. C A sketch of some of the main elements of Levenson's (1994) model of emotions

may have been. In this chapter, we can offer only a few brief remarks on these issues.

- Norms typically characterize a range of actions or behaviours that are *appropriate*, or *required*, or *prohibited* in various circumstances.
- Norms are typically '*common knowledge*' in the relevant community. More or less everyone knows them and more or less everyone knows that everyone knows them.
- Norms typically include an *affective component.* Doing an appropriate or required action typically generates a positive affect in others which may be accompanied by praise, admiration, or support. Doing inappropriate or prohibited actions typically generates negative affect in those who know about it, and this will often be accompanied by sanctions. It is important to note that norm violations (in contrast with actions which merely provoke anger but do not violate norms) typically provoke negative affect and an inclination to sanction or punish among members of the community who are not directly involved or harmed.

Along with Axelrod (1986), Boyd and Richerson (1992) and Sober and Wilson (1998), we think that norms and the psychological mechanisms subserving them may have evolved because they give communities a powerful and relatively inexpensive way of altering the 'pay-offs' for various sorts of behaviours that may benefit an individual to the detriment of the group (or vice versa).

6 Science is a norm-governed activity

What makes the preceding discussion of norms relevant to our current topic is that science is a norm-governed activity. It is a commonplace that norms govern many aspects of science, including: the sorts of evidence that must be offered in support of a claim; the sorts of arguments that may (and may not) be offered in support of a claim; the sorts of information that scientists are expected to report (and the sorts they are allowed or expected *not* to report); the procedures required for getting work accepted for presentation at scientific meetings and for publication in scientific publications; the ways in which researchers claim credit for work and the ways in which they share credit with other researchers and with their own collaborators and students. These norms not only affect what scientists do, they also affect what scientists *believe* – they affect which theories are accepted, which theories are taken seriously and which theories are rejected. Moreover, while it is possible that *some* of the norms that play a role in science are innate, many others certainly are not. Most scientific norms are neither universal nor unchanging; they are not a fixed or permanent feature of human minds or of human cultures. Quite the opposite. The norms of science have a history (indeed, in most cases a relatively short history) and they are subject to ongoing modification.

6.1 *Some examples of the role of changing norms in the history of science*

This is not the place to attempt a systematic survey of the role that norms have played in the history of science. However, a few examples should serve to make the point that they have been of enormous importance.

Nullius in verba ('On no man's word'). This was the motto of the Royal Society, founded in 1660, and it captures a fundamental norm that early scientists saw as setting them apart from the university based 'schoolmen' on whom they regularly poured scorn. The norm that the slogan reflected was that in order to understand how the world works, one ought to look at the 'testimony' of nature rather than reading and interpreting Aristotle and other ancient authors. The spread of this new norm marked a sea change in the intellectual history of the West – a change without which the emergence of science would obviously have been impossible.

Why did the norm emerge and spread? As in most cases of transformations in prevailing norms, the reasons were many and complex. We will mention just two parts of the story that historians of science have stressed. The first is that fathers of modern science were hardly alone in rebelling against established authorities. In the two centuries prior to the founding of the Royal Society, the questioning of intellectual and political authorities was becoming far more common in many areas of society. In religion, for example, Martin Luther famously asked people to focus on scripture rather than church doctrine.

Second, as a number of historians have emphasized, some of the earliest advocates of looking at nature were not saying 'give up copying and look at the facts'. Rather, they were actually saying that we should study better books, namely 'those which [God] wrote with his own fingers' which are found everywhere (Nicholas of Cusa (1401–64) quoted in Hacking, 1975, p. 41). It may seem that such claims are only metaphorical, but Ian Hacking argues that many claims like this one were meant literally. One piece of evidence that Hacking cites is the 'doctrine of signatures' that suggested that physicians should attempt to find natural signs which were considered to be like linguistic testimony (Hacking, 1975, pp. 41–3).

While there is much more to be said about the emergence of the new norm according to which scientists (or natural philosophers as they preferred to be called) should ignore the Ancients and look at nature, we think these two fragments of the story provide a nice illustration of fact that the evolution of culture is *partially autonomous*. Though the structure of the mind may impose important constraints on the evolution of norms, local and surprisingly idiosyncratic historical factors also play a crucial role in the process.

The management of testimony. Though their motto discouraged relying on the testimony of the Ancients, the members of the Royal Society relied heavily on testimony from each other, from scientists in other countries and from travellers who had been to distant lands and seen things that could not be seen in England. The members of the Society actively debated which principles they ought to apply in accepting such testimony. They wanted to avoid excessive scepticism toward strange tales from afar,[3] but they also did not want to give uncritical credence to stories from travellers.

In *A Social History of Truth*, Steven Shapin (1994) offers an extended discussion of how the Royal Society 'managed' testimony. On Shapin's account, seventeenth-century beliefs and norms regarding truthfulness had a major effect on the standards that these early scientists adopted. One belief that played a central role was that 'gentlemen' were more trustworthy than artisans, servants, merchants and other folk. This belief was intertwined with the prevailing norm according to which a gentleman's reputation for truthfulness was a matter of major importance. 'Giving the lie' or accusing someone of dishonesty was a very serious matter.[4]

Shapin argues that these attitudes had important effects on the members of the Royal Society, and thus on the development of science in the seventeenth century. One rather curious consequence was that quite different standards were applied in deciding whether to trust testimony from people of different social status. Thus, for example, Boyle refused to accept the testimony of 'common' divers when it contradicted his theories about water pressure. At the same time Boyle attempted to accommodate testimony from gentlemen travellers to the arctic even when their testimony regarding icebergs contradicted his theories. (See Shapin, 1994, pp. 253–66.) Testimony from non-gentlemen was sometimes accepted but it needed to be vouched for by someone of good standing. When Leeuwenhoek, a haberdasher by trade and a chamberlain for the sheriffs of Delft, provided the Society with strange reports of the abundance of microscopic creatures, he felt the need to have no less than eight 'local worthies' vouch for his claims, even though none had any relevant technical experience (Shapin, 1994, pp. 305–7).

For our purposes, a more important consequence was that members of the Royal Society typically took great pains to avoid being seen as anything other than disinterested investigators. They wished to avoid appearing like merchants or trades-people for then they would suffer a considerable drop in credibility (Shapin, 1994, pp. 175–92). As a result, the 'gentlemen scientists' of the Royal

[3] Locke and Boyle both discussed how people from warmer climates were sceptical of stories about ice. They wished to avoid making similar mistakes. See Shapin (1994, pp. 243, 249).

[4] Shapin makes a very strong case that these were forceful social norms in the seventeenth century (at least among those who had the time and resources for science).

Society tended to focus on what we would describe as 'pure science' rather than on potential applications or the development of technology. This stands in stark contrast to the pattern of inquiry to be found in China at that time and earlier. In China, inquiry was almost always conducted with some practical goal in mind; the ideals of pure inquiry and theory for its own sake were never embraced. The importance of this fact will emerge more clearly in sections 8 and 9.

Contemporary debates about scientific norms. The emergence of new norms regulating the conduct of inquiry is not a phenomenon restricted to the early history of science. Quite the opposite, it is a process that has operated throughout the history of science and continues in contemporary science. In experimental psychology, for example, there has recently been a heated debate about experimental designs in which the participants are deceived about some aspect of the experimental situation. Some of those who urge that the use of deceptions should be curtailed are motivated by moral concerns, but others have argued that deception should be avoided because of the 'methodological consequences of the use of deception on participants' attitudes, expectations, and in particular, on participants' behaviour in experiments' (Hertwig and Ortmann, forthcoming a). The behaviour revealed in experiments involving deception, these critics maintain, particularly when it is widely known that psychologists practice deception, may not give us reliable information about how participants would behave outside an experimental setting (Hertwig and Ortmann, forthcoming b). One fact that has come to play an important role in this debate is that the norms that obtain in experimental psychology are quite different from the norms that prevail in the closely related field of experimental economics. In the latter discipline the use of deception is strongly discouraged, and the use of financial incentives – which are rarely used in psychology – is mandatory. 'Experimental economists who do not use them at all can count on not getting their work published' (Hertwig and Ortmann, forthcoming a).

6.2 *A first modification to Gopnik's picture*

The conclusion that we want to draw from the discussion in this section is that Gopnik's picture of scientific cognition is in an important way incomplete. While it *may* be the case that the pattern of theory revision in early childhood is entirely determined by the three factors mentioned earlier – the nature of the innate theory revision devices, the current theory, and the available evidence – there is another factor that plays a crucial role in theory revision in adults: *norms* – more specifically norms governing how one ought to go about the process of inquiry and how one ought to revise one's beliefs about the natural world. Moreover, the emergence of new norms played a crucial role in

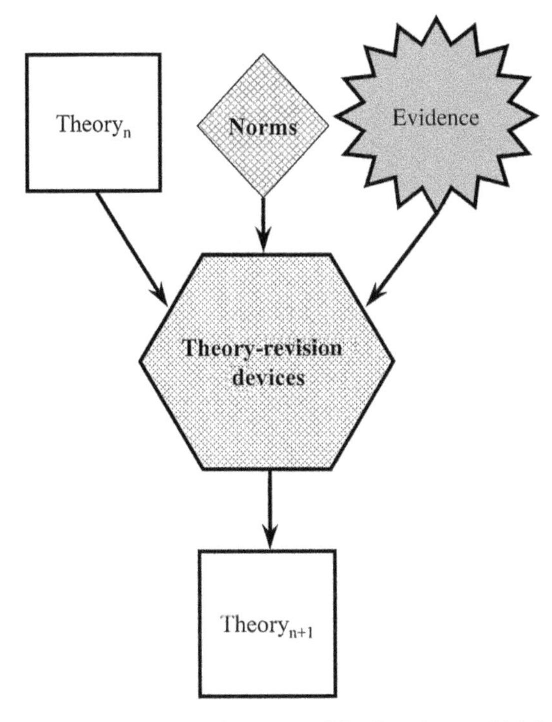

Fig. A* A sketch of theory revision in science which improves on Gopnik's picture in figure A, by indicating that norms play an important role

the emergence of science. So, while figure A *may* offer a plausible picture of theory revision in early childhood, something more along the lines of figure A* does a better job of capturing theory revision in science. And thus Gopnik's Continuity Thesis, depicted in figure B, must be replaced by something more like figure B*.

7 The cultural transmission of norms and theories (and two more modifications to Gopnik's picture)

In the previous section we argued that norms play an important role in science and we emphasized the emergence of new norms and the role that these new norms played in making science possible. But, as the literature in anthropology and social psychology makes abundantly clear, though norms do change they are also often remarkably stable. Within a culture, the norms governing many aspects of behaviour can be highly resistant to change. In traditional

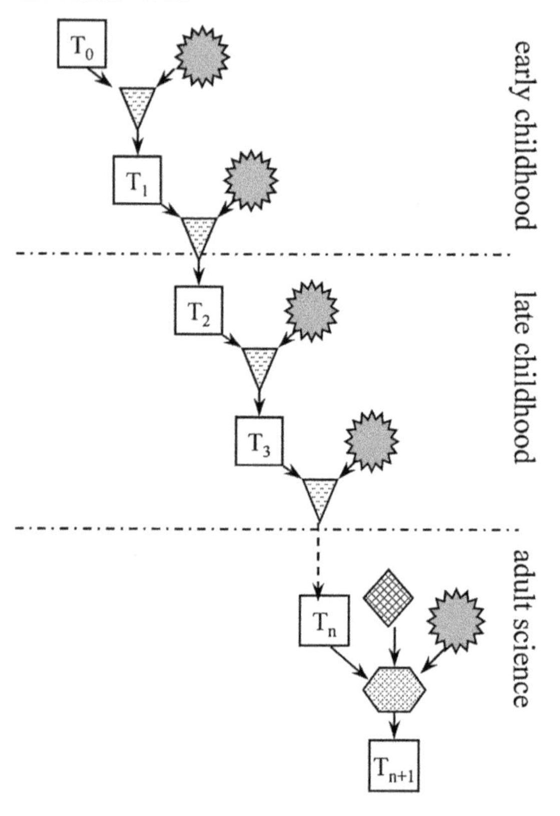

Fig. B* An alternative to Gopnik's Continuity Thesis indicating that norms play a role in scientific theory revision even though they may not play a role in theory revision in early or late childhood

societies with relatively homogenous norms, children and adolescents almost always end up sharing the norms of their culture.

To the best of our knowledge, relatively little is known about the details of the cognitive processes subserving the acquisition of norms, and even less is known about the processes that result in the emergence of new norms. However, a bit of reflection on familiar experience suggest that norm transmission *cannot* be subserved by the sort of evidence-driven theory revision process that, according to Gopnik, subserves cognitive development in infants (figure A). Rather, norm transmission is a process of *cultural transmission*.

Children acquire their norms from older members of their society and this process is not plausibly viewed as a matter of accumulating evidence indicating that one set of norms is better than another. Our point here is not that evidence is irrelevant to norm acquisition, but rather that to the extent that it is relevant it plays a very different role from the one suggested in figure A. An anthropologist will use evidence of various sorts to discover the norms that obtain in a culture. And it is entirely possible that children in that culture will use some of the same evidence to discover what norms their cultural parents embrace.[5] But this alone is not enough to explain the *transmission* of norms. For it does not explain the crucial fact that children typically *adopt* the norms of their cultural parents while anthropologists do not typically adopt the norms of the culture they are studying. Much the same, of course, is true for the acquisition of scientific norms which (we would speculate) typically occurs somewhat later in life. Evidence of various sorts may be necessary to figure out what norms of inquiry obtain in a given branch of science. But *knowing what the norms are* and *accepting them* are importantly different phenomena.

If this is right, it indicates that there is yet another mental mechanism (or cluster of mental mechanisms) that must be added to our evolving sketch of the cognitive underpinnings of scientific cognition. In addition to the mechanisms in figure A*, we need a mechanism (or cluster of mechanisms) whose function it is to subserve the process of cultural transmission, via which norms are acquired. Adding this to figure A* gives us figure A**.

Once mechanisms subserving social transmission have been added to the picture, a natural question to ask is whether these mechanisms might play any other role in helping to explain scientific cognition. And the answer, we think, is 'yes'. For there is an interesting *prima facie* parallel between the process by which norms are transmitted and the process by which children and adolescents acquire much of their basic knowledge of science. Learning science, like learning norms, appears to be largely a process of cultural transmission in which younger members of a culture are instructed and indoctrinated by appropriate senior members. In the case of science, of course, teachers are among the more important 'cultural parents'.

What makes this analogy important for our current concerns is that in the cultural transmission of both norms and science, evidence takes a back-seat to *authority*. When children are taught the basics of the heliocentric theory of the solar system, Newtonian physics, the atomic theory of matter or Watson

[5] It is also possible that children go about discovering the norms of their society in ways that are quite different from those that an anthropologist would use. Given the importance of norms in human cultures, it is entirely possible that children have special mechanisms whose function is to facilitate the process of norm acquisition, much as they have special mechanisms whose function is to facilitate the acquisition of language. And these mechanisms may play little or no role in anthropologists' attempts to come up with an explicit statement of the norms in the cultures they are studying, just as a linguist's 'language acquisition device' plays little or no role in her attempt to come up with an explicit statement of the grammar of the language she is studying. For more on this possibility, see Stich (1993) and Harman (1999).

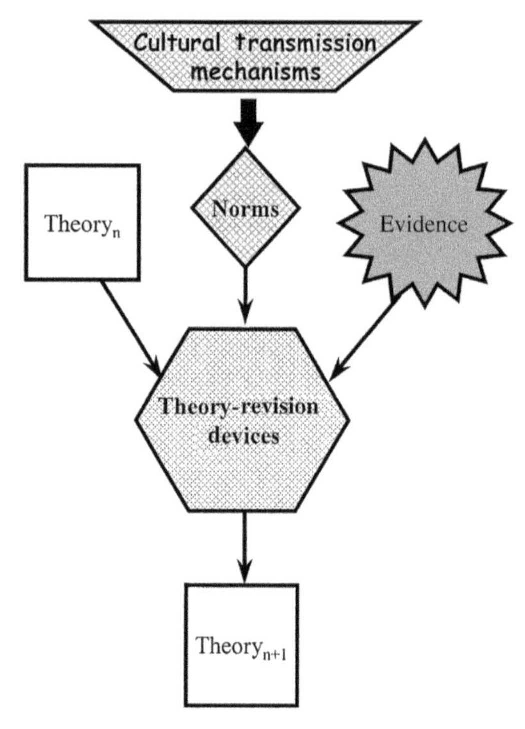

Fig. A** A sketch of theory revision in science which improves on figure A* by indicating that cultural transmission mechanisms play a role in the acquisition of norms

and Crick's theory about the structure of DNA, the challenge is to get them to *understand* the theory. Models (often real, physical models made of wood or plastic), diagrams and analogies play an important role. But *evidence* does not. Most people (including several of the authors of this chapter) who know that genes are made of DNA and that DNA molecules have a double helical structure haven't a *clue* about the evidence for these claims. If students can be got to understand the theory, the fact that it is endorsed by parents, teachers and textbooks is typically more than sufficient to get them to *accept* it. We teach science to children in much the same way that people in traditional cultures teach traditional wisdom. Adults explain what they think the world is like, and (by and large) children believe them. If this is right, it highlights another way in which culture plays an important role in scientific cognition. It is *cultural transmission* that lays the groundwork on which the adult scientist builds.

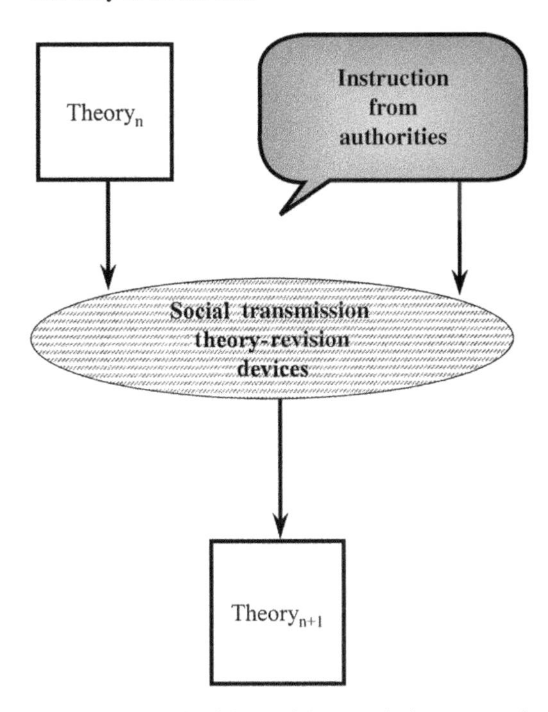

Fig. D A sketch of the social transmission process by which older children and adolescents acquire much of their knowledge of scientific theories; in contrast with Gopnik's account, in figure A, authority plays an important role and evidence plays little or no role

All of this, of course, is bad news for theories like Gopnik's. For, on her Continuity Thesis, sketched in figure B, the process of theory revision is much the same in young children, older children and in adult scientists. But if we are right, the acquisition of new scientific theories by older children and adolescents is subserved, in most instances at least, by a quite different process of social transmission in which authority looms large and evidence plays little or no role.[6] That process is sketched in figure D. So, if we continue to grant, for argument's sake, that Gopnik is right about young children, figure B, which was superseded by figure B*, must now be replaced by figure B**.

[6] It is, we think, a remarkable fact, little noted by cognitive scientists, that theories or other knowledge structures which people acquire via cultural transmission and authority can (in some cases at least) be later modified by what are apparently quite different evidence-driven theory revision processes. And if Gopnik is right about the role of evidence in early childhood, then theories acquired via an evidence-driven process can also be modified or replaced by cultural transmission.

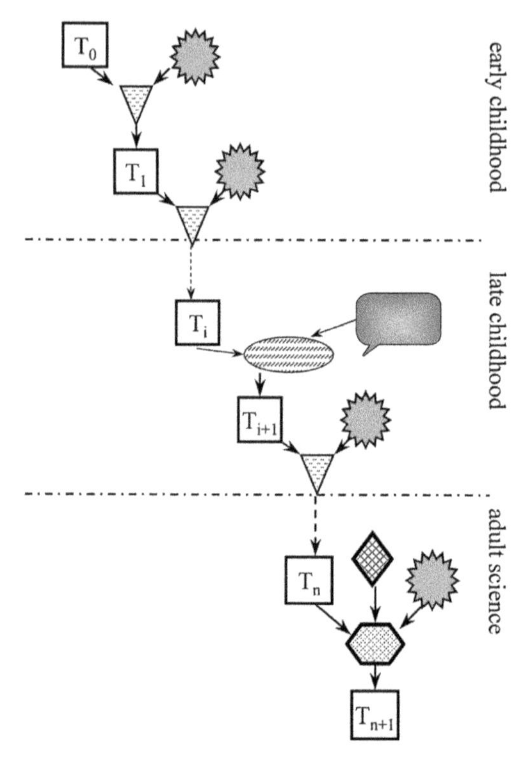

Fig. B** A second alternative to Gopnik's continuity Thesis that indicates the role played by norms in science and includes the social transmission process by which older children acquire much of their knowledge of scientific theories

8 Some surprising examples of the ways in which culture affects cognition

The examples of cultural transmission that we've focused on so far are hardly surprising. It comes as no news that both norms and theories are acquired from one's culture. What is surprising is that Gopnik and her collaborators have offered an account of scientific cognition that ignores the role of culturally transmitted norms and theories, and that there has been so little discussion of the cultural transmission of norms and theories in the cognitive science literature. In this section, our focus will be on some further examples of cultural transmission – examples which many people *do* find surprising. The examples we'll discuss are drawn primarily from the recent work of Richard Nisbett and his colleagues (Nisbett *et al.*, 2001). It is our belief that their findings have important implications for the study of the cognitive basis of science and that they provide an important part of the solution to the '1492 problem'. More importantly, this work indicates that a systematic re-examination of some of the more fundamental assumptions of cognitive science may be in order.

The work of Nisbett and his colleagues was inspired, in part, by a long tradition of scholarship which maintains that there were systematic cultural differences between ancient Greek and ancient Chinese societies, and that many of these differences have endured up to the present in the cultures that have been most deeply influenced by ancient Greek and Chinese cultures. This scholarship also suggests that these cultural differences were correlated with different 'mentalities' – that people in Greek-influenced Western cultures perceive and think about the world around them in very different ways from people in Chinese-influenced cultures, and that these differences are reflected in the way they describe and explain events and in the beliefs and theories they accept. What is novel, and startling, in the work of Nisbett and his colleagues is that they decided to explore whether these claims about differences in mentalities could be experimentally verified, and they discovered that many of them could.

One of the more important aspects of ancient Greek culture, according to the scholars that Nisbett and his colleagues cite was people's 'sense of personal *agency*' (Nisbett *et al.* 2001, pp. 292, emphasis in original). Ordinary people took control of their lives, and their daily lives reflected a sense of choice and an absence of constraint that had no counterpart in the ancient Chinese world. One indication of this was the Greek tradition of debate, which was already well established at the time of Homer, who repeatedly emphasizes that next to being a good warrior, the most important skill for a man to have is that of the debater. Even ordinary people could participate in the debates in the market place and the political assembly, and they could and did challenge even a king.

Another aspect of Greek civilization, one that Nisbett and his colleagues suggest may have had the greatest effect on posterity, was the Greeks' sense of curiosity about the world and their conviction that it could be understood by the discovery of rules or principles:

> The Greeks speculated about the nature of the objects and events around them and created models of them. The construction of these models was done by categorizing objects and events and generating rules about them for the purpose of systematic description, prediction and explanation. This characterized their advances in, some have said invention of, the fields of physics, astronomy, axiomatic geometry, formal logic, rational philosophy, natural history, history, ethnography and representational art. Whereas many great civilizations . . . made systematic observations in many scientific domains, only the Greeks attempted to model such observations in terms of presumed underlying physical causes. (Nisbett *et al.*, 2001, p. 292)

The ancient Chinese, according to scholars, present a radical contrast to this picture. Their civilization was much more technologically sophisticated than the

Greeks', and they have been credited with the original or independent invention of 'irrigation systems, ink, porcelain, the magnetic compass, stirrups, the wheelbarrow, deep drilling, the Pascal triangle, pound-locks on canals, fore-and-aft sailing, watertight compartments, the sternpost rudder, the paddle-wheel boat, quantitative cartography, immunization techniques, astronomical observations of novae, seismographs, and acoustics' (see Logan, 1986, p. 51; Nisbett *et al.*, 2001, p. 293). But most experts hold that this technological sophistication was not the result of scientific investigation or theory. Rather, it reflects the Chinese emphasis on (and genius for) *practicality*; they had little interest in knowledge for its own sake. Indeed, in the Confucian tradition, according to Munro 'there was no thought of *knowing* that did not entail some consequence for action' (1969, p. 55; quoted in Nisbett *et al.*, 2001: p. 293, emphasis in original).

The Chinese made good use of intuition and trial-and-error methods, but, many scholars insist, they never developed the notion of a law of nature, in part because they did not have the concept of *nature* as something distinct from human or spiritual entities. Neither debate nor a sense of agency played a significant role in Chinese culture. Rather, there was an emphasis on harmony and obligation. Individuals felt very much a part of a large and complex social system whose behavioural prescriptions and role obligations must be adhered to scrupulously. And no one could contradict another person without fear of making an enemy. (See Nisbett *et al.*, 2001, pp. 292–3.)

According to Nisbett and his colleagues, many of these 'aspects of Greek and Chinese life had correspondences in the mentalities or systems of thought in the two cultures' and these differences have left an important contemporary residue (Nisbett *et al.*, 2001, p. 293). In their account of the differences between ancient Greek and ancient Chinese mentalities, Nisbett *et al.* stress five themes:

- *Continuity vs. Discreteness.* The Chinese 'held the view that the world is a collection of overlapping and interpenetrating stuffs or substances' while the Greeks saw the world as 'a collection of discrete objects which could be categorized by reference to some subset of universal properties that characterized the object' (Nisbett *et al.*, 2001, p. 293).
- *Field vs. Object.* 'Since the Chinese were oriented toward continuities and relationships, the individual object was not a primary conceptual starting point... The Greeks, in contrast, were inclined to focus primarily on the central object and its attributes' (Nisbett *et al.*, 2001, p. 293).
- *Relationships and similarities vs. Categories and rules.* Because of the Chinese emphasis on continuity and the Greek emphasis on discreteness, 'the Chinese were concerned with relationships among objects and events. In contrast, the Greeks were more inclined to focus on the categories and rules that would help them to understand the behaviour of the object independent of

its context (Nakamura, 1964/1985, pp. 185–6). The Chinese were convinced of the fundamental relatedness of all things and the consequent alteration of objects and events by the context in which they were located. It is only the whole that exists; and the parts are linked relationally, like the ropes in a net' (Nisbett *et al.*, 2001, pp. 293–4).

- *Dialectics vs. Foundational principles and logic.* 'The Chinese seem not to have been motivated to seek for first principles underlying their mathematical procedures or scientific assumptions . . . The Chinese did not develop any formal systems of logic or anything like an Aristotelian syllogism . . . In place of logic, the Chinese developed a dialectic, . . . which involved reconciling, transcending or even accepting apparent contradiction' (Nisbett *et al.*, 2001, p. 294).
- *Experience-based knowledge vs. Abstract analysis.* '"The Chinese . . . sought intuitive instantaneous understanding through direct perception" (Nakumara 1964/1985, p. 171). This resulted in a focus on particular instances and concrete cases in Chinese thought . . .' By contrast, 'many Greeks favoured the epistemology of logic and abstract principles, and many Greek philosophers, especially Plato and his followers, actually viewed concrete perception and direct experiential knowledge as unreliable and incomplete at best, and downright misleading at worst . . . Ironically, important as the Greek discovery of formal logic was for the development of science, it also impeded it in many ways. After the 6th-century Ionian period, the empirical tradition in Greek science was greatly weakened. It was countered by the conviction on the part of many philosophers that it ought to be possible to understand things through reason alone, without recourse to the senses (Logan, 1986, pp. 114–15)' (Nisbett *et al.*, 2001, p. 294).

Nisbett and his colleagues suggest that these differences can be loosely grouped together under the heading of *holistic* vs. *analytic* thought – where holistic thought is defined as 'involving an orientation to the context or field as a whole, including attention to relationships between a focal object and the field, and a preference for explaining and predicting events on the basis of such relationships' and analytic thought is defined as 'involving detachment of the object from its context, a tendency to focus on attributes of the object in order to assign it to categories, and a preference for using rules about the categories to explain and predict the object's behaviour' (Nisbett *et al.*, 2001, p. 293). And, as noted earlier, they claim that these differences persist in the thought of contemporary cultures that have been influenced by China (including modern China, Japan and Korea) and by Greece (including Europe and North America). In support of this claim they assemble an impressive catalogue of experimental findings showing that there are indeed differences between Americans and East Asians in perception, attention and memory, and in the way they go about predicting, explaining, categorizing and revising beliefs in the face of new arguments and evidence. While this is not the place to offer a systematic review of these experimental results, we will very briefly sketch a few of the more remarkable findings.

- *Attention and perception.* In one task used to test the extent to which people can perceptually isolate an object from the context in which it embedded, a square frame was rotated independently of a rod located inside it. Subjects were asked to report when the rod appeared to be vertical. East Asian subjects were significantly more influenced by the position of the frame and made many more errors than American subjects (Nisbett *et al.*, 2001, p. 297).
- *Attention and memory.* In another experiment, subjects viewed realistic animated cartoons of fish and other underwater life and were then asked to report what they had seen. In each cartoon there was a focal fish or group of fish that was larger and moved more rapidly than anything else on the screen. In describing what they had seen, Japanese and American subjects were equally likely to refer to the focal fish, but the Japanese participants were much more likely to refer to background aspects of the environment. Later the subjects were tested on how well they could recognize the focal fish from the displays they had seen. Some were shown fish with the original background, others with a background the subjects had never seen. Japanese recognition was significantly harmed by showing the focal fish with the wrong background, but American recognition was unaffected (Nisbett *et al.*, 2001, p. 297).
- *Attention, curiosity and surprise.* As Nisbett *et al.* note, the Asian tendency to attend to a broad range of factors may make it too easy to come up with explanations of unexpected events:

  > If a host of factors is attended to, and if naïve metaphysics and tacit epistemology support the view that multiple, interactive factors are usually operative in any given outcome, then any outcome may seem to be understandable, even inevitable, after the fact . . . An advantage of the more simplistic, rule-based stance of the Westerner may be that surprise is a frequent event. *Post hoc* explanations may be relatively difficult to generate, and epistemic curiosity may be piqued. The curiosity, in turn, may provoke a search for new, possibly superior models to explain events. In contrast, if Eastern theories about the world are less focused, and a wide range of factors are presumed to be potentially relevant to any given outcome, it may be harder to recognize that a particular outcome could not have been predicted. (Nisbett *et al.*, 2001, p. 299)

  One prediction that this suggests is that Easterners might be more susceptible to what Fischoff (1975) has called *hindsight bias* – the tendency to assume that one knew all along that a given outcome was likely. And in a series of experiments Koreans subjects did indeed show less surprise than Americans when told of the results in Darley and Batson's famous 'Good Samaritan' experiment (1973), in which seminary students in a bit of a hurry refused to help a man lying in a doorway pleading for help. Importantly, Korean and American subjects make the same predictions about the likelihood of the seminary student helping when they have not been told about the results of the experiment.

- *Detection of co-variation.* If Asians are more attentive to relationships among objects then we might expect that they do a better job at detecting co-variation. And indeed they do. In one experiment requiring subjects to judge the degree of association between arbitrary objects presented together on opposite sides of a computer screen, Chinese subjects reported a higher degree of co-variation than American subjects and were more confident about their judgements. Their greater confidence was quite appropriate, since their judgements calibrated better with actual co-variation (Ji, Peng and Nisbett, 1999; Nisbett *et al.*, 2001, p. 297). In a similar experiment, American subjects showed a strong primacy effect. Their predictions of future co-variation was more influenced by the first pairing they saw than by the overall degree of co-variation to which they had been exposed. Chinese subjects showed no primacy effect at all (Yates and Curley, 1996).
- *Spontaneous categorization.* Norenzayan *et al.* (1999) showed subjects a series of stimuli on a computer screen in which a simple target object was displayed below two groups of four similar objects. The subjects were asked to say which group the target was most similar to. The groups were constructed so that those in one group had a close family resemblance to one another and to the target object. The objects in the other group had a different family resemblance structure, one that was dissimilar to the target object. However, all the objects in the second group could be characterized by a simple 'rule', like 'has a curved stem' which also applied to the target object. A majority of the East Asians said that the target was more similar to the 'family resemblance' group while a majority of European Americans said it was more similar to the 'rule' group (Nisbett *et al.*, 2001, p. 300).
- *Plausibility vs. logic.* It has long been known that the performance of subjects who are asked to assess the logical validity of an argument is affected by a 'belief bias' – an invalid argument is more likely to be judged valid if it has a plausible conclusion. However, Norenzayan *et al.* (1999) have shown that Korean subjects show a much stronger belief bias than Americans. As Nisbett *et al.* note, 'The results indicate that when logical structure conflicts with everyday belief, American students are more willing to set aside empirical belief in favour of logic than are Korean students' (Nisbett *et al.*, 2001, p. 301).
- *Dealing with conflicting arguments.* In a particularly striking experiment, Korean and American subjects were presented with a variety of arguments for and against funding a particular scientific project. The arguments were

independently rated for plausibility by separate groups of Korean and American subjects, and Americans and Koreans agreed on how strong the arguments were. Korean subjects presented with a strong argument for funding and a weak argument against funding, were less in favour of funding than Korean subjects presented with just the strong pro argument. However, for American subjects the results were just the opposite. Those presented with both a strong pro argument and a weak con argument were *more* favourable to funding than those presented with just the strong pro argument (Davis, 1999; Davis, Nisbett and Schwarz, 1999; Nisbett *et al.*, 2001, p. 302)!

9 Conclusions

What conclusions can we draw from this large body of work that are relevant to our concerns in this chapter? First, we think that the sorts of differences that Nisbett and his colleagues have found between East Asians and Westerners are *bound* to affect the processes of belief change and theory change in adults of those two groups – including adults who happen to be scientists. Since all of the studies that Nisbett and his colleagues cite were done on adult subjects, we can only speculate about when the differences emerge. However, it is very likely that many of them would be found in teenagers and in younger children as well. Moreover, since the differences between East Asians and Westerners begin to disappear in Asians whose families have been in America for several generations, there is no reason to think that there is any significant genetic component to these differences. The differences are acquired, not innate, and the process by which they are acquired is another example of social transmission. So we need to make yet another addition to our evolving picture of theory revision to indicate that, to some significant extent, *the theory revision mechanisms themselves are a product of culture* – figure A** is superseded by figure A***, and figure B** is superseded by figure B***. At this point neither Gopnik's account of adult theory revision (in figure A) nor her Continuity Thesis (in figure B) appear to be even remotely plausible.

Nisbett's work, along with our earlier discussion of the emergence of new norms, also suggests what we take to be a much richer and more plausible solution to the '1492 problem' than the one that Gopnik suggests. On her account, the engine that drove the emergence of science was simply the availability of new evidence produced by new technology, along with better communication and more leisure. Although we would not deny that this is part of the story, we maintain that it is only a small part. On Nisbett's account, the Western tradition, with its emphasis on reason and theory, its assumption that the important properties of things were not accessible to the senses and its greater likelihood to be surprised by and curious about unexpected events, was in a much better

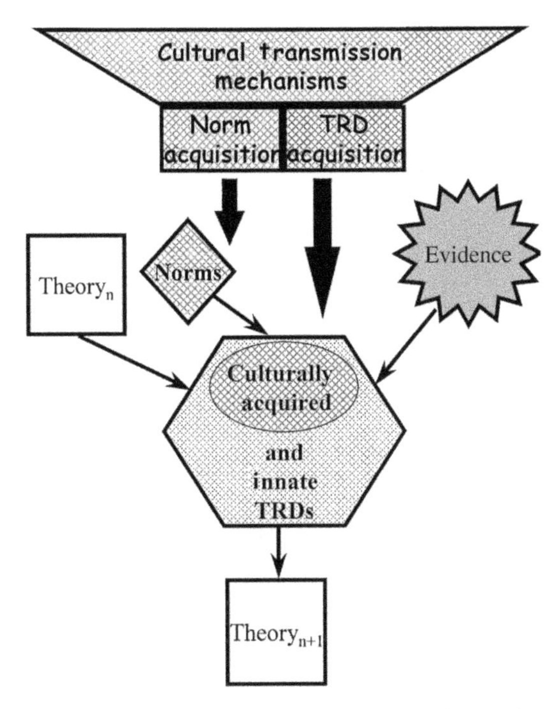

Fig. A*** A sketch of theory revision in science which improves on figure A** by indicating that, to some significant extent, the theory mechanisms themselves are a product of culture

position to generate and embrace scientific theories about invisible processes that give rise to observable phenomena. But, as Nisbett and his colleagues note, after the sixth century BC, the conviction that the world could be understood without appeal to highly misleading sensory input stifled experimental inquiry. This reluctance to observe the world was reversed by the new norm that insisted one ought to look at nature rather than accept the pronouncements of the Ancients. And as cultural transmission spread that new norm, something resembling the Ionian experimental tradition re-emerged and modern science was born. This is, we recognize, only a small fragment of the full story. But in the full story, we believe that the four factors we have emphasized – (i) norms, (ii) the mechanisms subserving the social transmission of norms, (iii) the mechanisms subserving the cultural transmission of theory, and (iv) culturally acquired differences in the cognitive mechanisms that subserve theory revision – will all play an important part, even if it is true, as Gopnik maintains, that none of these plays a significant role in the theory revision processes in

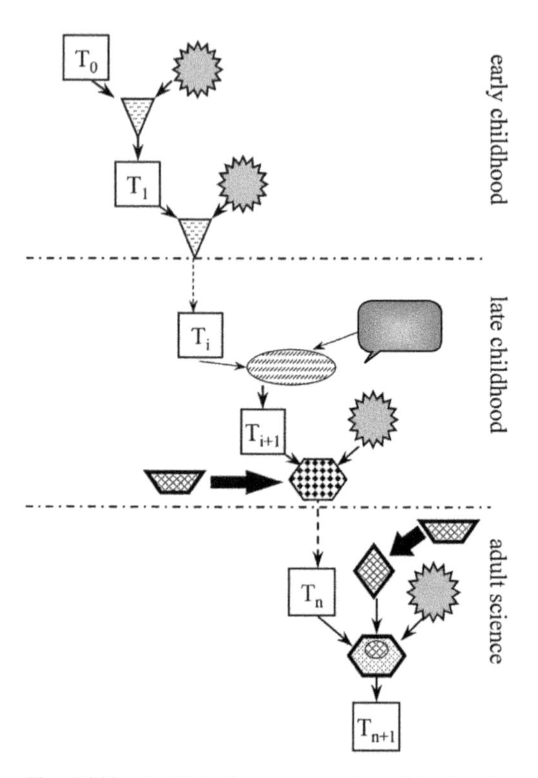

Fig. B*** A third alternative to Gopnik's Continuity Thesis that includes the fact that theory revision mechanisms are also, in part, a product of culture; the quite radical difference between figure B and figure B*** indicates how seriously mistaken Gopnik's Continuity Thesis is

young children. Though these four factors have received little attention from either cognitive or developmental psychologists, we believe that they will play an important role in understanding not only the cognitive basis of science but cognition in other domains as well. Gopnik's baby in the lab-coat is not an adequate model for understanding scientific cognition because there are more cognitive mechanisms involved in understanding heaven and earth than are dreamt of in Gopnik's psychology.

References

Adler, J. (1984). Abstraction is uncooperative. *Journal for the Theory of Social Behavior*, 14

AHG/APESA (1992). *Plan de desarollo integrado de Petén: Inventario forestal del Departamento del Petén* (Convenio Gobiernos Alemania y Guatemala). Santa Elena, Petén. SEGEPLAN

Ahn, W. and Bailenson, J. (1996). Causal attribution as a search for underlying mechanisms: an explanation of the conjunction fallacy and the discounting principle. *Cognitive Psychology*, 31

Ahn, W., Kalish, C., Medin, D. and Gelman, S. (1995). The role of covariation versus mechanism information in causal attribution. *Cognition*, 54

Aiello, L. and Wheeler, P. (1995). The expensive tissues hypothesis. *Current Anthropology*, 36

Amado, G. and Deumie, C. (1991). Pratiques magiques et régressives dans la gestion des ressources humaines. *Revue de Gestion des Ressources Humaines*, 1

Anderson, J. (1990). *The Adaptive Character of Thought*. Lawrence Erlbaum

Armstrong, D. (1968). *A Materialist Theory of the Mind*. Routledge

(1973). *Belief, Truth and Knowledge*. Cambridge University Press

Arsuaga, J.-L., Martinez, I., Gracia, A., Carretero, J.-M. and Carbonell, E. (1993). Three new human skulls from the Sima de los Huesos Middle Pleistocene site in Sierra de Atapuerca. *Nature*, 362

Atran, S. (1985). The nature of folk-botanical life forms. *American Anthropologist*, 87

(1987). Origins of the species and genus concepts. *Journal of the History of Biology*, 20

(1990). *Cognitive Foundations of Natural History: Towards an Anthropology of Science*. Cambridge University Press

(1998). Folk biology and the anthropology of science. *Behavioral and Brain Sciences*, 21

(1999). Itzaj Maya folk-biological taxonomy. In D. Medin and S. Atran (eds.), *Folk Biology*. MIT Press

Atran, S. and Sperber, D. (1991). Learning without teaching: its place in culture. In L. Tolchinsky-Landsmann (ed.), *Culture, Schooling and Psychological Development*. Norwood, NJ: Ablex

Atran, S. and Ucan Ek', E. (1999). Classification of useful plants among the Northern Peten Maya. In C. White (ed.), *Reconstructing Maya Diet*. University of New Mexico Press

Atran, S., Estin, P., Coley, J. and Medin, D. (1997). Generic species and basic levels: essence and appearance in folk biology. *Journal of Ethnobiology*, 17

Atran, S., Medin, D., Lynch, E., Vapnarsky, V., Ucan Ek', E. and Sousa, P. (2001). Folk biology doesn't come from folk psychology: evidence from Yukatek Maya in cross-cultural perspective. *Journal of Cognition and Culture*, 1

Atran, S., Medin, D., Ross, N., Lynch, E., Coley, J., Ucan Ek', E. and Vapnarsky, V. (1999). Folk ecology and commons management in the Maya Lowlands. *Proceedings of the National Academy of Sciences USA*, 96

Au, T., Romo, L. and DeWitt, J. (1999). Considering children's folk biology in health education. In M. Siegal and C. Peterson (eds.), *Children's Understanding of Biology and Health*. Cambridge University Press

Au, T., Sidle, A. and Rollins, K. (1993). Developing an intuitive understanding of conservation and contamination: invisible particles as a plausible mechanism. *Developmental Psychology*, 29

Axelrod, R. (1986). An evolutionary approach to norms. *American Political Science Review*, 80(4)

Ayer, A. (1946). *Language, Truth and Logic*. Gollancz

Bahrick, L. and Watson, J. (1985). Detection of intermodal proprioceptive–visual contingency as a potential basis of self-perception in infancy. *Developmental Psychology*, 21(6)

Baillargeon, R. (1995). Physical reasoning in infancy. In M. Gazzaniga (ed.), *The Cognitive Neurosciences*. MIT Press

Baker, L. and Dunbar, K. (2001). Experimental design heuristics for scientific discovery: the use of 'baseline' and 'known' standard controls. *International Journal of Computer Studies*

Ball, L., Evans, J. and Dennis, I. (1994). Cognitive processes in engineering design: a longitudinal study. *Ergonomics*, 37

Ball, L., Evans, J., Dennis, I. and Oremord, T. (1997). Problem solving strategies and expertise in engineering design. *Thinking and Reasoning*, 3

Banning, E. (1998). The Neolithic period: triumphs of architecture, agriculture and art. *Near Eastern Archaeology*, 61

Barkow, J., Cosmides, L. and Tooby, J. (eds.) (1992). *The Adapted Mind*. Oxford University Press

Barnes, A. and Thagard, A. (1997). Empathy and analogy. *Dialogue: Canadian Philosophical Review*, 36

Baron-Cohen, S. (1995). *Mindblindness: An Essay on Autism and Theory of Mind*. MIT Press

Barrett, J. and Keil, F. (1996). Conceptualizing a nonnatural entity: anthropomorphism in God concepts. *Cognitive Psychology*, 31

Barrett, J., Richert, R. and Driesenga, A. (2001). God's beliefs versus mother's: the development of non-human agent concepts. *Child Development*, 72

Barsalou, L. (1983). Ad hoc categories. *Memory and Cognition*, 11

(1991). Deriving categories to achieve goals. In G. Bower (ed.), *The Psychology of Learning and Motivation: Advances in Research and Theory*. Academic Press

(1999). Perceptual symbol systems. *Behavioral and Brain Systems*, 22

Bartlett, H. (1936). A method of procedure for field work in tropical American phytogeography based on a botanical reconnaissance in parts of British Honduras and the Petén forest of Guatemala. *Botany of the Maya Area, Miscellaneous Papers I*. Carnegie Institution of Washington Publication, 461

(1940). History of the generic concept in botany. *Bulletin of the Torrey Botanical Club*, 47

Barwise, J. and Etchemendy, J. (1996). Heterogeneous logic. In G. Allwein and J. Barwise (eds.), *Logical Reasoning with Diagrams*. Oxford University Press

Bar-Yosef, O. (1998). On the nature of transitions: the Middle to the Upper Palaeolithic and the Neolithic revolutions. *Cambridge Archaeological Journal*, 8

Basso, A., Spinnler, H., Vallar, G. and Zanobio, M. (1982). Left hemisphere damage and selective impairment of auditory verbal short-term memory: a case study. *Neuropsychologia*, 20

Bechtel, W. (1996a). What should a connectionist philosophy of science look like?. In R. McCauley (ed.), *The Churchlands and Their Critics*. Blackwell

(1996b). What knowledge must be in the head in order to acquire knowledge?. In B. Velichkovsky and D. Rumbaugh (eds.), *Communicating Meaning: The Evolution and Development of Language*. Lawrence Erlbaum

Ben-Shakhar, G., Bar-Hillel, M., Blui, Y., Ben-Abba, E. and Flug, A. (1989). Can graphology predict occupational success? *Journal of Applied Psychology*, 71

Bentham, J. (1879). *Introduction to the Principles of Morals and Legislation*. Oxford University Press

Berk, L. (1994). Why children talk to themselves. *Scientific American*, November

Berlin, B. (1978). Ethnobiological classification. In E. Rosch and B. Lloyd (eds.), *Cognition and Categorization*. Lawrence Erlbaum

(1992). *Ethnobiological Classification*. Princeton University Press

(1999). How a folk biological system can be both natural and comprehensive. In D. Medin and S. Atran (eds.), *Folk Biology*. MIT Press

Berlin, B., Breedlove, D. and Raven, P. (1973). General principles of classification and nomenclature in folk biology. *American Anthropologist*, 74

Bickerton, D. (1995). *Language and Human Behavior*. University of Washington Press (UCL Press, 1996)

Binford, L. (1981). *Bones: Ancient Men and Modern Myths*. Academic Press

(1986). Comment on 'Systematic butchery by Plio/Pleistocene hominids at Olduvai Gorge', by H. T. Bunn and E. M. Kroll. *Current Anthropology*, 27

Blanchette, I. and Dunbar, K. (2001). How analogies are generated: the roles of structural and superficial similarity. *Memory and Cognition*, 29

Block, N. and Fodor, J. (1972). What psychological states are not. *Philosophical Review*, 81

Bloom, P. (2000). *How Children Learn the Meanings of Words*. MIT Press

Bloor, D. (1976). *Knowledge and Social Imagery*. Routledge

Blurton-Jones, N. and Konner, M. (1976). !Kung knowledge of animal behaviour. In R. Lee and I. DeVore (eds.), *Kalahari Hunter–Gatherers*. Cambridge University Press

Bock, W. (1973). Philosophical foundations of classical evolutionary taxonomy. *Systematic Zoology*, 22

Boden, M. (1990). *The Creative Mind: Myths and Mechanisms*. London: Weidenfeld & Nicolson (expanded edn., Abacus, 1991)

Boëda, E. (1988). Le concept laminaire: rupture et filiation avec le concept Levallois. In J. Kozlowski (ed.), *L'Homme Neanderthal, Vol.8: La Mutation*. Liège, Belgium: ERAUL

Boesch, C. and Boesch, H. (1984). Mental maps in wild chimpanzees: an analysis of hammer transports for nut cracking. *Primates*, 25

Bohner, G., Bless, H., Schwarz, N. and Strack, F. (1988). What triggers causal attributions? The impact of valence and subjective probability. *European Journal of Social Psychology*, 18

Bond, R. and Smith, P. (1996). Culture and conformity: a meta-analysis of studies using Asch's (1952b, 1956) line-judgement task. *Psychological Bulletin*, 119

Boster, J. (1991). The information economy model applied to biological similarity judgement. In L. Resnick, J. Levine and S. Teasley (eds.), *Perspectives on Socially Shared Cognition*. American Psychological Association

Botterill, G. and Carruthers, P. (1999). *The Philosophy of Psychology*. Cambridge University Press

Bower, G. (1981). Mood and memory. *American Psychologist*, 36

Boyd, R. (1989). What Realism implies and what it does not. *Dialectica*, 43(1–2)

Boyd, R. and Richerson, P. (1992). Punishment allows the evolution of co-operation (or anything else) in sizeable groups. *Ethology and Sociobiology*, 13

Boyd, R. and Silk, J. (1997). *How Humans Evolved*. Norton

Boyer, P. and Walker, S. (2000). Intuitive ontology and cultural input in the acquisition of religious concepts. In K. Rosengren, C. Johnson and P. Harris (eds.), *Imagining the Impossible: Magical, Scientific and Religious Thinking in Children*. Cambridge University Press

Brown, C. (1984). *Language and Living Things: Uniformities in Folk Classification and Naming*. Rutgers University Press

Brown, D. and Boysen, S. (2000). Spontaneous discrimination of natural stimuli by chimpanzees (*Pan troglodytes*). *Journal of Comparative Psychology*, 114

Brown, R. and Fish, D. (1983). The psychological causality implicit in language. *Cognition*, 14

Brown, R. and van Kleeck, R. (1989). Enough said: three principles of explanation. *Journal of Personality and Social Psychology*, 57

Bullock, M. and Gelman, R. (1979). Preschool children's assumptions about cause and effect: temporal ordering. *Child Development*, 50(1)

Bulmer, R. (1974). Folk biology in the New Guinea Highlands. *Social Science Information*, 13

Bunn, H. and Kroll, E. (1986). Systematic butchery by Plio-Pleistocene hominids at Olduvai Gorge, Tanzania. *Current Anthropology*, 27

Bush, V. (1945). *Science: The Endless Frontier*. Washington DC: National Science Foundation

Bussey, K. and Bandura, A. (2000). Social cognitive theory of gender development and differentiation. *Psychological Review*, 106

Butterfield, H. (1957). *The Origins of Modern Science 1300–1800* (revised edn.). London: Bell

Butterworth, G., Siegal, M., Newcombe, P. and Dorfmann, M. (2002). Models and methodology in children's cosmology. Manuscript submitted for publication. University of Sheffield

Campbell, J. (1995). *Past, Space and Self*. MIT Press

Carbonell, E., Bermúdez de Castro, J.-M., Arsuaga, J.-M., Diez, J., Rosas, A., Cuenca-Bercos, G., Sala, R., Mosquera, M. and Rodriguez, X. (1995). Lower

Pleistocene hominids and artefacts from Atapuerca – TD-6 (Spain). *Science*, 269
Carey, S. (1985). *Conceptual Change in Childhood*. MIT Press
(1995). On the origin of causal understanding. In D. Sperber, D. Premack and A. Premack (eds.), *Causal Cognition*. Oxford University Press
(1996). Cognitive domains as modes of thought. In D. Olson and N. Torrance (eds.), *Modes of Thought*. Cambridge University Press
(1999). Sources of conceptual change. In E. Scholnick, K. Nelson, S. Gelman and P. Miller (eds.), *Conceptual Development: Piaget's Legacy*. Lawrence Erlbaum
(2000a). Science education as conceptual change. *Journal of Applied Developmental Psychology*, 21
(2000b). Whorf versus continuity theorists: bringing data to bear on the debate. In M. Bowerman, and S. Levinson (eds.), *Language Acquisition and Conceptual Development*. Cambridge University Press
Carey, S. and Spelke, E. (1994). Domain-specific knowledge and conceptual change. In L. Hirschfeld and S. Gelman (eds.), *Mapping the Mind: Domain-Specificity in Culture and Cognition*. Cambridge University Press
(1996). Science and core knowledge. *Philosophy of Science*, 63
Carnap, R. (1950). *Logical Foundations of Probability*. Chicago University Press
(1967). *The Logical Structure of the World* (trans. R. George). Routledge & Kegan Paul
Carruthers, P. (1992). *Human Knowledge and Human Nature*. Oxford University Press
(1996a). *Language, Thought and Consciousness*. Cambridge University Press
(1996b). Autism as mind-blindness: an elaboration and partial defence. In P. Carruthers and P. Smith (eds.), *Theories of Theories of Mind*. Cambridge University Press
(1998). Thinking in language: evolution and a modularist possibility. In P. Carruthers and J. Boucher (eds.), *Language and Thought*. Cambridge University Press
(2002). Human creativity: its cognitive basis, its evolution, and its connections with childhood pretence. *The British Journal for the Philosophy of Science*, 53
Carruthers, P. and Chamberlain, A. (eds.) (2000). *Evolution and the Human Mind*. Cambridge University Press
Cartwright, N. (1983). *How the Laws of Physics Lie*. Oxford University Press
(1989). *Nature's Capacities and Their Measurement*. Oxford University Press
Cerella, J. (1979). Visual classes and natural categories in the pigeon. *Journal of Experimental Psychology, Human Perception and Performance*, 5
Chandrasekaran, B., Glasgow, J. and Narayanan, N. (1995). *Diagrammatic Reasoning: Cognitive and Computational Perspectives*. MIT Press
Chater, N. (1999). The search for simplicity: a fundamental cognitive principle? *Quarterly Journal of Experimental Psychology*, 52A
Cheney, D. and Seyfarth, R. (1990). *How Monkeys See the World*. Chicago University Press
Cheng, P. (1993). Separating causal laws from casual facts: pressing the limits of statistical relevance. In D. Medin (ed.), *The Psychology of Learning and Motivation*. Academic Press
(1997). From covariation to causation: a causal power theory. *Psychological Review*, 104(2)

(1999). Causal reasoning. In R. Wilson and F. Keil (eds.), *The MIT Encyclopedia of the Cognitive Sciences*. MIT Press

Cheng, P. and Holyoak, K. (1985). Pragmatic reasoning schemas. *Cognitive Psychology*, 17

Cheng, P. and Novick, L. (1992). Convantion in natural causal induction. *Psychological Review*, 99

Chi, M., Feltovich, P. and Glaser, R. (1981). Categorization and representation of physics problems by experts and novices. *Cognitive Science*, 5

Chinn, C. and Brewer, W. (1993). Factors that influence how people respond to anomalous data. *Proceedings of the Fifteenth Annual Conference of the Cognitive Science Society*

(1998). An empirical test of a taxonomy of responses to anomalous data in science. *Journal of Research in Science Teaching*, 35(6)

(2000). Knowledge change in response to science, religion and magic. In K. Rosengren, C. Johnson and P. Harris (eds.), *Imagining the Impossible: Magical, Scientific and Religious Thinking in Children*. Cambridge University Press

Chomsky, N. (1988). *Language and Problems of Knowledge: The Managua Lectures*. MIT Press

Churchland, P. (1981). Eliminative materialism and the propositional attitudes. *Journal of Philosophy*, 78

(1989). *A Neurocomputational Perspective: The Nature of Mind and the Structure of Science*. MIT Press

(1990). On the nature of explanation: A PDP approach. *Physica*, 42

Clark, A. (1997). *Being There: Putting Brain, Body, and World Together Again*. MIT Press

(1998). Magic words: how language augments human computation. In P. Carruthers and J. Boucher (eds.), *Language and Thought: Interdisciplinary Themes*. Cambridge University Press

Clement, J. (1989). Learning via model construction and criticism. In G. Glover, R. Ronning and C. Reynolds (eds.), *Handbook of Creativity: Assessment, Theory, and Research*. Plenum

Coady, C. (1992). *Testimony*. Oxford University Press

Cohen, L. (1981). Can human irrationality be experimentally demonstrated? *Behavioral and Brain Sciences*, 4

Coley, J., Medin, D. and Atran, S. (1997). Does rank have its privilege? Inductive inferences in folkbiological taxonomies. *Cognition*, 63

Coley, J., Medin, D., Lynch, E., Proffitt, J. and Atran, S. (1999). Inductive reasoning in folk-biological thought. In D. Medin and S. Atran (eds.), *Folk Biology*. MIT Press

Cosmides, L. (1989). The logic of social exchange: has natural selection shaped how humans reason? Studies with Wason Selection Task. *Cognition*, 31

Cosmides, L. and Tooby, J. (1992). Cognitive adaptations for social exchange. In J. Barkow, L. Cosmides and J. Tooby (eds.), *The Adapted Mind*. Oxford University Press

(1994). Origins of domain specificity: the evolution of functional organization. In L. Hirschfeld and S. Gelman (eds.), *Mapping the Mind: Domain-Specificity in Culture and Cognition*. Cambridge University Press

Craik, K. (1943). *The Nature of Explanation*. Cambridge University Press

Crick, F. (1988). *What Mad Pursuit: A Personal View of Scientific Discovery*. Basic Books

Crombie, A. (1994). *Styles of Scientific Thinking in the European Tradition: The History of Argument and Explanation Especially in the Mathematical and Biomedical Sciences*. London: Duckworth

Cunningham, A. and Williams, P. (1993). De-centering the 'big' picture: the origins of modern science and the modern origins of science. *British Journal of the History of Science*, 26

Curtiss, S. (1977). *Genie: A Psycholinguistic Study of a Modern-day 'Wild Child'*. Academic Press

D'Errico, F. (1995). A new model and its implications for the origin of writing: the La Marche antler revisited. *Cambridge Archaeological Journal*, 5

D'Errico, F. and Nowell, A. (2000). A new look at the Berekhat Ram figurine: implications for the origins of symbolism. *Cambridge Archaeological Journal*, 10

D'Errico, F., Zilhao, J., Julien, M., Baffier, D. and Pelegrin, J. (1998). Neanderthal acculturation in Western Europe? A critical review of the evidence and its interpretation. *Current Anthropology*, 39

Daly, M. and Wilson, M. (1988). *Homicide*. Hawthorne, NY: Aldine de Gruyter

Dama, M. and Dunbar, K. (1996). Distributed reasoning. When social and cognitive worlds fuse. *Proceedings of the Eighteenth Annual Meeting of the Cognitive Science Society*

Damasio, A. (1994). *Descartes' Error: Emotion, Reason and the Human Brain*. Putnam Books (Picador, 1995)

Darden, L. (1980). Theory construction in genetics. In T. Nickles (ed.), *Scientific Discovery: Case Studies*. Dordrecht: Reidel

(1991). *Theory Change in Science: Strategies from Mendelian Genetics*. Oxford University Press

Darley, J. and Batson, C. (1973). From Jerusalem to Jericho: a study of situational and dispositional variables in helping behavior. *Journal of Personality and Social Psychology*, 27

Darwin, C. (1859). *On the Origins of Species by Means of Natural Selection*. London: Murray

(1883). *On the Origins of Species by Means of Natural Selection* (6th edn.). New York: Appleton (originally published 1872)

Davidson, D. (1970). Mental events. In L. Foster and J. Swanson (eds.), *Experience and Theory*. Duckworth

Davis, M. (1999). *Response to Weak Argument on the Part of Asians and Americans*. University of Michigan

Davis, M., Nisbett, R. and Schwarz, N. (1999). *Responses to Weak Arguments by Asians and Americans*. University of Michigan

Dawes, R. (1994). *House of Cards: Psychology and Psychotherapy Built on Myth*. Free Press

de Sousa, R. (1987). *The Rationality of Emotion*. MIT Press

de Villiers, J. and de Villiers, P. (1999). Linguistic determinism and the understanding of false beliefs. In P. Mitchell and K. Riggs (eds.), *Children's Reasoning and the Mind*. Psychology Press

Dehaerne, S. (1997). *The Number Sense*. Penguin Press

DeKleer, J. and Brown, J. (1983). Assumptions and ambiguities in mechanistic mental models. In D. Stevens (ed.), *Mental Models*. Lawrence Erlbaum

Dennett, D. (1991). *Consciousness Explained*. London: Allen Lane

(1993). *Consciousness Explained* (2nd edn.). Penguin

(1995). *Darwin's Dangerous Idea*. Penguin Press

Diamond, J. (1966). Zoological classification system of a primitive people. *Science*, 15

Diamond, J. and Bishop, D. (1999). Ethno-ornithology of the Ketengban people, Indonesian New Guinea. In D. Medin and S. Atran (eds.), *Folk Biology*. MIT Press

Dias, M. and Harris, P. (1988). The effect of make-believe play on deductive reasoning. *British Journal of Developmental Psychology*, 6

Dickinson, A. (1980). *Contemporary Animal Learning Theory*. Cambridge University Press

Dickinson, A., Shanks, D. and Evenden, J. (1984). Judgement of act–outcome contingency: the role of selective attribution. *Quarterly Journal of Experimental Psychology*, 37B

Diver, C. (1940). The problem of closely related species living in the same area. In J. Huxley (ed.), *The New Systematics*. Oxford University Press

Dixon, E. (2001). Human colonization of the Americas: timing, technology and process. *Quaternary Science Reviews*, 20

Doherty, M., Mynatt, C., Tweney, R. and Schiavo, M. (1979). Pseudodiagnosticity. *Acta Psychologica*, 43

Donaldson, M. (1978). *Children's Minds*. Fontana

Donnellan, K. (1971). Necessity and criteria. In J. Rosenberg and C. Travis (eds.), *Readings in the Philosophy of Language*. Prentice-Hall

Dougherty, J. (1979). Learning names for plants and plants for names. *Anthropological Linguistics*, 21

Dretske, F. (1969). *Seeing and Knowing*. Chicago University Press

(1981). *Knowledge and the Flow of Information*. MIT Press

Dunbar, K. (1993). 'In vivo cognition: knowledge representation and change in real-world scientific laboratories.' Paper presented at the *Society for Research in Child Development*. New Orleans

(1995). How scientists really reason: scientific reasoning in real-world laboratories. In R. Sternberg and J. Davidson (eds.), *The Nature of Insight*. MIT Press

(1997). How scientists think: online creativity and conceptual change in science. In T. Ward, S. Smith and S. Vaid (eds.), *Conceptual Structures and Processes: Emergence, Discovery and Change*. APA Press

(1999a). The scientist in vivo: how scientists think and reason in the laboratory. In L. Magnani, N. Nersessian and P. Thagard (eds.), *Model-Based Reasoning in Scientific Discovery*. Plenum Press

(1999b). Science. In M. Runco and S. Pritzker (eds.), *The Encyclopedia of Creativity*. Academic Press

(1999c). Cognitive development and scientific thinking. In R. Wilson and F. Keil (eds.), *The MIT Encyclopedia of Cognitive Science*. MIT Press

(2000). How scientists think and reason: implications for education. *Journal of Applied Developmental Psychology*, 21

(2001a). What scientific thinking reveals about the nature of cognition. In K. Crowley, C. D. Scheenn and T. Okada (eds.), *Designing for Science: Implications from Everyday, Classroom, and Professional Settings*. Hillsdale, NJ: LEA

(2001b). The analogical paradox: why analogy is so easy in naturalistic settings, yet so difficult in the psychology laboratory. In D. Gentner, K. Holyoak and B. Kokinov (eds.), *The Analogical Mind: Perspective from Cognitive Science*. MIT Press

Dunbar, K. and Dama, M. (in preparation). Why groups sometimes work and sometimes fail: the representational change cycle

Dunbar, K. and Klahr, D. (1989). Developmental differences in scientific discovery processes. In D. Klahr and K. Kotovsky (eds.), *Complex Information Processing: The Impact of Herbert A. Simon*. Brighton: Lawrence Erlbaum

Dunbar, R. (1995) *The Trouble with Science*. Faber & Faber

(1996). *Gossip, Grooming and the Evolution of Language*. Faber & Faber

Dupré, J. (1993). *The Disorder of Things*. Harvard University Press

Ehreshefsky, M. (1992). Eliminative pluralism. *Philosophy of Science*, 59

Einhorn, H. and Hogarth, R. (1978). Confidence in judgement: persistence of the illusion of validity. *Psychological Review*, 85

(1986). Judging probable cause. *Psychological Bulletin*, 99

Ekman, P. (1972). Universals and cultural differences in facial expressions of emotion. In J. Cole (ed.), *Nebraska Symposium on Motivation 1971, Vol. 4*. University of Nebraska Press

(1992). An argument for basic emotions. *Cognition and Emotion*, 6

Eldredge, N. (1986). Information, economics and evolution. *Annual Review of Ecology and Systematics*, 17

Ellen, R. (1993). *The Cultural Relations of Classification*. Cambridge University Press

Elman, J., Bates, E., Johnson, M., Karmiloff-Smith, A., Parisi, D. and Plunkett, K. (1997). *Rethinking Innateness*. MIT Press

Ericsson, K. A. and Simon, H. A. (1984). *Protocol Analysis: Verbal Reports as Data* (revised edn., 1993). MIT Press

Evans, E. (2001). Cognitive and contextual factors in the emergence of diverse belief systems: creation versus evolution. *Cognitive Psychology*, 42

Evans, J. (1989). *Bias in Human Reasoning: Causes and Consequences*. Brighton: Erlbaum

(1998). Matching bias in conditional reasoning: do we understand it after 25 years?. *Thinking and Reasoning*, 4

(1999). Hypothetical thinking in reasoning and decision making. *ESRC Workshop on Reasoning and Thinking*, London Guildhall University

(2001). Thinking and believing. In J. Garcia-Madruga, N. Carriedo and M. Gonzales-Labra (eds.), *Mental Models on Reasoning*. Madrid: UNED

Evans, J. and Handley, S. (1997). Necessary and possible inferences: a test of the mental model theory of reasoning. *Report to the Economic and Social Research Council* (Grant R000221742)

Evans, J. and Over, D. (1996). *Rationality and Reasoning*. Psychology Press

(1997). Rationality in reasoning: the case of deductive competence. *Current Psychology of Cognition*, 16

Evans, J., Allen, J. L., Newstead, S. E. and Polland, P. (1994). Debiasing by instruction: the case of belief bias. *European Journal of Cognitive Psychology*, 6

Evans, J., Handley, S., Harper, C. and Johnson-Laird, P. (1999). Reasoning about necessity and possibility: a test of the mental model theory of deduction. *Journal of Experimental Psychology: Learning, Memory and Cognition*, 25

Evans, J., Newstead, S. and Byrne, R. (1993). *Human Reasoning: The Psychology of Deduction*. Lawrence Erlbaum

Evans-Pritchard, E. (1976). *Witchcraft, Oracles and Magic Among the Azande*. Oxford University Press (original work published 1937)

Faraday, M. (1839). *Experimental Researches in Electricity*, vol. 1, plate I. London: Taylor (reprinted New York: Dover, 1965)

Feeney, A. (1996). Information selection and belief updating in hypothesis evaluation. Unpublished PhD thesis. University of Plymouth, UK

Feeney, A., Evans J. St B. T. and Clibbens, J. (2000). Background beliefs and evidence interpretation. *Thinking and Reasoning*, 6

Feyerabend, P. (1970). Consolations for the specialist. In I. Lakatos and A. Musgrave (eds.), *Criticism and the Growth of Knowledge*. Cambridge University Press

(1975). *Against Method*. London: Verso

Feynman, R. (1999). *The Pleasure of Finding Things Out*. Perseus Books

Fischhoff, B. (1975). Hindsight ≠ foresight: the effect of outcome knowledge on judgement under uncertainty. *Journal of Experimental Psychology: Human Perception and Performance*, 1

(1982). Debiasing. In D. Kahneman, P. Slovic and A. Tversky (eds.), *Judgement Under Uncertainty: Heuristics and Biases*. Cambridge University Press

Flavell, J. (1986). The development of children's knowledge about the appearance–reality distinction. *American Psychologist*, 41

(1993). The development of children's understanding of false belief and the appearance–reality distinction. *International Journal of Psychology*, 28

(1999). Cognitive development: children's knowledge about the mind. *Annual Review of Psychology*, 50

Flavell, J., Flavell, E. and Green, F. (1983). Development of the appearance–reality distinction. *Cognitive Psychology*, 15

Flavell, J., Green, F. and Flavell, E. (1986). Development of the appearance–reality distinction. *Monographs of the Society for Research in Child Development*, 51, serial no. 212

Fodor, J. (1975). *The Language of Thought*. Thomas Y. Crowell Company

(1981). *RePresentations*. Harvester Press

(1983). *The Modularity of Mind*. MIT Press

(1987). *Psychosemantics*. MIT Press

(2000). *The Mind Doesn't Work that Way*. MIT Press

Gabunia, L., Vekua, A., Lordkipanidze, D., Swisher III, C., Ferring, R., Justus, A., Nioradze, M., Tvalchrelidze, M., Antón, S., Bosinski, G., Jöris, O., de Lumley, M., Majsuradze, G. and Mouskhelishvili, A. (2000). Earliest Pleistocene hominid cranial remains from Dmanisi, Republic of Georgia. *Science*, 288

Gallistel, C. (1990). *The Organization of Learning*. MIT Press

Gallistel, R. and Gelman, R. (1992). Preverbal and verbal counting and computation. *Cognition*, 44

Gamble, C. (1993). *Timewalkers: The Prehistory of Global Colonization*. Stroud, UK: Alan Sutton

Gargett, R. (1989). Grave shortcomings: the evidence for Neanderthal burial. *Current Anthropology*, 30

Garrett, M. (1982). Production of speech: observations from normal and pathological language use. In A. Ellis (ed.), *Normality and Pathology in Cognitive Functions*. Academic Press

Gauvain, M. and Greene, J. (1994). What do children know about objects? *Cognitive Development*, 9

Gazzaniga, M., Ivry, R. and Mangun, G. (1998). *Cognitive Neuroscience: The Biology of the Mind*. W.W. Norton

Geary, D. (1995). Reflections on evolution and culture in children's cognition: Implications for mathematical development and instruction. *American Psychologist*, 50

Gelman, R. (1991). Epigenetic foundations of knowledge structures: initial and transcendent constructions. In S. Carey and R. Gelman (eds.), *The Epigenesis of Mind: Essays on Biology and Cognition*. Lawrence Erlbaum

(2000). The epigenesis of mathematical thinking. *Journal of Applied Developmental Psychology*, 21

Gelman, S. and Bloom, P. (2000). Young children are sensitive to how an object was created when deciding how to name it. *Cognition*, 76

Gelman, S. and Wellman, H. (1991). Insides and essence: early understandings of the non-obvious. *Cognition*, 38(3)

Gelman, S., Collman, P. and Maccoby, E. (1986). Inferring properties from categories versus inferring categories from properties: the case of gender. *Child Development*, 57

Gentner, D. (1983). Structure-mapping: a theoretical framework for analogy. *Cognitive Science*, 7

(1989). The mechanisms of analogical learning. In S. Vosniadou and A. Ortony (eds.), *Similarity and Analogical Reasoning*. Cambridge University Press

Gentner, D. and Gentner, D. R. (1983). Flowing waters and teeming crowds: mental models of electricity. In D. Gentner and A. Stevens (eds.), *Mental Models*. Lawrence Erlbaum

Gentner, D., Brem, S., Ferguson, R., Markman, A., Levidow, B., Wolff, P. and Forbus, K. (1997). Analogical reasoning and conceptual change: a case study of Johannes Kepler. *The Journal of the Learning Sciences*, 6(1)

Gentner, D., Holyoak, K. and Kokinov. B. (eds.) (2001). *The Analogical Mind: Perspective from Cognitive Science*. MIT Press

George, C. (1995). The endorsement of the premises: assumption-based or belief-based reasoning. *British Journal of Psychology*, 86

German, T. and Leslie, A. (2001). Children's inferences from *knowing* to *pretending* and *believing* . *British Journal of Developmental Psychology*, 19

Gick, M. and Holyoak, K. (1980). Analogical problem solving. *Cognitive Psychology*, 12

(1983). Schema induction and analogical transfer. *Cognitive Psychology*, 15

Giere, R. (1988). *Explaining Science: A Cognitive Approach*. University of Chicago Press

(1989). Computer discovery and human interests. *Social Studies of Science*, 19

(1992). Cognitive models of science. *Minnesota Studies in the Philosophy of Science, Vol. 15*. University of Minnesota Press

(1994). The cognitive structure of scientific theories. *Philosophy of Science*, 61

(1996a). Visual models and scientific judgement. In B. Baigrie (ed.), *Picturing Knowledge: Historical and Philosophical Problems Concerning the Use of Art in Science*. University of Toronto Press

(1996b). The scientist as adult. *Philosophy of Science*, 63

(1999a). *Science without Laws*. University of Chicago Press

(1999b). Using models to represent reality. In L. Magnani, N. Nersessian and P. Thagard (eds.), *Model-Based Reasoning in Scientific Discovery*. Kluwer

Gigerenzer, G. (2000). *Adaptive Thinking: Rationality in the Real World*. Oxford University Press

Gigerenzer, G., Todd, P. and the ABC Research Group (1999). *Simple Heuristics That Make Us Smart*. Oxford University Press

Giménez, M. and Harris, P. (2001). Understanding the impossible: intimations of immortality and omniscience in early childhood. Paper presented at *Biennial Meeting of the Society for Research in Child Development*, Minneapolis, MI, 19–22 April

Girotto, V., Evans, J. and Legrenzi, P. (2000). Pseudodiagnosticity in hypothesis testing: a focussing phenomenon. Unpublished manuscript. University of Provence

Glymour, C. (2000). Bayes-Nets as psychological models. In F. Keil and R. Wilson (eds.), *Cognition and Explanation*. MIT Press

(2001). *The Mind's Arrows. Bayes Nets and Graphical Causal Models Psychology*. MIT Press

Glymour, C. and Cooper, G. (eds.) (1999). *Computation, Causation, and Discovery*. AAAI/MIT PRESS

Goldin-Meadow, S. and Zheng, M.-Y. (1998).Thought before language: the expression of motion events prior to the impact of a conventional language system. In P. Carruthers and J. Boucher (eds.), *Language and Thought: Interdisciplinary Themes*. Cambridge University Press

Goldman, A. (1976). Discrimination and perceptual knowledge. *Journal of Philosophy*, 73

(1979). What is justified belief? In G. Pappas (ed.), *Justification and Knowledge*. Reidel

(1986). *Epistemology and Cognition*. Harvard University Press

(1993). The psychology of folk psychology. *Behavioral and Brain Sciences*, 16

(1999). *Knowledge in a Social World*. Oxford University Press

Gooding, D. (1981). Final steps to the field theory: Faraday's study of electromagnetic phenomena, 1845–1850. *Historical Studies in the Physical Sciences*, 11

(1990). *Experiment and the Making of Meaning: Human Agency in Scientific Observation and Experiment*. Kluwer

(1992). The procedural turn: or why did Faraday's thought experiments work? In R. Giere (ed.), *Cognitive Models of Science*. University of Minnesota Press

Gopnik, A. (1988). Conceptual and semantic development as theory change. *Mind and Language*, 3

(1993). The illusion of first-person knowledge of intentionality. *Behavioral and Brain Sciences*, 16

(1996a). The scientist as child. *Philosophy of Science*, 63

(1996b). A reply to commentators. *Philosophy of Science*, 63

(1998). Explanation as orgasm. *Minds and Machines*, 8(1)

Gopnik, A. and Meltzoff, A. (1997). *Words, Thoughts and Theories*. MIT Press

Gopnik, A. and Sobel, D. (2000). Detecting blickets: how young children use information about novel causal powers in categorization and induction. *Child Development*, 71(5)

Gopnik, A., Sobel, O. M., Schulz, L. and Glymour, C. (2001). Causal learning mechanisms in very young children. Two, three, and four-year-olds inter causal relations from patterns of dependent and independent probability. *Developmental Psychology*, 37(5)

Gopnik, A. and Wellman, H. (1992). Why the child's theory of mind really is a theory. *Mind and Language*, 7

(1994). The theory-theory. In L. Hirschfeld and S. Gelman (eds.), *Mapping the Mind: Domain Specificity in Cognition and Culture*. Cambridge University Press

Gopnik, A., Meltzoff, A. and Kuhl, P. (1999a). *The Scientist in the Crib: Minds, Brains and How Children Learn*. William Morrow

(1999b). *How Babies Think*. Weidenfeld & Nicolson

Gorman, M. (1995). Confirmation, disconfirmation and invention: the case of Alexander Graham Bell and the telephone. *Thinking and Reasoning*, 1

Goswami, U. (1998). *Cognition in Children*. Psychology Press

Gould, S. and Lewontin, R. (1979). The spandrels of San Marco and the Panglossian paradigm: a critique of the adaptationist programme. In E. Sober (ed.), *Conceptual Issues in Evolutionary Biology*. MIT Press

Gowlett, J. (1984). Mental abilities of early man: a look at some hard evidence. In R. Foley (ed.), *Hominid Evolution and Community Ecology*. Academic Press

Green, D. and Over, D. (2000). Decision theoretic effects in testing a causal conditional. *Cahiers de Psychologie Cognitive/Current Psychology of Cognition*, 19

Greenberg, J. and Baron, R. (1997). *Behavior in Organizations*. (6th edn.). Prentice-Hall

Greeno, J. (1998). The situativity of knowing, learning, and research. *American Psychologist*, 53

Grice, H. (1961). The causal theory of perception. *Proceedings of the Aristotelian Society*, supplementary volume 35

(1975). Logic and conversation. In P. Cole and J. Morgan (eds.), *Syntax and Semantics 3: Speech Acts*. New York: Wiley

Griesemer, J. (1991a). Material models in biology. *PSA 1990*. East Lansing, MI: PSA

(1991b). Must scientific diagrams be eliminable? The case of path analysis. *Biology and Philosophy*, 6

Griesemer, J. and Wimsatt, W. (1989). Picturing Weismannism: a case study of conceptual evolution. In M. Ruse (ed.), *What the Philosophy of Biology is: Essays for David Hull*. Kluwer

Griffith, T. (1999). A computational theory of generative modeling in scientific reasoning. PhD thesis. College of Computing, Georgia Institute of Technology, Atlanta

Griffith, T., Nersessian, N. and Goel, A. (1996). The role of generic models in conceptual change. *Proceedings of the Cognitive Science Society*, 18. Lawrence Erlbaum

(2001). Function-follows-form transformations in scientific problem solving. *Proceedings of the Cognitive Science Society, Vol. 22*. Lawrence Erlbaum

Griggs, R. and Cox, J. (1982). The elusive thematic materials effect in the Wason selection task. *British Journal of Psychology*, 73

Guantao, J., Hongye, F. and Qingfeng, L. (1996). The structure of science and technology in history: on the factors delaying the development of science and technology in China in comparison with the west since the 17th century (Parts One and Two). In F. Dainian and R. Cohen (eds.), *Boston Studies in the Philosophy of Science (Vol. 179): Chinese Studies in the History and Philosophy of Science and Technology*. Kluwer

Gutmann, A. and Thompson, D. (1996). *Deliberative Democracy*. Harvard University Press

Hacking, I. (1975). *The Emergence of Probability*. Cambridge University Press

Hamilton, D. and Gifford, R. (1976). Illusory correlation in interpersonal perception: a cognitive basis of stereotypic judgements. *Journal of Experimental Social Psychology*, 12

Harman, G. (1999). Moral philosophy and linguistics. In K. Brinkmann (ed.), *Proceedings of the 20th World Congress of Philosophy, Vol. I: Ethics*. Bowling Green, OH: Philosophy Documentation Center

Harris, P. (2000a). On not falling down to earth: children's metaphysical questions. In K. Rosengren, C. Johnson and P. Harris (eds.), *Imagining the Impossible: Magical, Scientific, and Religious Thinking in Children*. Cambridge University Press

(2000b). *The Work of the Imagination*. Blackwell

Harris, P. L. and Giménez, M. (2001). Intimations of mortality and omniscience in early childhood. Paper presented at Biennial Meeting of the Society for Research in Child Development, Minneapolis, MI, 19–22 April

Harris, P., German, T. and Mills, P. (1996). Children's use of counterfactual thinking in causal reasoning. *Cognition*, 61(3)

Hart, H. and Honoré, A. (1985). *Causation in the Law* (2nd edn.). Oxford University Press

Hastie, R. (1984). Causes and effects of causal attributions. *Journal of Personality and Social Psychology*, 46

Hatano, G. and Inagaki, K. (1994). Young children's naive theory of biology. *Cognition*, 50

(1999). A developmental perspective on informal biology. In D. Medin and S. Atran (eds.), *Folk Biology*. MIT Press

Hauser, M. (2000). *Wild Minds: What Animals Really Think*. New York: Henry Holt

Hays, T. (1983). Ndumba folkbiology and general principles of ethnobotanical classification and nomenclature. *American Anthropologist*, 85

Heaton, R., Chelune, G., Talley, J., Kay, G. and Curtiss, G. (1993). *Wisconsin Card Sorting Test*. Bowling Green, FL: Psychological Assessment Resources

Hegarty, M. (1992). Mental animation: inferring motion from static diagrams of mechanical systems. *Journal of Experimental Psychology: Learning, Memory, and Cognition*, 18(5)

Hegarty, M. and Just, M. (1994). Constructing mental models of machines from text and diagrams. *Journal of Memory and Language*, 32

Hegarty, M. and Sims, V. (1994). Individual differences in mental animation from text and diagrams. *Journal of Memory and Language*, 32

Heider, F. (1958). *The Psychology of Interpersonal Relations*. New York: Wiley

Hejmadi, A., Rozin, P. and Siegal, M. (2002). Children's understanding of contamination and purification in India and the United States. Unpublished manuscript. University of Pennsylvania

Hempel, C. (1965). *Aspects of Scientific Explanation and other Essays in the Philosophy of Science*. London: Collier Macmillan/New York: Free Press

(1966). *The Philosophy of Natural Science*. Prentice-Hall

Hermer-Vazquez, L., Spelke, E. and Katsnelson, A. (1999). Sources of flexibility in human cognition: dual-task studies of space and language. *Cognitive Psychology*, 39

Herrnstein, R. (1984). Objects, categories, and discriminative stimuli. In H. Roitblat (ed.), *Animal Cognition*. Lawrence Erlbaum

Hertwig, R. and Ortmann, A. (forthcoming a). Experimental practices in economics: a challenge for psychologists? *Behavioral and Brain Sciences*

(forthcoming b). Does deception impair experimental control? A review of the evidence

Hesslow, G. (1988). The problem of causal selection. In D. Hilton (ed.), *Contemporary Science and Natural Explanation: Common Sense Conceptions of Causality*. Harvester Press

Hickling, A. and Gelman, S. (1995). How does your garden grow? Evidence of an early conception of plants as biological kinds. *Child Development*, 66(3)

Hilton, D. (1990). Conversational processes and causal explanation. *Psychological Bulletin*, 107

(1991). A conversational model of causal explanation. In W. Stroebe and M. Hewstone (eds.), *European Review of Social Psychology*, 2

(1995a). Logic and language in causal explanation. In D. Sperber, D. Premack and A. Premack (eds.), *Causal Cognition: A Multidisciplinary Debate*. Oxford University Press

(1995b). The social context of reasoning: conversational inference and rational judgement. *Psychological Bulletin*, 118

Hilton, D. and Erb, H.-P. (1996). Mental models and causal explanation: judgements of probable cause and explanatory relevance. *Thinking and Reasoning*, 2

Hilton, D. and Jaspars, J. (1987). The explanation of occurrences and non-occurrences: a test of the inductive logic model of causal attribution. *British Journal of Social Psychology*, 26

Hilton, D. and Neveu, J.-P. (1996). Induction and utility in managerial decision making: the enduring value of John Stuart Mill's perspective. Paper presented at the Colloque *L'Utilitarisme: analyse et histoire*. Faculté des Sciences Economiques et Sociales de l'Université Lille I, 25–26 January 1996

Hilton, D. and Slugoski, B. (1986). Knowledge-based causal attribution: the abnormal conditions focus model. *Psychological Review*, 93

(2000). Discourse processes and rational inference: judgement and decision-making in a social context. In T. Connolly, H. Arkes and K. Hammond (eds.), *Judgement and Decision-making: A Reader* (2nd edn.). Cambridge University Press

Hilton, D., Mathes, R. and Trabasso, T. (1992). The study of causal explanation in natural language: Analysing reports of the Challenger disaster in the 'New York Times'. In M. McLaughlin, S. Cody and S. Read (eds.), *Explaining One's Self to Others: Reason-Giving in a Social Context*. Lawrence Erlbaum

Hirschfeld, L. (1995). Do children have a theory of race? *Cognition*, 54

(1996). *Race in the Making*. MIT Press

Holland, J., Holyoak, K., Nisbett, R. and Thagard, P. (1986). *Induction: Processes of Inference, Learning, and Discovery*. MIT Press

Holmes, F. (1981). The fine structure of scientific creativity. *History of Science*, 19

(1985). *Lavoisier and the Chemistry of Life: An Exploration of Scientific Creativity*. University of Wisconsin Press

Holyoak, K. and Thagard, P. (1989). Analogical mapping by constraint satisfaction: a computational theory. *Cognitive Science*, 13

(1996). *Mental Leaps: Analogy in Creative Thought*. MIT Press

Hookway, C. (1993). Mimicking foundationalism: on sentiment and self-control. *European Journal of Philosophy*, 1

(1999). Doubt: affective states and the regulation of inquiry. *Canadian Journal of Philosophy*, Supplementary Volume 24

(2001). Epistemic akrasia and epistemic virtue. In A. Fairweather and L. Zagzebski (eds.), *Virtue Epistemology*. Oxford University Press

Howson, C. and Urbach, P. (1993). *Scientific Reasoning* (2nd edn.). Chicago: Open Court

Hrdy, S. (1999). *Mother Nature: A History of Mothers, Infants and Natural Selection*. Pantheon

Hughes, C. and Plomin, R. (2000). Individual differences in early understanding of mind: genes, non-shared environment and modularity. In P. Carruthers and A. Chamberlain (eds.), *Evolution and the Human Mind*. Cambridge University Press

Hull, D. (1997). The ideal species definition and why we can't get it. In M. Claridge, H. Dawah and M. Wilson (eds.), *Species: The Units of Biodiversity*. Chapman & Hall

Hume, D. (1739). *A Treatise of Human Nature* (Oxford, 1888)

Hunn, E. (1976). Toward a perceptual model of folk biological classification. *American Ethnologist*, 3

(1977). *Tzeltal Folk Zoology*. Academic Press

(1982). The utilitarian factor in folk biological classification. *American Anthropologist*, 84

Hutchins, E. (1995). *Cognition in the Wild*. MIT Press

Inagaki, K. (1990). The effects of raising animals on children's biological knowledge. *British Journal of Developmental Psychology*, 8

Inagaki, K. and Hatano, G. (1991). Constrained person analogy in young children's biological inference. *Cognitive Development*, 6

(1993). Young children's understanding of the mind–body distinction. *Child Development*, 64

Inhelder, B. and Piaget, J. (1958). *The Growth of Logical Thinking from Childhood to Adolescence*. Routledge & Kegan Paul

Ippolito, M. and Tweney, R. (1995). The inception of insight. In R. Sternberg and J. Davidson (eds.), *The Nature of Insight*. MIT Press

Isaac, G. (1978). The food-sharing behaviour of proto-human hominids. *Scientific American*, 238

(1984). The archaeology of human origins: studies of the Lower Pleistocene in East Africa. 1971–1981. *Advances in World Archaeology*, 3

Ishihara, S. (1983). *Ishihara's Tests for Colour-Blindness*. Kanehara

Jacob, F. (1988). *The Statue Within* (trans. F. Philip). Basic Books

Janis, I. (1971). Groupthink. *Psychology Today*, 5

Ji, L., Peng, K. and Nisbett, R. (2000). Culture, control and the perception of relationships in the environment. *Journal of Personality and Social Psychology*, 78

Job, R. and Surian, L. (1998). A neurocognitive mechanism for folk biology? *Behavioral and Brain Sciences*, 21

Johansen, D. and Edgar, B. (1996). *From Lucy to Language*. Weidenfeld & Nicolson

Johnson, C. (1990). If you had my brain, where would I be? Children's developing conceptions of the mind and brain. *Child Development*, 61

Johnson, C. and Wellman, H. (1982). Children's developing conception of the mind and brain. *Child Development*, 53

Johnson, S. and Solomon, G. (1997). Why dogs have puppies and cats have kittens: the role of birth in young children's understanding of biological origins. *Child Development*, 68

Johnson-Laird, P. (1982). The mental representation of the meaning of words. *Cognition*, 25

(1983). *Mental Models*. MIT Press

(1989). Mental models. In M. Posner (ed.), *Foundations of Cognitive Science*. MIT Press

Johnson-Laird, P. and Byrne, R. (1991). *Deduction*. Lawrence Erlbaum

Jordan, M. (ed.) (1998). *Learning in Graphical Models*. MIT Press

Judson, H. (1979). *The Eighth Day of Creation: Makers of the Revolution in Biology*. Simon & Schuster

Kahneman, D. (1999). Objective happiness. In D. Kahneman, E. Diener and N. Schwarz (eds.), *Well-Being: Foundations of Hedonic Psychology*. Russell Sage Foundation

Kahneman, D. and Miller, D. (1986). Norm theory: comparing reality to its alternatives. *Psychological Review*, 93

Kahneman, D., Slovic, P. and Tversky, A. (eds.) (1982). *Judgement Under Uncertainty: Heuristics and Biases*. Cambridge University Press

Kalish, C. (1996). Preschoolers' understanding of germs as invisible mechanisms. *Cognitive Development*, 11

(1997). Preschoolers' understanding of mental and bodily reactions to contamination: what you don't know can hurt you, but cannot sadden you. *Developmental Psychology*, 33

(1999). What young children's understanding of contamination and contagion tells us about their concepts of illness. In M. Siegal and C. Peterson (eds.), *Children's Understanding of Biology and Health*. Cambridge University Press

Kant, I. (1951). *Critique of Judgement* (trans. J. Bernard). New York: Hafner Press (originally published in German in 1790)

Karmiloff-Smith, A. (1992). *Beyond Modularity*. MIT Press

Kay, J., Lesser, R. and Coltheart, M. (1992). *Psycholinguistic Assessment of Language Processing in Aphasia*. Psychology Press

Keil, F. (1979). *Semantic and Conceptual Development: An Ontological Perspective*. Harvard University Press

(1989). *Concepts, Kinds, and Cognitive Development*. MIT Press

(1994). The birth and nurturance of concepts by domains: the origins of concepts of living things. In L. Hirschfeld and S. Gelman (eds.), *Mapping the Mind: Domain-Specificity in Culture and Cognition*. Cambridge University Press

(1995). The growth of causal understandings of natural kinds. In D. Sperber, D. Premack and A. Premack (eds.), *Causal Cognition: A Multidisciplinary Debate*. Oxford University Press

Kelley, H. (1967). Attribution in social psychology. *Nebraska Symposium on Motivation*, 15

(1973). The processes of causal attribution. *American Psychologist*, 28(2)
Kempton, W. (1986). Two theories of home heat control. *Cognitive Science*, 10
Key, C. (2000). The evolution of human life-history. *World Archaeology*, 31
Kingery, D. W., Vandiver, P. B. and Pickett, M. (1988). The beginnings of pyrotechnology. Part II: production and use of lime and gypsum plaster in the pre-pottery Neolithic Near East. *Journal of Field Archaeology*, 15
Kitcher, P. (1985). *Vaulting Ambition*. MIT Press
(1990). The division of cognitive labor. *Journal of Philosophy*, 87
(1993). *The Advancement of Science*. Oxford University Press
Klahr, D. with K. Dunbar, A. Fay, D. Pennes and C. Schunn (2000). *Exploring Science: The Cognition and Development of Discovery Processes*. MIT Press
Klahr, D. and Simon, H. (1999). Studies of scientific discovery: complementary approaches and convergent findings. *Psychological Bulletin*, 125(5)
Klayman, J. (1995). Varieties of confirmation bias. *The Psychology of Learning and Motivation*, 32
Klayman, J. and Burt, R. (1999). Individual differences in confidence and experiences in social networks. *Working Paper*, Centre for Decision Research, University of Chicago
Klayman, J. and Ha, Y.-W. (1987). Confirmation, disconfirmation and information in hypothesis testing. *Psychological Review*, 94
Klayman, J., Soll, J. B., González-Vallejo and Barlas, S. (1999). Overconfidence: it depends on how, what and whom you ask. *Organizational Behavior and Human Decision Processes*, 79
Knorr-Cetina, K. (1999). *Epistemic Cultures: How the Sciences Make Knowledge*. Harvard University Press
Kohlberg, L. (1966). A cognitive–developmental analysis of children's sex-role concepts and attitudes. In E. Maccoby (ed.), *The Development of Sex Differences*. Stanford University Press
Koslowski, B. (1996). *Theory and Evidence: The Development of Scientific Reasoning*. MIT Press
Koslowski, B. and Masnick, A. (2001) The development of causal reasoning. In U. Goswami, (ed.), *Handbook of Cognitive Development*. Blackwell
Koslowski, B. and Winsor, A. (1981). Preschool children's spontaneous explanations and requests for explanations: a non-human application of the child-as-scientist metaphor. Unpublished manuscript. Department of Human Development, Ithaca, NY: Cornell University
Kosslyn, S. (1980). *Image and Mind*. Harvard University Press
(1994). *Image and Brain*. MIT Press
Kripke, S. (1972). Naming and necessity. In G. Harman and D. Davidson (eds.), *Semantics of Natural Language*. Reidel
Kubovy, M. (1999). On the pleasures of the mind. In D. Kahneman, E. Diener and N. Schwarz (eds.), *Well-Being: Foundations of Hedonic Psychology*. Russell Sage Foundation
Kuhn, D. and Lao, J. (1996). Effects of evidence on attitudes: is polarization the norm? *Psychological Science*, 7
Kuhn, D., Amsel, E. and O'Loughlin, M. (1988). *The Development of Scientific Thinking Skills*. Academic Press
Kuhn, T. (1962). *The Structure of Scientific Revolutions*. University of Chicago Press

(1979). Metaphor in science. In A. Ortony (ed.), *Metaphor and Thought*. Cambridge University Press
Kulkarni, D. and Simon, H. (1988). The processes of scientific discovery: the strategy of experimentation. *Cognitive Science*, 12
Kurz, E. and Tweney, R. (1998). The practice of mathematics and science: from the calculus to the clothesline problem. In M. Oaksford and N. Chater (eds.), *Rational Models of Cognition*. Oxford University Press
Labandeira, C. and Seposki, J. (1993). Insect diversity in the fossil record. *Science*, 261
Lahr, M. and Foley, R. (1998). Towards a theory of modern human origins: geography, demography and diversity in recent human evolution. *Yearbook of Physical Anthropology*, 41
Lakatos, I. (1970). The methodology of scientific research programmes. In I. Lakatos and A. Musgrave (eds.), *Criticism and the Growth of Knowledge*. Cambridge University Press
Langley, P., Simon, H., Bradshaw, G. and Zytkow, J. (1987). *Scientific Discovery: Computational Explorations of the Creative Processes*. MIT Press
Larkin, J. (1983). The role of problem representation in physics. In D. Gentner and A. Stevens (eds.), *Mental Models*. Lawrence Erlbaum
Latour, B. (1986). Visualization and cognition: Thinking with eyes and hands. *Knowledge and Society: Studies in the Sociology of Culture, Past and Present*, 6
(1987). *Science in Action*. Harvard University Press
Latour, B. and Woolgar, S. (1986). *Laboratory Life: The Construction of Scientific Facts*, 2nd edn. Princeton University Press
Lazarus, R. (1994). Universal antecedents of the emotions. In P. Ekman and R. Davidson (eds.), *The Nature of Emotion: Fundamental Questions*. Oxford University Press
Legrenzi, P., Butera, F., Mugny, G. and Perez, J. (1991). Majority and minority influence in inductive reasoning: a preliminary study. *European Journal of Social Psychology*, 21
Leslie, A. (1994a). Pretending and believing: issues in the theory of ToMM. *Cognition*, 50
(1994b). ToMM, ToBY and Agency: core architecture and domain specificity. In L. Hirschfeld and S. Gelman (eds.), *Mapping the Mind: Domain-Specificity in Culture and Cognition*. Cambridge University Press
Leslie, A. and Keeble, S. (1987). Do six-month-old infants perceive causality? *Cognition*, 25(3)
Levenson, R. (1994). Human emotion: a functional view. In P. Ekman and R. Davidson (eds.), *The Nature of Emotion: Fundamental Questions*. Oxford University Press
Lévi-Strauss, C. (1963). The bear and the barber. *The Journal of the Royal Anthropological Institute*, 93
Lewis, D. (1966). An argument for the identity theory. *Journal of Philosophy*, 63
(1970). How to define theoretical terms. *Journal of Philosophy*, 67
(1980). Mad pain and Martian pain. In N. Block (ed.), *Readings in Philosophy of Psychology, Vol. 1*. Methuen
Leyens, J.-P. and Scaillet, N. (2000). The Wason selection task as an opportunity to improve social identity. Unpublished manuscript
Leyens, J.-P., Dardenne, B., Yzerbyt, V., Scaillet, N. and Snyder, M. (1999). Confirmation and disconfirmation: their social advantages. In W. Stroebe and M. Hewstone (eds.), *European Review of Social Psychology*, 10

Liebenberg, L. (1990). *The Art of Tracking: The Origin of Science*. Cape Town: David Philip Publishers

Linnaeus, C. (1738). *Classes Plantarum*. Leiden: Wishoff

(1751). *Philosophia Botanica*. Stockholm: G. Kiesewetter

Lloyd, G. (1990). *Demystifying Mentalities*. Cambridge University Press

Lober, K. and Shanks, D. (1999). Is causal induction based on causal power? Critique of Cheng (1997). *Psychological Review*, 107

Locke, D. (1971). *Memory*. Macmillan

Loewenstein, G. (1994). The psychology of curiosity: a review and reinterpretation. *Psychological Bulletin*, 116

Logan, R. (1986). *The Alphabet Effect*. Morrow

López, A., Atran, S., Coley, J., Medin, D. and Smith, E. (1997). The tree of life: universals of folk-biological taxonomies and inductions. *Cognitive Psychology*, 32

Lord, C., Ross, L. and Lepper, M. (1979). Biased assimilation and attitude polarization: the effects of prior theories on subsequently considered evidence. *Journal of Personality and Social Psychology*, 37(11)

Lorenz, K. (1966). The role of gestalt perception in animal and human behavior. In L. White (ed.), *Aspects of Form*. Indiana University Press

Lutz, C. (1988). *Unnatural Emotions: Everyday Sentiments on a Micronesian Atoll and Their Challenge to Western Theory*. The University of Chicago Press

Lynch, M. and Woolgar, S. (eds.) (1990). *Representation in Scientific Practice*. MIT Press

Mackie, J. (1980). *The Cement of the Universe* (2nd edn.). Oxford University Press

Macnamara, J. (1982). *Names for Things*. MIT Press

Mandel, D. and Lehman, D. (1998). Integration of contingency information in judgements of cause, covariation and probability. *Journal of Experimental Psychology: General*, 127

Mandler, J. and McDonough, L. (1996). Drinking and driving don't mix: inductive generalization in infancy. *Cognition*, 59

Mandler, J., Bauer, P. and McDonough, L. (1991). Separating the sheep from the goats: differentiating global categories. *Cognitive Psychology*, 23

Mani, K. and Johnson-Laird, P. (1982). The mental representation of spatial descriptions. *Memory and Cognition*, 10

Mania, D. and Mania, U. (1988). Deliberate engravings on bone artefacts by *Homo erectus*. *Rock Art Research*, 5

Manktelow, K. (1999). *Reasoning and Thinking*. Psychology Press

Manktelow, K. and Over, D. (1991). Social roles and utilities in reasoning with deontic conditionals. *Cognition*, 39

Manktelow, K., Fairley, N., Kilpatrick, S. and Over, D. (2000). Pragmatics and strategies for practical reasoning. In W. Schaeken, G. De Vooght, A. Vandierendonck and G. d'Ydewalle (eds.), *Deductive Reasoning and Strategies*. Lawrence Erlbaum

Marshall, B. and Warren, J. (1984). Unidentified curved bacilli in the stomach of patients with gastritis and peptic ulceration, *Lancet*, 1(8390)

Martin, A., Ungerleider, L. and Haxby, J. (2000). Category specificity and the brain: the sensory–motor model of semantic representations of objects. In M. Gazzaniga (ed.), *The New Cognitive Sciences*. MIT Press

Martin, C. and Deutscher, M. (1966). Remembering. *Philosophical Review*, 75

Martin, R. (1993). Short-term memory and sentence processing: evidence from neuropsychology. *Memory and Cognition*, 21

Martin, R. and Feher, E. (1990). The consequences of reduced memory span for the comprehension of semantic versus syntactic information. *Brain and Language*, 38

Marx, R., Stubbart, C., Traub, V. and Cavanaugh, M. (1987). The NASA space shuttle disaster: a case study. *Journal of Management Case Studies*, 3

Masnick, A., Barnett, S., Thompson, S. and Koslowski, B. (1998). Evaluating explanations in the context of a web of information. Paper presented at the *Twentieth Annual Meeting of the Cognitive Science Society*. Madison, WI

Massey, C. and Gelman, R. (1988). Preschoolers' ability to decide whether a pictured unfamiliar object can move itself. *Developmental Psychology*, 24

Maxwell, J. C. (1890). On physical lines of force. In *The Scientific Papers of James Clerk Maxwell*, D. Niren (ed.), Cambridge University Press, vol. 1

McAllister, J. (1996). *Beauty and Revolution in Science*. Ithaca, NY: Cornell University Press

McArthur, L. (1972). The how and what of why: some determinants and consequences of causal attributions. *Journal of Personality and Social Psychology*, 22

McClelland, J., Rumelhart, D. and the PDP Research Group (eds.) (1986). *Parallel Distributed Processing: Explorations in the Microstructure of Cognition*. MIT Press

McClure, J. (1998). Discounting causes of behavior: are two reasons better than one? *Journal of Personality and Social Psychology*, 47

McGill, A. (1989). Context effects in causal judgement. *Journal of Personality and Social Psychology*, 57

McGill, A. and Klein, J. (1995). Counterfactual and contrastive reasoning in explanations for performance: implications for gender bias. In N. Roese and J. Olson (eds.), *What Might Have Been: The Social Psychology of Counterfactual Thinking*. Lawrence Erlbaum

McKenzie, C. and Mikkelsen, L. (in press). The psychological side of Hempel's paradox of confirmation. *Psychonomic Bulletin and Review*,

McKenzie, C., Mikkelsen, L., McDermott, K. and Skrable, R. (2000). Are hypotheses phrased in terms of rare events? Unpublished manuscript

McNamara, T. and Sternberg, R. (1983). Mental models of word meaning. *Journal of Verbal Learning and Verbal Behavior*, 22

Medin, D., Lynch, E. and Solomon, K. (2000). Are there kinds of concepts? *Annual Review of Psychology*, 51

Medin, D., Lynch, E., Coley, J. and Atran, S. (1997). Categorization and reasoning among tree experts: do all roads lead to Rome? *Cognitive Psychology*, 32

Mellars, P. (1996). *The Neanderthal Legacy*. Princeton University Press

Michotte, A. (1962). *Causalité, Permanence et Réalité Phénoménales: Etudes de Psychologie Expérimentale*. Louvain: Publications Universitaires

Mill, J. (1872/1973). System of logic. In J. Robson (ed.), *Collected Works of John Stuart Mill* (8th edn., Vols. 7 and 8). University of Toronto Press

Miller, G. (2000). *The Mating Mind: How Sexual Choice Shaped the Evolution of Human Nature*. Pantheon

Miller, R. (1987). *Fact and Method*. Princeton University Press

Miller, R. and Matute, H. (1996). Biological significance in forward and backward blocking: resolution of a discrepancy between animal conditioning and human causal judgement. *Journal of Experimental Psychology: General*, 125(4)

Mithen, S. (1988). Looking and learning: Upper Palaeolithic art and information gathering. *World Archaeology*, 19

(1990). *Thoughtful Foragers: A Study of Prehistoric Decision Making*. Cambridge University Press

(1996). *The Prehistory of the Mind*. Thames & Hudson

(1998). The supernatural beings of prehistory and the external storage of religious ideas. In C. Renfrew and C. Scarre (eds.), *Cognition and Material Culture: The Archaeology of Symbolic Storage*. Cambridge: McDonald Institute

(2000). Palaeoanthropological perspectives on the theory of mind. In S. Baron-Cohen, H. Talgar-Flusberg and D. Cohen (eds.), *Understanding Other Minds*. Oxford University Press

Mitroff, I. (1974). *The Subjective Side of Science*. Elsevier

Morris, M. and Larrick, R. (1995).When one cause casts doubt on another: a normative analysis of discounting in causal attribution. *Psychological Review*, 102(2)

Morris, S., Taplin, J. and Gelman, S. (2000). Vitalism in naïve biological thinking. *Developmental Psychology*, 36

Morrow, D., Bower, G. and Greenspan, S. (1989). Updating situation models during narrative comprehension. *Journal of Memory and Language*, 28

Munro, D. (1969). *The Concept of Man in Early China*. Stanford University Press

Murphy, G. and Medin, D. (1985). The role of theories in conceptual coherence. *Psychological Review*, 92(3)

Mynatt, C., Doherty, M. and Dragan, W. (1993). Information relevance, working memory and the consideration of alternatives. *Quarterly Journal of Experimental Psychology*, 46A

Mynatt, C., Doherty, M. and Tweney, R. (1977). Confirmation bias in a simulated research environment: an experimental study of scientific inference. *Quarterly Journal of Experimental Psychology*, 29

Nagel, E. (1961). *The Structure of Science*. Harcourt & Brace

Nagel, T. (1974). What is it like to be a bat? *Philosophical Review*, 83

Nakamura, H. (1964/1985). *Ways of Thinking of Eastern Peoples*. University of Hawaii Press

Nazzi, T. and Gopnik, A. (2000). A shift in children's use of perceptual and causal cues to categorization. *Developmental Science*, 3

Needham, J. (1954). *Science and Civilization in China I*. Cambridge University Press

Nemeroff, C. and Rozin, P. (1994). The contagion concept in adult thinking in the United States: transmission of germs and of interpersonal influence. *Ethos: Journal of Psychological Anthropology*, 22

Nersessian, N. J. (1984a). Aether/or: The creation of scientific concepts. *Studies in the History & Philosophy of Science*, 15

(1984b). *Faraday to Einstein: Constructing Meaning in Scientific Theories*. Dordrecht: Martinus Nijhoff/Kluwer Academic Publishers

(1985). Faraday's field concept. In D. Gooding and F. James (eds.), *Faraday Rediscovered: Essays on the Life & Work of Michael Faraday*. Macmillan

(1988). Reasoning from imagery and analogy in scientific concept formation. In A. Fine and J. Leplin (eds.), *PSA 1988*. Philosophy of Science Association

(1992a). How do scientists think? Capturing the dynamics of conceptual change in science. In R. Giere (ed.), *Minnesota Studies in the Philosophy of Science, Vol. 15*. University of Minnesota Press

(1992b). In the theoretician's laboratory: thought experimenting as mental modeling. In D. Hull, M. Forbes and K. Okruhlik (eds.), *PSA 1992*. Philosophy of Science Association

(1995). Opening the black box: cognitive science and the history of science. *Osiris*, 10

(1999). Model-based reasoning in conceptual change. In L. Magnani, N. Nersessian and P. Thagard (eds.), *Model-Based Reasoning in Scientific Discovery*. Kluwer Academic/Plenum Publishers

(2001a). Abstraction via generic modeling in concept formation in science. In M. Jones and N. Cartwright (eds.), *Correcting the Model: Abstraction and Idealization in Science*. Amsterdam: Rodopi

(2001b). Maxwell and the method of physical analogy: model-based reasoning, generic abstraction, and conceptual change. In D. Malamet (ed.), *Reading Natural Philosophy: Essays in History and Philosophy of Science and Mathematics in Honor of Howard Stein on his 70th Birthday*. LaSalle, IL: Open Court

Neter, E. and Ben-Shakhar, G. (1989). Predictive validity of graphological inferences: a meta-analysis. *Personality and Individual Differences*, 10

Newcombe, P. and Siegal, M. (1996). Where to look first for suggestibility in children's memory. *Cognition*, 59

(1997). Explicitly questioning the nature of suggestibility in preschoolers' memory and retention. *Journal of Experimental Child Psychology*, 67

Newell, A. and Simon, H. (1972). *Human Problem Solving*. Prentice-Hall

Nichols, S. and Stich, S. (2000). A cognitive theory of pretence. *Cognition*, 74

Nisbett, R. and Ross, L. (1980). *Human Inference: Strategies and Shortcomings of Social Judgement*. Prentice-Hall

Nisbett, R. and Wilson, T. (1977). Telling more than we can know: verbal reports on mental processes. *Psychological Review*, 84

Nisbett, R., Peng, K., Choi, I. and Norenzayan, A. (2001). Culture and systems of thought: holistic vs. analytic cognition. *Psychological Review*, 108(2)

Norenzayan, A., Nisbett, R., Smith, E. and Kim, B. (1999). *Rules vs. Similarity as a Basis for Reasoning and Judgement in East and West*. University of Michigan

O'Connell, J., Hawkes, K. and Blurton-Jones, N. (1999). Grandmothering and the evolution of *Homo erectus*. *Journal of Human Evolution*, 36

Oakes, L. and Cohen, L. (1990). Infant perception of a causal event. *Cognitive Development*, 5

(1994). Infant causal perception. In C. Rovee-Collier and L. Lipsitt (eds.), *Advances in Infancy Research, vol. 9*. Norfolk, NJ: Ablex

Oakhill, J. and Garnham, A. (eds.) (1996). *Mental Models in Cognitive Science: Essays in Honor of Philip Johnson-Laird*. Psychology Press

Oaksford, M. and Chater, N. (1994). A rational analysis of the selection task as optimal data selection. *Psychological Review*, 101

(1995). Information gain explains relevance which explains the selection task. *Cognition*, 57

(1998). *Rationality in an Uncertain World*. Psychology Press

Oatley, K. (1992). *Best Laid Schemes: The Psychology of Emotions*. Cambridge University Press

Occhipinti, S. and Siegal, M. (1994). Reasoning about food and contamination. *Journal of Personality and Social Psychology*, 66

(1996). Cultural evolution and divergent rationalities in human reasoning. *Ethos*, 24
O'Keefe, J. and Nadel, L. (1978). *The Hippocampus as a Cognitive Map*. Oxford University Press
Olby, R. (1974). *The Path to the Double Helix*. Macmillan
Osherson, D., Smith, E., Wilkie, O., López, A. and Shafir, E. (1990). Category-based induction. *Psychological Review*, 97
Palmer, S. (1999). *Vision Science: From Photons to Phenomenology*. MIT Press
Palmerino, C., Rusiniak, K. and Garcia, J. (1980). Flavor–illness aversions: The peculiar roles of odor and taste in memory for poison. *Science*, 208(4445)
Panksepp, J. (1998). *Affective Neuroscience: The Foundations of Human and Animal Emotions*. Oxford University Press
Papineau, D. (2000). The evolution of knowledge. In P. Carruthers and A. Chamberlain (eds.), *Evolution and the Human Mind*. Cambridge University Press
Parker, G. (1978). Searching for mates. In J. Krebs and N. Davies (eds.), *Behavioural Ecology: An Evolutionary Approach*. Blackwell
Payne, J. (1997). The scarecrow's search: a cognitive psychologist's perspective on organizational decision-making. In Z. Shapira (ed.), *Organizational Decision-making*. Cambridge University Press
Payne, J., Bettman, J. and Johnson, E. (1993). *The Adaptive Decision-Maker*. Cambridge University Press
Pearce, J. and Hall, G. (1980). A model of Pavlovian learning: variations in the effectiveness of conditioned but not of unconditioned stimuli. *Psychological Review*, 87
Pearl, J. (1988). *Probabilistic Reasoning in Intelligent Systems*. Morgan Kaufman
(2000). *Causality*. Oxford University Press
Pearl, J. and Verma, T. (1991). A theory of inferred causation. *Second Annual Conference on Principles of Knowledge Representation and Reasoning*. Morgan Kaufmann
Perner, J. (1991). *Understanding the Representational Mind*. MIT Press
Perrig, W. and Kintsch, W. (1985). Propositional and situational representations of text. *Journal of Memory and Language*, 24
Piaget, J. (1928). *Judgement and Reasoning in the Child*. Routledge & Kegan Paul
(1929). *The Child's Conception of the World*. Routledge & Kegan Paul
(1930). *The Child's Conception of Physical Causality*. Routledge & Kegan Paul
(1952). *The Child's Conception of Number*. Routledge & Kegan Paul
(1954). *The Construction of Reality in the Child*. Basic Books
(1962). *Play, Dreams, and Imitation*. Routledge & Kegan Paul
Pickering, D. (1992). *Science as Practice and Culture*. University of Chicago Press
Pinker, S. (1984). *Language Learnability and Language Development*. Harvard University Press
(1994). *The Language Instinct*. William Morrow
(1997). *How the Mind Works*. Allen Lane
Place, U. (1956). Is consciousness a brain process? *British Journal of Psychology*, 47
Plato (1961). *The Collected Dialogues*. Princeton University Press
Polak, A. and Harris, P. (1999). Deception by young children following non-compliance. *Developmental Psychology*, 35
Polanyi, M. (1958). *Personal Knowledge*. University of Chicago Press
Pollard, P. and Evans, J. (1987). On the relationship between content and context effects in reasoning. *American Journal of Psychology*, 100

Popper, K. (1935). *Logik der Forschung*. Vienna. (trans. as *The Logic of Scientific Discovery*. Hutchinson, 1959)
(1945). *The Open Society and its Enemies*. Routledge
(1959). *The Logic of Scientific Discovery*. Hutchinson
(1963). *Conjectures and Refutations*. Routledge & Kegan Paul
(1972). *Objective Knowledge*. Oxford University Press
Potts, R. (1988). *Early Hominid Activities at Olduvai Gorge*. New York: Aldine de Gruyter
Povinelli, D. (2000). *Folk Physics for Apes?* Oxford University Press
Premack, D. (1995). Forward to part IV: causal understanding in naïve biology. In S. Sperber, D. Premack and A. Premack (eds.), *Causal Cognition*. Oxford University Press
Putnam, H. (1960). Minds and machines. In S. Hook (ed.), *Dimensions of Mind*. Harvard University Press
(1962). The analytic and the synthetic. In H. Feigland and G. Maxwell (eds.), *Minnesota Studies in the Philosophy of Science, Vol. III*. University of Minnesota Press
(1967). The nature of mental states. In W. Capitan and D. Merrill (eds.), *Art, Mind and Religion*. University of Pittsburgh Press
(1975). The meaning of 'meaning'. *Minnesota Studies in Philosophy of Science*, 7
Quine, W. (1951). Two dogmas of empiricism. *Philosophical Review*, 60
(1960). *Word and Object*. MIT Press
Rakison, D. and Poulin-Dubois, D. (2001). Developmental origins of the animate–inanimate distinction. *Psychological Bulletin*, 127
Ramón y Cajal, S. (1999). *Advice for a Young Investigator* (trans. N. Swanson and L. Swanson). MIT Press
Rapoport, J. (1989). *The Boy Who Couldn't Stop Washing*. Harper Collins
Read, H. and Varley, R. (in preparation). A dissociation between music and language in severe aphasia
Reber, A. (1993). *Implicit Learning and Tacit Knowledge*. Oxford University Press
Reichenbach, H. (1938). *Experience and Prediction*. University of Chicago Press
(1956). *The Direction of Time*. University of California Press
Reid, T. (1764/1997). *An Inquiry into the Human Mind on the Principles of Common Sense*. Edinburgh University Press
Rescorla, R. and Wagner, A. (1972). A theory of Pavlovian conditioning: variations in the effectiveness of reinforcement and non-reinforcement. In A. Black and W. Prokasy (eds.), *Classical Conditioning II: Current Theory and Research*. New York: Appleton-Century-Croft
Resnick, L., Levine, J. and Teasley, S. (eds.) (1991). *Perspectives on Socially Shared Cognition*. American Psychological Association
Richards, C. and Sanderson, J. (1999). The role of imagination in facilitating deductive reasoning in 2-, 3- and 4-year-olds. *Cognition*, 72
Richmond, B. G. and Strait, D. (2000). Evidence that humans evolved from a knuckle walking ancestor. *Nature*, 404
Rieff, C., Koops, W., Terwogt, M., Stegge, H. and Oomen, A. (2001). Preschoolers' appreciation of uncommon desires and subsequent emotions. *British Journal of Developmental Psychology*, 19
Rips, L. (1989). Similarity, typicality, and categorization. In S. Vosniadou and A. Ortony (eds.), *Similarity and Analogical Reasoning*. Cambridge University Press

Roberts, M. (1997). Boxgrove. *Current Archaeology*, 153

Rorty, R. (1979). *Philosophy and the Mirror of Nature*. Princeton University Press

Rosch, E. (1975). Universals and cultural specifics in categorization. In R. Brislin, S. Bochner and W. Lonner (eds.), *Cross-Cultural Perspectives on Learning*. NY: Halstead

Rosch, E., Mervis, C., Grey, W., Johnson, D. and Boyes-Braem, P. (1976). Basic objects in natural categories. *Cognitive Psychology*, 8

Rosen, A. and Rozin, P. (1993). Now you see it . . . now you don't: the preschool child's conception of invisible particles in the context of dissolving. *Developmental Psychology*, 29

Rosenberg, L. (1998). *Scientific Opportunities and Public Needs*. National Academy Press

Ross, B. and Murphy, G. (1999). Food for thought: cross-classification and category organization in a complex real-world domain. *Cognitive Psychology*, 38

Ross, L. (1977). The intuitive psychologist and his shortcomings: distortions in the attribution process. In L. Berkowitz (ed.), *Advances in Experimental Social Psychology, Vol. 14*. Academic Press

Ross, N., Medin, D., Coley, J. and Atran, S. (submitted). Cultural and experiential differences in the development of folkbiological induction

Rozin, P. (1976). The evolution of intelligence and access to the cognitive unconscious. In J. Sprague and A. Epstein (eds.), *Progress in Psychobiology and Physiological Psychology, Vol. 6*. Academic Press

Rozin, P. and Srull, J. (1988). The adaptive–evolutionary point of view in experimental psychology. In R. Atkinson, R. Herrnstein, G. Lindzey and R. Luce (eds.), *Handbook of Experimental Psychology*. New York: Wiley-Interscience

Rudwick, M. (1976). The emergence of a visual language for geological science. *History of Science*, 14

Ryle, G. (1949). *The Concept of Mind*. Hutchinson

Salmon, W. (1984). *Scientific Explanation and the Causal Structure of the World*. Princeton University Press.

Samuels, R. (forthcoming). Innateness in cognitive science

Sapp F., Lee, K. and Muir, D. (2000). Three-year-olds' difficulty with the appearance-reality distinction: is it real or is it apparent? *Developmental Psychology*, 36

Sartori, G. and Job, R. (1988). The oyster with four legs: a neuro-psychological study on the interaction of semantic and visual information. *Cognitive Neuropsychology*, 5

Savage-Rumbaugh, S., Murphy, J., Sevcik, R., Brakke, K., Williams, S. and Rumbaugh, D. (1993). Language comprehension in ape and child. *Monographs of the Society for Research in Child Development*, 58(3–4)

Scaillet, N. and Leyens, J.-P. (2000). From incorrect deductive reasoning to ingroup favouritism. In D. Capozza and R. Brown (eds.). *Social Identity Processes: Trends in Theory and Research*. Sage

Schaeken, W., De Vooght, G., Vandierendonck, A. and d'Ydewalle, G. (2000). *Deductive Reasoning and Strategies*. Lawrence Erlbaum

Schaller, S. (1995). *A Man Without Words*. University of California Press

Schank, R. and Abelson, R. (1977). *Scripts, Plans, Goals and Understanding: An Enquiry into Human Knowledge Structures*. Lawrence Erlbaum

Schauble, L. (1990). Belief revision in children: the role of prior knowledge and strategies for generating evidence. *Journal of Experimental Child Psychology*, 49

Scheffler, I. (1991). *In Praise of the Cognitive Emotions*. Routledge
Scheines, R., Spirtes, P., Glymour, C. and Meek, C. (1984). *TETRAD II*. Lawrence Erlbaum
Schepartz, L. (1993). Language and modern human origins. *Yearbook of Physical Anthropology*, 36
Schick, K., Toth, N., Garufi, G., Savage-Rumbaugh, S., Rumbaugh, D. and Sevcik, R. (1999). Continuing investigations into the stone tool making and tool-using capabilities of a Bonobo (*Pan piniscus*). *Journal of Archaeological Science*, 26
Schick, T. (1988). Nahal Hemar Cave: cordage, basketry and fibres. In O. Bar-Yosef and D. Alon (eds.), *Nahal Hemar Cave, Atiqot 18*. Jerusalem: Department of Antiquities and Museums
Schlottmann, A. (1999). Seeing it happen and knowing how it works: how children understand the relation between perceptual causality and underlying mechanism. *Developmental Psychology*, 35
Scholl, B. and Tremoulet, P. (2000). Perceptual causality and animacy. *Trends in Cognitive Sciences*, 4
Schustack, M. and Sternberg, R. (1981). Evaluation of evidence in causal inference. *Journal of Experimental Psychology: General*, 110
Schwartz, D. and Black, J. (1996). Analog imagery in mental model reasoning: depictive models. *Cognitive Psychology*, 30
Segal, G. (1996). Representing representations. In P. Carruthers and J. Boucher (eds.), *Language and Thought: Interdisciplinary Themes*. Cambridge University Press
Sen, A. (1985). *Choice, Welfare, and Measurement*. Harvard University Press
Shanks, D. (1985). Forward and backward blocking in human contingency judgement. *Quarterly Journal of Experimental Psychology*, 37B
Shanks, D. and Dickinson, A. (1987). Associative accounts of causality judgement. In G. Bower (ed.), *The Psychology of Learning and Motivation: Advances in Research and Theory, Vol. 21*. Academic Press
(1988). The role of selective attribution in causality judgement. In D. Hilton (ed.), *Contemporary Science and Natural Explanation: Common Sense Conceptions of Causality*. Harvester Press
Shapin, S. (1994). *A Social History of Truth: Civility and Science in Seventeenth-Century England*. University of Chicago Press
(1996). *The Scientific Revolution*. University of Chicago Press
Shelley, C. (1996). Visual abductive reasoning in archeology. *Philosophy of Science*, 63
Shennan, S. (in press). Demography and culture change. *Cambridge Archaeological Journal*
Shepard, R. and Cooper, L. (1982). *Mental Images and their Transformations*. MIT Press
Shultz, T. (1982). Rules of causal attribution. *Monographs of the Society for Research in Child Development*, serial no. 194
Siegal, M. (1988). Children's knowledge of contagion and contamination as causes of illness. *Child Development*, 59
(1995). Becoming mindful of food and conversation. *Current Directions in Psychological Science*, 6
(1996). Conversation and cognition. In R. Gelman and T. Au (eds.), *Handbook of Perception and Cognition: Perceptual and Cognitive Development* (2nd edn.). Academic Press

(1997). *Knowing Children: Experiments in Conversation and Cognition* (2nd edn.). Psychology Press

(1999). Language and thought: the fundamental significance of conversational awareness for cognitive development. *Developmental Science*, 2

Siegal, M. and Peterson, C. (1996). Breaking the mold: a fresh look at questions about children's understanding of lies and mistakes. *Developmental Psychology*, 32

(1998). Children's understanding of lies and innocent and negligent mistakes. *Developmental Psychology*, 34

(1999). Becoming mindful of biology and health: an introduction. In M. Siegal and C. Peterson (eds.), *Children's Understanding of Biology and Health*. Cambridge University Press

Siegal, M. and Robinson, J. (1987). Order effects in children's gender-constancy responses. *Developmental Psychology*, 23

Siegal, M. and Share, D. (1990). Contamination sensitivity in young children. *Developmental Psychology*, 26

Siegal, M., Surian, L., Nemeroff, C. and Peterson, C. (2001). Lies, mistakes and blessings: defining and characteristic features in conceptual development. *Journal of Cognition and Culture*, 1

Siegler, R. (1994). Cognitive variability: a key to understanding cognitive development. *Current Directions in Psychological Science*, 3

Simmons, D. and Keil, F. (1995). An abstract to concrete shift in the development of biological thought: the insides story. *Cognition*, 56

Simon, H. (1977). *Models of Discovery*. Dordrecht: Reidel

Slaughter, V., Jaakkola, R. and Carey, S. (1999). Constructing a coherent theory: children's biological understanding of life and death. In M. Siegal and C. Peterson (eds.), *Children's Understanding of Biology and Health*. Cambridge University Press

Slugoski, B. and Wilson, A. (1998). Contribution of conversational skills to the production of judgmental errors. *European Journal of Social Psychology*, 28

Slugoski, B., Lalljee, M., Lamb, R. and Ginsburg, J. (1993). Attribution in conversational context: effect of mutual knowledge on explanation-giving. *European Journal of Social Psychology*, 23

Slugoski, B., Sarson, D. and Krank, M. (1991). Cognitive load has paradoxical effects on the formation of illusory correlation. Unpublished manuscript

Smart, J. (1959). Sensations and brain processes. *Philosophical Review*, 68

Smedslund, J. (1963). The concept of correlation in adults. *Scandinavian Journal of Psychology*, 4

Smith, B. (1995). *The Emergence of Agriculture*. Washington: Smithsonian Press

Smith, E. and Medin, D. (1981). *Categories and Concepts*. Harvard University Press

Sober, E. (1993). *Philosophy of Biology*. Westview Press

Sober, E. and Wilson, D. (1998). *Unto Others: The Evolution and Psychology of Unselfish Behavior*. Harvard University Press

Sodian, B., Zaitchik, D. and Carey, S. (1991). Young children's differentiation of hypothetical beliefs from evidence. *Child Development*, 62(4)

Solomon, G. and Cassimatis, N. (1999). On facts and conceptual systems: young children's integration of their understanding of germs and contagion. *Developmental Psychology*, 35

Solomon, G. and Johnson, S. (2000). Conceptual change in the classroom: teaching young children to understand biological inheritance. *British Journal of Developmental Psychology*, 18

Solomon, G., Johnson, S., Zaitchik, D. and Carey, S. (1996). Like father, like son: young children's understanding of how and why offspring resemble their parents. *Child Development*, 67

Spelke, E. (1994). Initial knowledge: six suggestions. *Cognition*, 50

Spelke, E., Breinlinger, K., Macomber, J. and Jacobson, K. (1992). Origins of knowledge. *Psychological Review*, 99(4)

Spelke, E., Vishton, P. and von Hofsten, C. (1995). Object perception, object-directed action, and physical knowledge in infancy. In M. Gazzaniga (edn.), *The Cognitive Neurosciences*, MIT Press

Sperber, D. (1996). *Explaining Culture: A Naturalistic Approach*. Oxford: Blackwell

Sperber, D., Cara, F., and Girotto, V. (1995). Relevance theory explains the selection task. *Cognition*, 57

Spirtes, P., Glymour, C., and Scheines, R. (1993). *Causation, Prediction and Search*. NY: Springer-Verlag

(2000). *Causation, Prediction and Search* (rev. 2nd edn.). MIT Press

Springer, K. (1995). Acquiring a naïve theory of kinship through inference. *Child Development*, 66

(1999). How a naïve theory of biology is acquired. In M. Siegal and C. Peterson (eds.), *Children's Understanding of Biology and Health*. Cambridge University Press

Springer, K. and Belk, A. (1994). The role of physical contact and association in early contamination sensitivity. *Developmental Psychology*, 30

Springer, K. and Keil, F. (1989). On the development of biologically specific beliefs: the case of inheritance. *Child Development*, 60

(1991). Early differentiation of causal mechanisms appropriate to biological and non-biological kinds. *Child Development*, 62

Stanovich, K. (1999). *Who is Rational? Studies of Individual Differences in Reasoning*. Lawrence Erlbaum

Stanovich, K. and West, R. (1998). Cognitive ability and variation in selection task performance. *Thinking and Reasoning*, 4

(in press). Individual differences in reasoning: implications for the rationality debate? *Behavioral and Brain Sciences*

Stein, E. (1996). *Without Good Reason*. Oxford University Press

Sterman, J. (1994). Learning in and about complex systems. *Systems Dynamics Review*, 10

Stevenson, R. and Over, D. (1995). Deduction from uncertain premises. *Quarterly Journal of Experimental Psychology*, 48A

Stich, S. (1983). *From Folk Psychology to Cognitive Science*. MIT Press

(1993). Moral philosophy and mental representation. In M. Hechter, L. Nadel and R. Michod (eds.), *The Origin of Values*. New York: Aldine de Gruyter

Stich, S. and Nichols, S. (1998). Theory theory to the max. *Mind and Language*, 13(3)

Straus, L. (1990). The original arms race: Iberian perspectives on the Solutrean phenomenon. In J. Kozlowski (ed.), *Feuilles de Pierre: Les Industries Foliacées du Paléolithique Supérieur Européen*. Liège, Belgium: ERAUL

Stringer, C. and Gamble, C. (1993). *In Search of the Neanderthals*. Thames & Hudson

Stross, B. (1973). Acquisition of botanical terminology by Tzeltal children. In M. Edmonson (ed.), *Meaning in Mayan Languages*. The Hague: Mouton
Suchman, L. (1987). *Plans and Situated Actions*. Cambridge University Press
Susman, R. (1991). Who made the Oldowan tools? Fossil evidence for tool behaviour in Plio-Pleistocene hominids. *Journal of Anthropological Research*, 47
Swiderek, M. (1999). Beliefs can change in response to disconfirming evidence and can do so in complicated ways, but only if collateral beliefs are disconfirmed. Unpublished PhD Dissertation. Ithaca, NY: Cornell University
Swisher III, C., Curtis, G., Jacob, T., Getty, A., Suprijo, A. and Widiasmoro (1994). Age of the earliest known hominids in Java, Indonesia. *Science*, 263
Tetlock, P. (1992). The impact of accountability on judgement and choice: toward a social contingency model. In M. Zanna (ed.), *Advances in Experimental Social Psychology*, 25
Thagard, P. (1988). *Computational Philosophy of Science*. MIT Press
(1992). *Conceptual Revolutions*. Princeton University Press
(1996). The concept of disease: structure and change. *Communication and Cognition*, 29
(1999). *How Scientists Explain Disease*. Princeton University Press
(2000). *Coherence in Thought and Action*. MIT Press
(2001). How to make decisions: coherence, emotion, and practical inference. In E. Millgram (ed.), *Varieties of Practical Inference*. MIT Press
Thagard, P. and Shelley, C. (2001). Emotional analogies and analogical inference. In D. Gentner, K. Holyoak and B. Kokinov (eds.), *The Analogical Mind: Perspectives from Cognitive Science*. MIT Press
Thagard, P. and Zhu, J. (forthcoming). Acupuncture, incommensurability, and conceptual change. In G. Sinatra and P. Pintrich (eds.), *Intentional Conceptual Change*. Lawrence Erlbaum
Thagard, P., Holyoak, K., Nelson, G. and Gochfield, D. (1990). Analog retrieval by constraint satisfaction. *Artificial Intelligence*, 46
Thelan, E. and Smith, L. (1994). *A Dynamic Systems Approach to the Development of Cognition and Action*. MIT Press
Thieme, H. (1997). Lower Palaeolithic hunting spears from Germany. *Nature*, 385
Thompson, V. (1996). Reasoning from false premises: the role of soundness in making correct logical deductions. *Canadian Journal of Experimental Psychology*, 50
Tolman, E. (1932). *Purposive Behavior in Animals and Men*. New York: The Century Co.
Tomasello, M. (1996). The cultural roots of language. In B. Velichkovsky and D. Rumbaugh (eds.), *Communicating Meaning: The Evolution and Development of Language*. Lawrence Erlbaum
Tomasello, M. and Call, J. (1997). *Primate Cognition*. Oxford University Press
Tooby, J. and Cosmides, L. (1992). The psychological foundations of culture. In J. Barkow, L. Cosmides and J. Tooby (eds.), *The Adapted Mind*. Oxford University Press
Toth, N., Schick, K., Savage-Rumbaugh, S., Sevcik, R. and Rumbaugh, D. (1993). Pan the tool maker: investigations into the stone tool-making and tool-using capabilities of a bonobo (*Pan paniscus*). *Journal of Archaeological Science*, 20
Tournefort, J. (1694). *Elémens de Botanique*. Paris: Imprimerie Royale

Trumpler, M. (1997). Converging images: techniques of intervention and forms of representation of sodium-channel proteins in nerve cell membranes. *Journal of the History of Biology*, 20

Turq, A. (1992). Raw material and technological studies of the Quina Mousterian in Perigord. In H. Dibble and P. Mellars (eds.), *The Middle Palaeolithic: Adaptation, Behaviour and Variability*. The University Museum, University of Pennsylvania

Tversky, A. and Kahneman, D. (1974). Judgement under uncertainty: heuristics and biases. *Science*, 185

Tweney, R. (1985). Faraday's discovery of induction: a cognitive approach. In D. Gooding and F. James (eds.), *Faraday Rediscovered*. Stockton Press

(1987). What is scientific thinking? Unpublished manuscript

(1992). Stopping time: Faraday and the scientific creation of perceptual order. *Physis*, 29

Tweney, R. and Chitwood, S. (1995). Scientific reasoning. In S. Newstead and J. Evans (eds.), *Perspectives on Thinking and Reasoning: Essays in Honour of Peter Wason*. Lawrence Erlbaum

Ungerleider, L. and Mishkin, M. (1982). Two cortical visual systems. In D. Ingle, M. Goodale and R. Mansfield (eds.), *Analysis of Visual Behavior*. MIT Press

Van der Lely, H., Rosen, S. and McClelland, A. (1998). Evidence for a grammar-specific deficit in children. *Current Biology*, 8

Varela, F., Thompson E. and Rosch, E. (1993). *The Embodied Mind*. MIT Press

Varley, R. (1998). Aphasic language, aphasic thought: an investigation of propositional thinking in an a-propositional aphasic. In P. Carruthers and J. Boucher (eds.), *Language and Thought: Interdisciplinary Themes*. Cambridge University Press

Varley, R. and Siegal M. (2000). Evidence for cognition without grammar from causal reasoning and 'theory of mind' in an agrammatic aphasic patient. *Current Biology*, 10

Varley, R., Siegal M. and Want, S. (2001). Severe impairment in grammar does not preclude theory of mind. *Neurocase*, 7

Velichkovsky, B. and Rumbaugh, D. (eds.) (1996). *Communicating Meaning: The Evolution and Development of Language*. Lawrence Erlbaum

Vicente, K. and Brewer, W. (1993). Reconstructive remembering of the scientific literature. *Cognition*, 46

Vosniadou, S. and Brewer, W. (1992). Mental models of the earth: a study of conceptual change in childhood. *Cognitive Psychology*, 24

Vygotsky, L. (1962). *Thought and Language*. MIT Press

(1978). *Mind in Society: The Development of Higher Psychological Processes*. Harvard University Press

Waddington, C. (1959). Canalization of development and the inheritance of acquired characteristics. *Nature*, 183

Walker, A. and Leakey, R. (eds.) (1993). *The Nariokotome Homo erectus Skeleton*. Berlin: Springer-Verlag

Wallace, A. (1889/1901). *Darwinism* (3rd edn.). Macmillan. (1st edn. 1889)

Warburton, F. (1967) The purposes of classification. *Systematic Zoology*, 16

Warrington, E. and Shallice, T. (1984). Category specific impairments. *Brain*, 107

Wason, P. (1960). On the failure to eliminate hypotheses in a conceptual task. *Quarterly Journal of Experimental Psychology*, 12

(1966). Reasoning. In B. Foss (ed.), *New Horizons in Psychology I.* Penguin

(1968). On the failure to eliminate hypotheses in a conceptual task – a second look. In P. Watson and P. Johnson-Laird (eds.),*Thinking and Reasoning*. Cambridge University Press

Wason, P. and Evans, J. (1975). Dual processes in reasoning? *Cognition*, 3

Wason, P. and Johnson-Laird, P. (1970). A conflict between selecting and evaluating information in an inferential task. *British Journal of Psychology*, 61

Wasserman, E. and Berglan, L. (1998). Backward blocking and recovery from overshadowing in human causal judgement. The role of within-compound associations. *Quarterly Journal of Experimental Psychology: Comparative & Physiological Psychology*, 51

Watson, J. (1967). Memory and 'contingency analysis' in infant learning. *Merrill-Palmer Quarterly*, 13

(1979). Perception of contingency as a determinant of social responsiveness. In E. Tohman (ed.), *The Origins of Social Responsiveness*. Lawrence Erlbaum

Watson, J. D. (1969). *The Double Helix*. New York: New American Library

(2000). *A Passion for DNA: Genes, Genomes, and Society*. Cold Spring Harbor Laboratory Press

Watts, I. (1999). The origin of symbolic culture. In R. Dunbar, C. Knight and C. Power (eds.),*The Evolution of Culture*. Edinburgh University Press

Wechsler, D. (1981). *Wechsler Adult Intelligence Scale – Revised.* The Psychological Corporation

Weiner, B. (1995). 'Spontaneous' causal thinking. *Psychological Bulletin*, 109

Wellman, H. (1990). *The Child's Theory of Mind.* MIT Press

Wellman, H. and Estes, D. (1986). Early understanding of mental entities: a re-examination of childhood realism. *Child Development*, 57

Wellman, H. and Gelman, S. (1992). Cognitive development: foundational theories of core domains. *Annual Review of Psychology*, 43

(1997). Knowledge acquisition in foundational domains. In D. Kuhn and R. Siegler (eds.), *Handbook of Child Psychology* (5th edn.). Wiley

Wellman, H., Cross, D. and Watson, J. (2001). Meta-analysis of theory of mind development: the truth about false belief. *Child Development*, 72

Wertsch, J. (1985). *Culture, Communication, and Cognition: Vygotskian Perspectives.* Cambridge University Press

Wetherick, N. (1962). Eliminative and enumerative behaviour in a conceptual task. *Quarterly Journal of Experimental Psychology*, 14

Wheeler, P. (1994). The thermoregulatory advantages of heat storage and shade seeking behaviour to hominids foraging in equatorial savannah environments. *Journal of Human Evolution*, 26

White, R. (1989). Production, complexity and standardization in early Aurignacian bead and pendant manufacture: evolutionary implications. In P. Mellars and C. Stringer (eds.),*The Human Revolution*. Edinburgh University Press

Whiten, A. (1996). When does smart behavior become mind reading? In P. Carruthers and P. Smith (eds.),*Theories of Theories of Mind.* Cambridge University Press

Wilkins, J. (1675). *Of the Principles and Duties of Natural Religion.* London: Bonwicke

Wittgenstein, L. (1953). *Philosophical Investigations.* Blackwell

Wolpert, L. and Richards, A. (1997). *Passionate Minds: The Inner World of Scientists*. Oxford University Press
Wood, B. (1997). The oldest whodunnit in the world. *Nature*, 385
Wood, B. and Collard, M. (1999). The Human genus. *Science*, 284
Woodward, J. (1965).*Industrial Organization: Theory and Practice*. Oxford University Press
Wynn, K. (1990). Children's understanding of counting. *Cognition*, 36
(1995). Origins of mathematical knowledge. *Mathematical Cognition*, 1
(1998). Psychological foundations of number: numerical competence in human infants. *Trends in Cognitive Sciences*, 4
Wynn, T. (1991). Tools, grammar and the archaeology of cognition. *Cambridge Archaeological Journal*, 1
(2000). Symmetry and the evolution of the modular linguistic mind. In P. Carruthers and A. Chamberlain (eds.), *Evolution and the Human Mind*. Cambridge University Press
Wynn, T. and McGrew, B. (1989). An ape's eye view of the Oldowan. *Man*, 24
Yates, J. and Curley, S. (1996). Contingency judgement: primacy effects and attention decrement. *Acta Psychologica*, 62
Yellen, J., Brooks, A., Cornelissen, E., Mehlman, M. and Steward, K. (1995). A Middle stone age worked bone industry from Katanda, Upper Semliki Valley, Zaire. *Science*, 268
Zubin, D. and Köpcke, K.-M. (1986). Gender and folk taxonomy. In C. Craig (ed.), *Noun Classes and Categorization*. Amsterdam: John Benjamins
Zvelebil, M. (1984). Clues to recent human evolution from specialized technology. *Nature*, 307

Author index

Subject index

www.ingramcontent.com/pod-product-compliance
Ingram Content Group UK Ltd.
Pitfield, Milton Keynes, MK11 3LW, UK
UKHW040702180125
453697UK00010B/344